Crumbling Genome

Crumbling Genome

The Impact of Deleterious Mutations on Humans

Alexey S. Kondrashov

WILEY Blackwell

This edition first published 2017

Registered Office
John Wiley & Sons, Inc., 111 River Street, Hoboken, NJ 07030, USA

Editorial Office
111 River Street, Hoboken, NJ 07030, USA

For details of our global editorial offices, customer services, and more information about Wiley products visit us at www.wiley.com.

Wiley also publishes its books in a variety of electronic formats and by print-on-demand. Some content that appears in standard print versions of this book may not be available in other formats.

Library of Congress Cataloguing-in-Publication data applied for

ISBN: 9781118952115

Cover Image: A mosaic mutant fly strives to reach perfection. Image by Glafira Kolbasova.
Cover Design: Wiley

Set in 10/12pt Warnock by SPi Global, Pondicherry, India

10 9 8 7 6 5 4 3 2 1

Contents

Preface

Before he died, or rather passed into Mahaparinirvana, Buddha gave final instructions to his disciples. Canonical text in Pali records his last sentence as: "vayadhammā saṅkhārā appamādena sampādetha", which can be translated as "Everything consisting of parts crumbles – but you will succeed through diligence".

Buddha's words provide a perfect summary of this book. Our subject is how complex and vulnerable our genes and bodies are, how they crumble under the relentless pressure of spontaneous mutation, and what can be done to mitigate the impact of deleterious mutations on individuals and populations.

Indeed, we are amazingly complex. The human genome consists of DNA molecules with a combined length of over 3 billion nucleotides and contains over 20 000 protein-coding genes, each encoding either one protein or a set of similar proteins. These genes and proteins interact with each other, forming staggering networks of mutual influences. The human body, which develops from a single cell on the basis of instructions provided by the genome, comprises many billions of interconnected cells. Verily, we consist of parts.

Thus, it is hardly surprising that we are very vulnerable to all kinds of insults. A mutation that replaces a single nucleotide within a protein-coding gene may lead to synthesis of a dysfunctional protein, and this can be inconsistent with life or cause a tragic disease. Chaos is common, but order is rare, and chances of a typo improving *Hamlet* are slim. Thus, deleterious mutations vastly outnumber beneficial mutations.

Far from being a moot point, vulnerability of our genome is constantly exposed by mutation. Despite of the all elaborate mechanisms that a cell employs to handle its DNA with the utmost care, a newborn human carries about 100 new (*de novo*) mutations, originated in the germline of their parents, about 10% of which are substantially deleterious. Several percent of even young people suffer from overt diseases that are caused, exclusively or primarily, by pre-existing and *de novo* mutations in their genomes, including both a wide variety of genetically simple Mendelian diseases and even more diverse complex diseases such as birth anomalies, diabetes, and schizophrenia. Milder, but still substantial, negative effects of mutations are harder to detect, but are even more pervasive. As of now, we possess no means of reducing the rate at which mutations appear spontaneously.

In the course of the long evolution of life, natural selection eliminated deleterious mutations with cruel efficiency, establishing what is known as the mutation–selection equilibrium. Still, even genotypes that exist at this equilibrium are far from being perfect. In the average human genotype, over 100 genes are dysfunctional or altogether missing, and over 1000 genes are substantially impaired.

Moreover, the Industrial Revolution radically improved the environments of many human populations. Although it is not yet clear how strong selection against deleterious mutations is in developed countries, it can hardly remain as comprehensive as it used to be when a newborn was expected to live only 25 years. If so, populations in these countries cannot be at the mutation–selection equilibrium and are accumulating deleterious mutations.

After the basic laws of heredity were discovered in the early 20th century, a number of geneticists, including Hermann J. Muller and James F. Crow, developed a conceptual framework for studying negative effects of deleterious mutations on individual humans and on human populations. Throughout this book, I will refer to these effects as the "Main Concern". It is clear that both pre-existing mutations, which appeared some generations ago and have not yet been purged from the population, and *de novo* mutations, which are happening now, do a lot of harm. However, until very recently, estimates of the scale of this problem were rather fuzzy.

The ongoing avalanche of genome-level data is rapidly changing this situation. Now we know much more than even 10 years ago about the current contribution of deleterious mutations to human suffering and imperfection, and about potential consequences of accumulation of such mutations in the future. Substantial uncertainties remain, but the post-genomic biology will likely eliminate many of them in the next 20–30 years.

My wish is to enable a motivated lay person to thoroughly comprehend the Main Concern. With this goal in mind, I provide about equal coverage to concepts, data, and unsolved problems in this book. This book is intended for three types of readers: those who are simply fascinated by the life sciences and want to learn more about genetics and evolution; those who are worried about challenges that deleterious mutations pose to humanity; and those who are personally touched by a disease with a large genetic component and need a background for understanding it.

Only a general scientific literacy is expected from a reader, and the necessary biological basics are presented in the first three chapters. After this, three chapters deal with different facets of mutation, and the next four with different facets of selection against mutations. The last five chapters directly address the Main Concern. I tried to follow the sage advice commonly attributed to Einstein: "Everything should be made as simple as possible, but not simpler". You will succeed through diligence.

I would like to thank Raquel Assis, Charlie Baer, Laura Eidietis, Eugene Koonin, Michael Lynch, Sergey Mirkin, Vladimir Seplyarsky, Mashaal Sohail and Shamil Sunyaev whose comments on the manuscript have spared me a number of embarrassments. I am most thankful to Glafira Kolbasova for producing almost all illustrations.

Alexey S. Kondrashov
University of Michigan

1

Genotypes and Phenotypes

I grew my own body... Nobody else did it for me. So if I grew it, I must have known how to grow it.

Jerome D. Salinger, Teddy, 1953.

Hereditary information of a cell is stored in double-stranded DNA molecules which, for a geneticist, are texts written in a four-letter alphabet {A, T, G, C}. Recent technological advances have made it possible to efficiently read long DNA texts. Before dividing, a cell produces two copies of its hereditary information by DNA replication. From each parent, an individual human receives a haploid genotype, a 3.2-billion-letter-long DNA text which is subdivided into 23 molecules, chromosomes. Also, tiny mitochondrial chromosomes are received from the mother only. The consensus of genotypes within a species is called the genome of the species. Some segments of the genome are transcribed into RNAs, and some segments of these RNAs are translated into proteins. Parental genotypes recombine in the course of meiosis, a form of cell division which halves the amount of DNA in a cell. In the course of sexual reproduction, a multicellular diploid organism develops from a single cell, zygote. Together with the environment, the genotype of the zygote determines the phenotype of the organism.

1.1 DNA is a Text

Inheritance is a salient phenomenon. Dogs beget dogs, and Beagles beget Beagles. Because an instruction on how to make a unique individual Beagle dog must be a lengthy one, there must be something, transmitted from parents to offspring, which carries an enormous amount of information. In the course of sexual reproduction, a multicellular organism develops from a single cell, called a zygote, which is produced by fertilization, fusion of two gametes, an egg and a sperm. A mammalian egg is just ~0.2 mm in diameter, and the head of a mammalian sperm is still much smaller, being only ~0.005 mm across. And yet even this tiny sperm is large enough for the job: mammals are about as similar to their fathers as to their mothers (Figure 1.1).

Thus, hereditary information must be packed at the molecular scale. Which molecule carries instructions from parents to offspring? In 1944, DNA emerged as a likely answer when it was shown that bacteria may acquire traits of other bacteria after ingesting their DNA. Then, in 1953, came the discovery, by Raymond Gosling, Rosalind Franklin,

Crumbling Genome: The Impact of Deleterious Mutations on Humans, First Edition. Alexey S. Kondrashov.

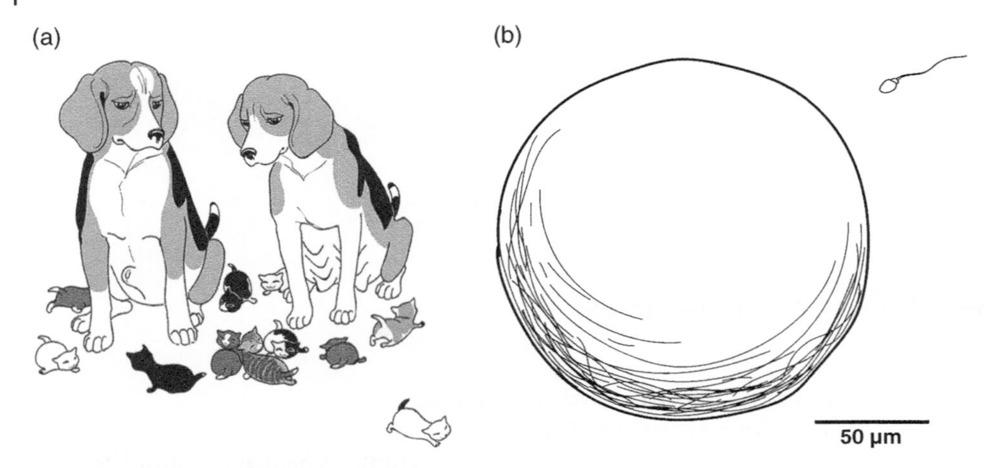

Figure 1.1 Phenomenon of heredity. (a) Mother and Father Beagle dogs stare with dismay at their kittens (joke). (b) A tiny mammalian sperm approaching an egg.

Maurice Wilkins, James Watson, and Francis Crick, of three properties of DNA which make it exquisitely suitable for storing and propagating information (Figure 1.2).

First, DNA is a linear polymer. A strand of DNA consists of a regular alternation of phosphates (residues of phosphoric acid, H_3PO_4) and sugars called deoxyribose, to each of which one of the four possible bases (known as A, T, G, and C, for adenine, thymine, guanine, and cytosine) is attached at a side. A and G are bigger molecules, called purines, and T and C are smaller molecules, called pyrimidines. The only thing we care about in deoxyribose is that its two ends, by which it is attached to two adjacent phosphates, are different, and one is called 5' and the other 3' (both "ends" of a phosphate are identical). All deoxyriboses within a DNA strand have the same orientation, which provides direction to the whole strand. Traditionally, a single DNA strand is shown in the 5' > 3' direction. Together, a phosphate, a deoxyribose, and a base are called a nucleotide, and nucleotides are denoted by the same letters as bases. Chemical details are not too important for a geneticist: I do not remember the exact structures of A, T, G, or C.

Secondly, there is no firm limit on the length of a DNA strand, which can consist of hundreds of millions of nucleotides. Moreover, there are also no restrictions on the order of nucleotides within a strand. Thus, if we think of a sequence of nucleotides as a text, written in a four-letter alphabet {A, T, G, C}, any message (such as ACCATCATCGATGACT...) is chemically possible. A four-letter alphabet is perfectly sufficient to store information: computers are content with a two-letter alphabet {0, 1}, and a 26-letter English alphabet is just a luxury.

Thirdly, two DNA strands can nicely fit each other, if (i) they are arranged side-by-side in the opposite (antiparallel) directions, 5' end to 3' end and vice versa, and (ii) their nucleotide sequences are complementary to each other. Complementarity means that A's in one strand are opposed by T's in the other (and vice versa), and G's are opposed by C's (and vice versa). Overall shapes of A:T and G:C nucleotide pairs (referred to as complementary, or Watson–Crick pairs) are very similar, leading to a nearly perfect fit of strands with complementary sequences (called complementary

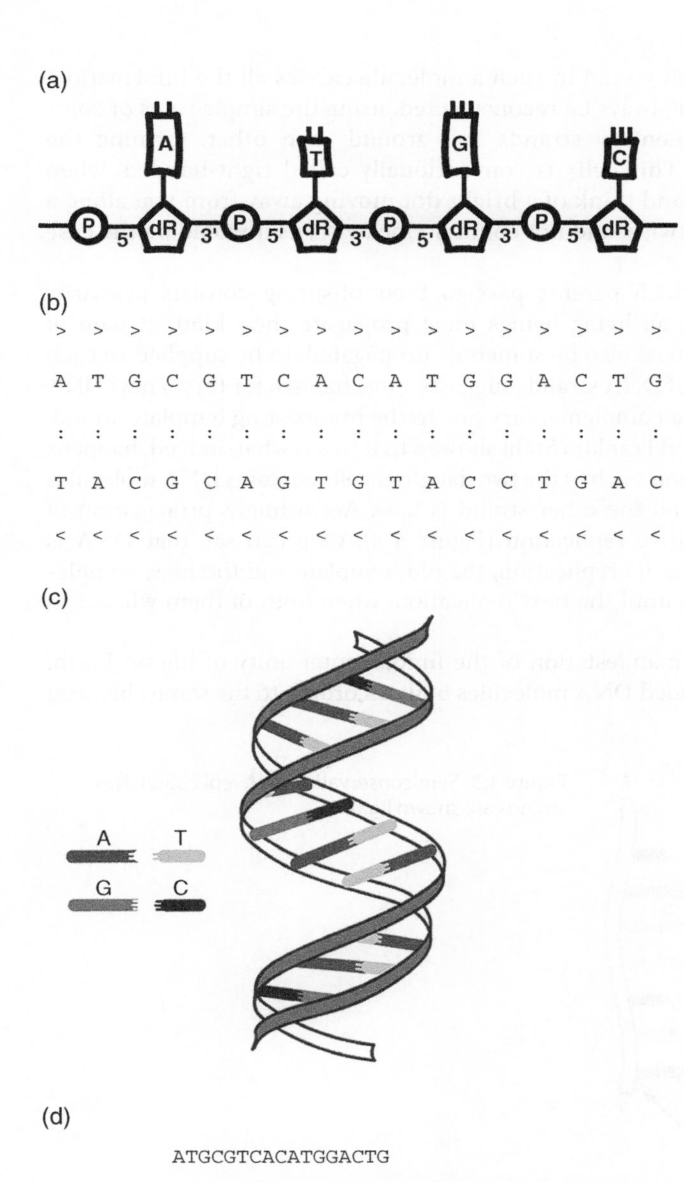

Figure 1.2 The fundamentals of DNA. (a) A single DNA strand with bases A, T, G, and C attached to it. P stands for a phosphate, and dR for a deoxyribose. (b) A scheme of double-stranded DNA, consisting of two complementary strands (">" shows the 5' > 3' direction of a strand, and ":" shows weak bounds connecting the two strands to each other). (c) A double-stranded DNA shown in its actual shape of a right-handed helix. (d) A double-stranded DNA shown as a sequence of nucleotides in one of its strands, or a text written in a four-letter alphabet. For a geneticist, this simplistic representation is usually enough.

strands). Within a Watson–Crick nucleotide pair, A (large) and T (small), or G (large) and C (small), are attached to each other by weak chemical bonds, known as hydrogen bonds (two of them in A:T pairs, and three in G:C pairs). DNA in all cells exists not as single-stranded but as double-stranded molecules, consisting of two complementary strands. Each strand in such a molecule carries all the information, because the other strand can always be reconstructed, using the simple rules of complementarity. Two complementary strands coil around each other, forming the famous DNA double helix. This helix is conventionally called right-handed: when you look at either end of it, and think of a bright dot moving away from you along a strand, this dot rotates clockwise, and not counter clockwise, as it would be the case for a left-handed helix.

Hereditary information which parents pass to their offspring consists primarily of DNA sequences. Because all living beings must propagate their kind, at pain of extinction, DNA molecules must also be somehow propagated, to be supplied to each offspring. Complementarity of DNA strands suggests a mechanism for this: a new DNA strand can be synthesized as a complementary one to the pre-existing template strand. In 1958 Matthew Meselson and Franklin Stahl showed that this is what, indeed, happens in living cells: after a cell division, each of the two daughter cells contains DNA molecules in which one strand is old and the other strand is new. Accordingly, propagation of DNA is called semiconservative replication (Figure 1.3). One can say that DNA is double-stranded because, after its replication, the old, template and the new, complementary strand stay together until the next replication, when both of them will act as templates.

DNA is the most striking manifestation of the fundamental unity of life on Earth. All cells contain double-stranded DNA molecules built according to the same chemical

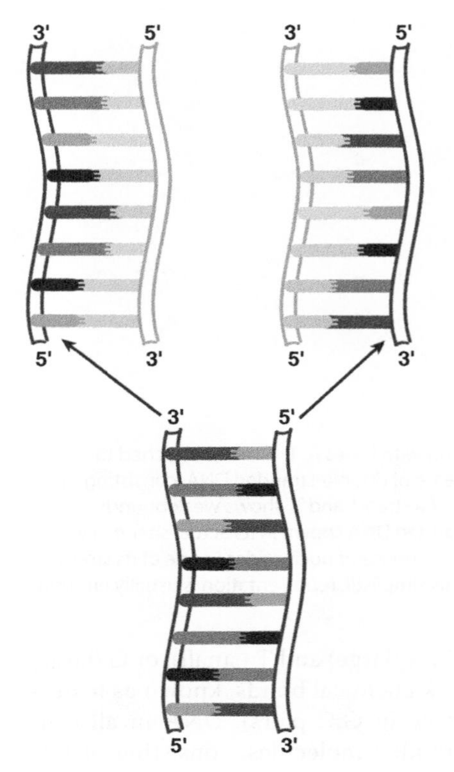

Figure 1.3 Semiconservative DNA replication. New strands are shown light.

rules, with only occasional minor secondary modifications. Moreover, some segments of DNA from all kinds of living beings, including bacteria, protists, plants, and animals, have rather similar sequences. This is the case, for example, for many DNA segments which carry instructions on how to make ribosomes, protein-synthesizing molecular machines.

Of course, there are also a lot of differences between DNA sequences from individuals that belong to different species. These differences are responsible for one individual being a dog and another one being the dog's owner. Even different individuals of the same species possess slightly different DNA sequences. These differences are responsible for one dog being a Beagle and another a Collie, as well as for hereditary variation within a breed. The DNA sequence present in every cell of an individual is called its genotype (there could be minor differences even between DNA sequences from different cells of an individual, see Chapters 4 and 5).

In order to study similarity and differences between DNA sequences, one first needs to align them. Aligning means introducing gaps into sequences in such a way that insertions and deletions of nucleotides which occurred in the course of evolution of these sequences from their common ancestor do not obscure their pattern of similarity. If a sequence segment was deleted, a gap must be placed in its stead, and if a sequence segment was inserted, a gap must be placed against it in all other sequences (Figure 1.4). For each position in an alignment, the consensus letter can be determined, by the majority rule (ties need to be processed somehow).

```
1    tcccttgtttggtctcttttttgttcttcactgat-ttttgtacag
2    tcccttgtttgctctatctttt--atcccttcactaat-ttctctacag
3    tcccttgcttgctctatttttgaatcctt actgatgttttatacag
4    tcccttgtttggtcttcttttttgttcttcactgat-tttggtacag
5    tcccttgtttggtcttcttttttgttcttcactgat-tttttacag
6    tcccttatttgcccatttctctaatccctactga--ttttatacag
7    tcccttgtttggtcttctttcttgttcttcactgat-ttttgtacag
8    tcccttgtttggtcttctttttttcttcactgat-tcttatacag
9    tcctttgtttgctctatcttt--actcctcactgat-ttttatacag
10   tcccttgtttggtcttcttttttgttcttcactaat-ttttgtacag
11   tcccttacttgctccatttctctaatcccagttgat-tttcatacag
12   tcccttgtttggtcttctttcttgttcttcgctgat-ttttgtacag
CONS TCCCTTGTTTGGTCTTCTTTTTTGTTCTTCACTGAT-TTTTGTACAG
```

Figure 1.4 Alignment of short pieces of 12 genotypes (shown in lower case) of the fungus *Schizophyllum commune*, the most genetically variable species known (see Chapter 2). Gaps of length 2 were inserted into genotypes 2 and 9, a gap of length 1 into genotype 6, instead of lost segments of the corresponding lengths. Gaps of length 1 were inserted into all genotypes, except genotype 3, because genotype 3 gained nucleotide g at the corresponding location. Consensus of these genotypes (a piece of *S. commune* genome) is shown in upper case at the bottom. Deviations of genotypes from the genome are shown in grey.

The consensus of genotypes of different individuals which belong to a particular species is called the genome of the species. Sometimes, the whole range of genetic variation within a species is also informally included into the concept of its genome. Thus, each human possesses a genotype (often, it is convenient to think that in humans and many other diploid species an individual possesses two genotypes, maternal and paternal), but no human has a genotype which exactly coincides with the human genome. There is no fundamental difference between aligning genotypes of different individuals from the same species and genomes of different species, although genotypes are typically much more similar to each other than genomes.

1.2 Genomes Small and Large

For some time after the key role of DNA in inheritance had been revealed, this fundamental discovery was hard to apply, because there were no practical means to determine nucleotide sequences of actual genotypes, or, in a professional slang, to sequence DNA. Do not forget that physically DNA is just a molecule, and a nucleotide occupies only 0.33 nm along its length, so that ~3.2 billion nucleotides that comprise a human genotype fit into about 1 m. Although it is possible to directly read a DNA sequence using a transmission electron microscope, after labeling different nucleotides by different heavy atoms, this approach began to take shape only very recently and is not yet used widely.

Instead, sequencing is performed by indirect chemical methods. We do not need to go into the complex and fascinating details; a very general outline will suffice. The first substantial genome segment, a 77-nucleotide-long piece called alanine tRNA gene from yeast, was sequenced in 1965, after over 2 years of hard work by the laboratory of Robert Holley. In 1977, two reasonably efficient methods, capable of sequencing DNA segments ("reads") up to ~1000 nucleotides long, were invented. A longer DNA molecule can be sequenced by breaking it randomly into many short reads, sequencing each of them, and assembling the complete sequence from these reads, using their overlaps. This shotgun method (Figure 1.5) works best when a DNA molecule does not contain any long repeats, so that almost every segment longer than, say, 20 nucleotides has a unique sequence. Repeats, which are common in many genomes, can interfere with assembling, obscuring the correct order of reads. However, modern computer programs that perform assembly often work well even when a lot of repeats are present.

The shotgun method requires the sequencing of enough reads so that each nucleotide is covered, on average, many times, as otherwise a substantial proportion of nucleotides

```
actacggactactgg
    cggactactggtacg
         tactggtacgaccaa
                    ccaagcatactgagt
ACTACGGACTACTGGTACGACCAAGCATACTGAGT
11112222233333222212222111111111111
```

Figure 1.5 Assembly of a long DNA sequence from relatively short overlapping reads. Reads are in lower case, the assembled sequence is in upper case, and the numbers show how many times each nucleotide of the sequence is covered by reads.

would remain uncovered and, thus, unknown. A 20-fold average coverage may be enough, although the higher the coverage, the better. The first 1 830 140-nucleotide complete genome of a free-living organism, a bacterium *Haemophilus influenzae*, was sequenced in this way in 1995. A rough draft of the human genome, containing about 90% of all its sequences, was published in 2001, and most of the missing pieces were added in 2004. The cost of this huge undertaking was several billion dollars.

More recently, several next-generation sequencing (NGS) methods have been invented. They retain the key shortcoming of older methods: only relatively short reads can be sequenced. However, NGS is massively parallel, being capable of sequencing very many reads at once. One run of a cutting-edge NGS instrument Illumina NextSeq 2000 produces 3 billion 200-nucleotide-long reads, which is enough to sequence 5 human genomes with coverage 40, at a cost of US$ 4000 each. As a result, genomes of hundreds of species have been sequenced in the last decade. Even more importantly, from our perspective, NGS made it possible to sequence (almost) complete genotypes of many individuals from the same species. In particular, tens of thousands of human genotypes are currently known, and this number is increasing rapidly.

Methods of DNA sequencing continue to improve, and their cost keeps declining. Hopefully, the length of a read will reach ~1000 nucleotides in the not-too-distant future, which will make it possible to assemble each human genotype individually, instead of relying on the human genome as a reference – a common practice currently, which can occasionally lead to errors. Soon, precise sequencing of the genotype of a human will become trivial.

Genomes of prokaryotes, bacteria and archaea, which are basically unicellular organisms whose cells do not have a nucleus, often consist of one circular DNA molecule. By contrast, genomes of eukaryotes, unicellular and multicellular organisms, including ourselves, whose cells contain nuclei separated from cytoplasm by a nuclear membrane, mostly consist of linear DNA molecules. DNA molecules which constitute any genome are called chromosomes, although originally this term was applied only to eukaryotes. Human genome consists of 23 linear chromosomes that reside in the nucleus and a tiny (16 569 nucleotides) circular chromosome of bacterial origin, that resides in mitochondria. A normal human diploid genotype consists of 23 pairs of nuclear chromosomes and of multiple tiny mitochondrial chromosomes (Figure 1.6). Table 1.1 presents key properties of a small sample of genomes of prokaryotes (the first three species) and eukaryotes. Except for *Ostreococcus tauri*, all the eukaryotes listed in this table are diploid, so that every cell of their bodies carries two genotypes, of maternal and paternal origin.

1.3 Genes and Intergenic Regions

It is obvious from Table 1.1 that across genomes of different species numbers of protein-coding genes are much more uniform than genome sizes. Small genomes are lean, in the sense that protein-coding genes constitute most of them. By contrast, large genomes, including ours, are bloated, consisting mostly of non-protein-coding, and often repetitive, segments of the DNA. Of course, in order to fully appreciate the importance of these facts, we first need to know what protein-coding genes are and, more generally, how a genome works.

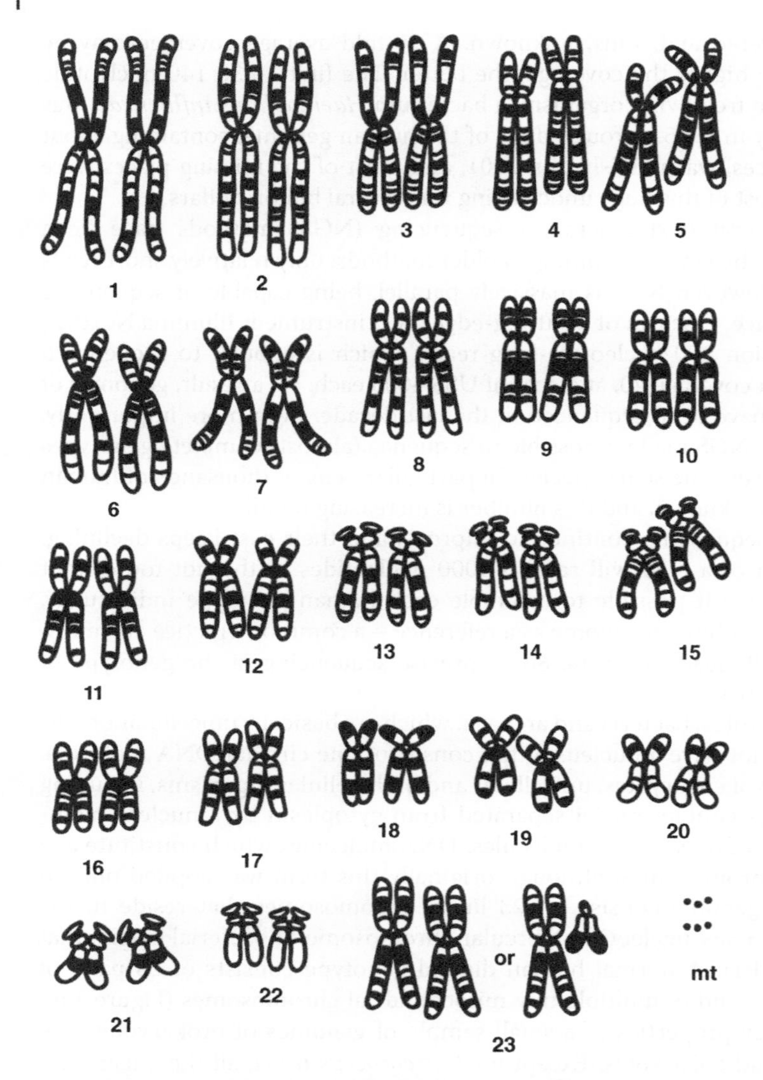

Figure 1.6 A normal human genotype. Each individual carries two copies of chromosomes 1–22, called autosomes, one inherited from the mother and the other from the father. A woman also carries two X chromosomes, one inherited from each parent, and a man carries one X chromosome, inherited from the mother, and one Y chromosome, inherited from the father. X and Y chromosomes are called sex chromosomes. Thus, a human genotype normally consists of 46 linear chromosomes, as well as of many small circular mitochondrial chromosomes, which are inherited only from the mother. Here, linear chromosomes are shown after DNA replication and before cell division, so that each of them consists of two identical copies, still joined to each other at one point. Bands on linear chromosomes roughly correspond to how they look under a light microscope. Mitochondrial chromosomes (mt) are too small to be seen in this way, and are shown disproportionally large.

On top of a physical subdivision of many genomes into chromosomes, all genomes can also be subdivided into segments performing different functions. Indeed, it is hard to imagine a whole ~100 000 000-nucleotide-long chromosome doing the same thing.

Table 1.1 Key properties of some representative genomes.

Species	Description	Total length	Number of chromosomes (not counting eukaryotic mitochondria)	Proportion of the genome that codes for proteins	Number of protein-coding genes	Comment
Carsonella ruddii	A bacterium which is an intracellular symbiont of insects, and cannot live on its own	159 662	1	0.9	182	The shortest known genome of a cellular organism
Pelagibacter ubique	A free-living bacterium	1 308 759	1	0.9	1354	The shortest genome of a free-living organism. In many lean bacterial genomes, there are ~1000 nucleotides per gene
Escherichia coli	A free-living bacterium which is the most-studied organism of all	4 639 221	1	0.85	4377	A typical bacterial genome. Different strains of bacteria which we all call "*E. coli*" possess rather different numbers of genes
Ostreococcus tauri	A unicellular green alga	12 560 000	20	0.7	8166	The shortest genome of a free-living eukaryote
Genlisea aurea	A carnivorous flowering plant	63 000 000	26	0.3	17 755	The shortest genome of a flowering plant. Some close relatives of this species have genomes ~20 times larger
Caenorhabditis elegans	A nematode worm	100 000 000	10	0.25	21 733	One of key model organisms in many fields of biology, which reproduces mostly by selfing
Drosophila melanogaster	A fruit fly	140 000 000	4	0.2	15 682	The most important model organism in genetics, possessing a small number of genes for a multicellular eukaryote
Takifugu rubripens	A puffer fish	400 000 000	22	0.1	~25 000	The smallest genome of a vertebrate
Homo sapiens	Human	3 200 000 000	23	0.01	20 051	Overall properties of genomes of all mammals are rather similar
Picea abies	Norwegian spruce	19 600 000 000	12	0.003	28 354	The longest genome sequenced so far
Paris japonica	A beautiful flowering plant	~150 000 000 000	20	0.001	~30 000 (guess)	The longest genome known, not yet sequenced

Any genome segment that performs a particular function can be loosely called a gene. Thus, understanding of genome functioning begins from subdividing it into genes or, in our professional slang, from annotating it. Three properties of many genomes can complicate this task. First, genes sometimes overlap – there is no reason why a particular DNA segment must always perform only one function. Secondly, bloated genomes contain many long functionless, or junk, segments which are not genes (Chapter 8). Thirdly, a gene can be split into several separate segments. Still, reliable, although incomplete, annotations are now available for many genomes.

Double-stranded DNA is itself a rather chemically passive molecule, which primarily serves as a repository for hereditary information and interacts with other molecules. Information encoded by DNA directs synthesis of other, more active, molecules. The first step along this way is copying of some DNA segments into RNA (ribonucleic acid) molecules, a process called transcription. DNA and RNA are close relatives, both being nucleic acids. There are three differences between them: (i) an RNA strand consists of phosphates alternating with riboses, instead of deoxyriboses (a ribose contains an extra oxygen atom, which does not matter much for us); (ii) RNA uses, instead of T, a related base U (uracil) (again, here this does not make much difference); and (iii) in living cells, RNA molecules are not double-stranded but single-stranded, although different parts of a single-stranded RNA with approximately complementary sequences may interact and form hairpin-like secondary structures with double-stranded stems and single-stranded loops (Figure 1.7). Chemically, however, double-stranded RNA molecules, as well as single-stranded DNA molecules, are perfectly possible and are both present in some viruses.

With some exceptions, cells do not perform RNA replication (which is essential for many viruses), so that almost all RNA molecules that are present in cells are produced by transcription. A DNA segment from which an RNA molecule is transcribed is called a coding gene. An RNA molecule is transcribed using one strand of double-stranded DNA as a template, with the same rules of complementarity as in the case of DNA replication. Transcription always starts from a 3' end of a template DNA strand, so that the nascent RNA molecule grows starting from its 5' end. Obviously, the

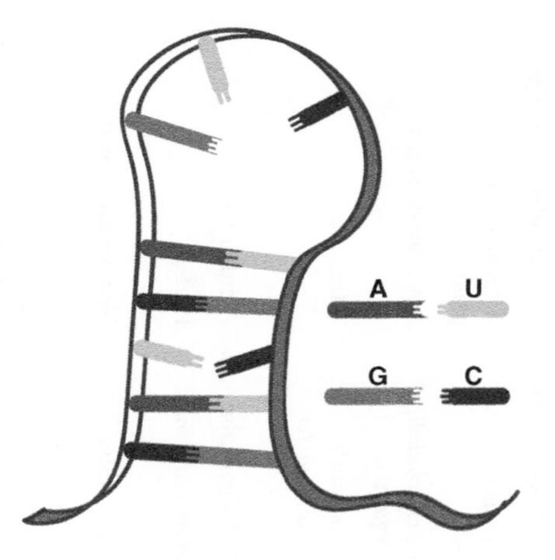

Figure 1.7 A single-stranded RNA molecule which forms a hairpin secondary structure, due to local self-complementarity.

resulting RNA molecule (a transcript) has the same sequence as the other DNA strand of the coding gene, if we equate U to T (Figure 1.8). In fact, the first sequence of a nucleic acid, determined in 1965, was that of an RNA molecule, called alanine tRNA, and the sequence of the gene which encodes it was inferred using the rules of complementarity.

Within a long chromosome, different coding genes may use segments of both DNA strands as templates (Figure 1.9). Together, coding genes cover over 90% of lean genomes, and over 50% of many bloated genomes. In eukaryotes, some encoded RNA molecules are over 1 million nucleotides long, but usually they are much shorter.

A freshly transcribed RNA molecule may undergo a process called splicing. In the course of splicing, some segments, called introns, are removed from the molecule and the remaining segments, called exons, are joined, to form a mature, spliced RNA (Figure 1.10). Bacteria usually do not have introns and splicing, and why this process evolved in eukaryotes (there are on average 7 introns within a human protein-coding gene, although some of genes are intronless) is still not clear. Often, different segments are removed from the same RNA molecule, for example, in different tissues. This phenomenon, called alternative splicing, allows one coding gene to encode two or more varieties of mature RNA.

RNA molecules can do a lot of things themselves. They constitute key parts of the machinery which performs the process of translation, to be discussed shortly, catalyze a variety of chemical reactions, regulate action of genes, and so on. Still, RNAs encoded by many coding genes act primarily as messengers, transmitting information from DNA to proteins. Such RNAs are called mRNAs, and the corresponding genes are regarded as protein-coding. Obviously, protein-coding genes represent a subset of all coding genes. There are at least ~90 000 different RNAs transcribed from the human genome that

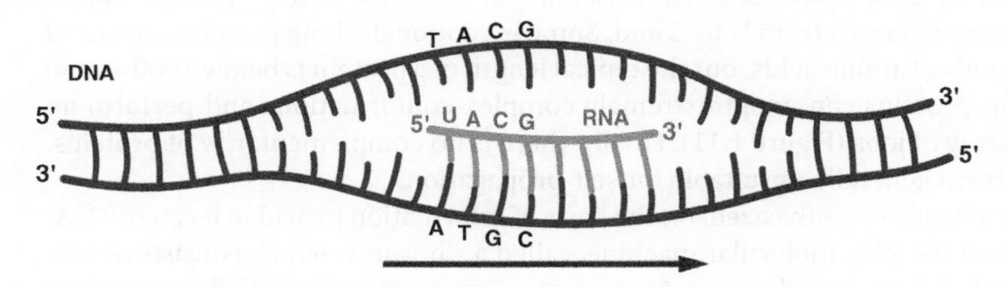

Figure 1.8 Scheme of DNA transcription.

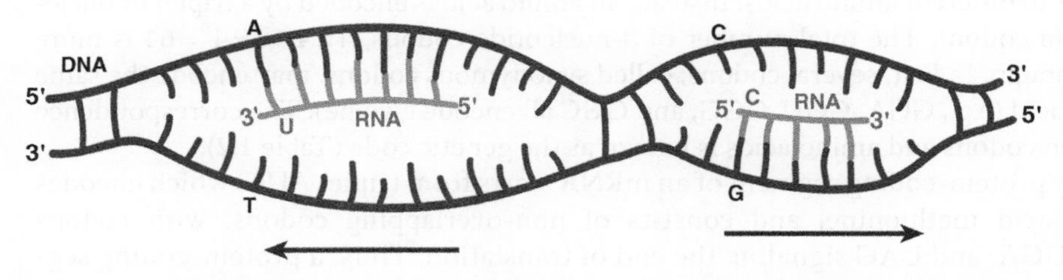

Figure 1.9 Two adjacent coding genes, arranged head-to-head, use different DNA strands as templates for transcription.

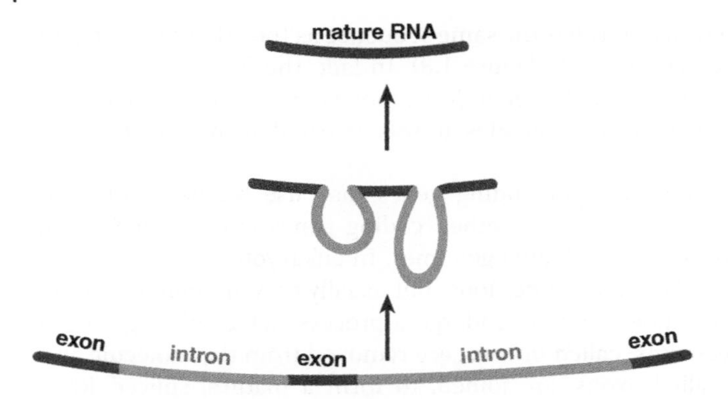

Figure 1.10 Scheme of RNA splicing.

apparently do not encode any proteins. However, it is not yet clear what proportion of them performs a function, as opposed to being produced by accident. Thus, the number of human genes that are coding, but not protein-coding, remains enigmatic.

Proteins is the other, on top of nucleic acids, key class of linear biopolymers. Proteins differ from nucleic acids in several ways. First, a protein strand is chemically very different from a nucleic acid strand, and consists of amino acids connected to each other by what is called a peptide bond. Thus, proteins are also called polypeptides. A protein strand, like a nucleic acid strand, has an orientation, having N (short for NH_2) and C (short for COOH) ends. Secondly, proteins use a 20-letter alphabet: residues of 20 different amino acids can be attached at the side of a polypeptide strand. For our purposes, there is no need to memorize the names of all these amino acids, which are usually denoted either by their three-letter abbreviations or by single letters. Traditionally, a protein sequence is shown from N to C end. Some exceptionally long proteins consist of tens of thousands of amino acids, but the typical length of a protein is below 1000 amino acids. Thirdly, proteins can adopt extremely complex conformations and perform an astonishing array of jobs (Figure 1.11). Finally, there is no complementarity of proteins, which make them generally unsuitable for self-propagation.

A protein molecule is synthesized, on the basis of information provided by an mRNA, by an incredibly complex molecular machine, called a ribosome, which consists of several ribosomal RNAs and dozens of proteins. This process is called translation. Obviously, not only 4 individual nucleotides, but even all their $4 \times 4 = 16$ possible pairs (AA, AU, AG, AC, UA, UU, UG, UC, GA, GU, GG, GC, CA, CU, CG, CC) cannot encode 20 different amino acids. Instead, an amino acid is encoded by a triplet of nucleotides (a codon). The total number of 3-nucleotide codons, $4 \times 4 \times 4 = 4^3 = 64$ is more than enough. In fact, several codons, called synonymous codons, may encode the same amino acid (e.g., GGA, GGU, GGG, and GGC all encode glycine). The correspondence between codons and amino acids is known as the genetic code (Table 1.2).

Every protein-coding segment of an mRNA starts from triplet AUG, which encodes amino acid methionine, and consists of non-overlapping codons, with codons UAA, UGA, and UAG signaling the end of translation. Thus, a protein-coding segment is three times longer that the protein it encodes. In eukaryotes, each mRNA usually contains just one protein-coding segment, flanked by a 5'UTR and a 3'UTR

(a)

```
glycine-arginine-proline-cysteine-asparagine-glutamine-
phenylalanine-tyrosine-cysteine
```

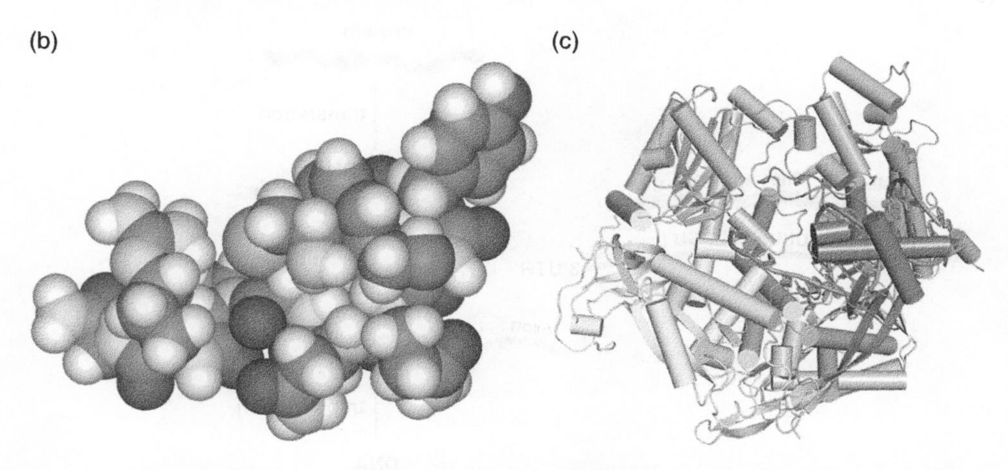

Figure 1.11 (a) An amino acid sequence of a very short protein, human hormone vasopressin, which, among other things, is involved in regulation of blood pressure. (b) A model of the spatial structure of vasopressin. (c) A scheme of the spatial structure of DNA polymerase ε, which plays a key role in DNA replication (see Chapter 5). https://en.wikipedia.org/wiki/Vasopressin#/media/File:Arginine_vasopressin3d.png. http://journals.plos.org/plosone/article?id=10.1371/journal.pone.0094835.

Table 1.2 Genetic code.

	Second nucleotide								
	U		C		A		G		
First nucleotide	Codon	Amino acid	Codon	Amino acid	Codon	Amino acid	Codon	Amino acid	Third nucleotide
U	UUU	Phe	UCU	Ser	UAU	Tyr	UGU	Cys	U
	UUC	Phe	UCC	Ser	UAC	Tyr	UGC	Cys	C
	UUA	Leu	UCA	Ser	UAA	Stop	UGA	Stop	A
	UUG	Leu	UCG	Ser	UAG	Stop	UGG	Trp	G
C	CUU	Leu	CCU	Pro	CAU	His	CGU	Arg	U
	CUC	Leu	CCC	Pro	CAC	His	CGC	Arg	C
	CUA	Leu	CCA	Pro	CAA	Gln	CGA	Arg	A
	CUG	Leu	CCG	Pro	CAG	Gln	CGG	Arg	G
A	AUU	Ile	ACU	Thr	AAU	Asn	AGU	Ser	U
	AUC	Ile	ACC	Thr	AAC	Asn	AGC	Ser	C
	AUA	Ile	ACA	Thr	AAA	Lys	AGA	Arg	A
	AUG	Met	ACG	Thr	AAG	Lys	AGG	Arg	G
G	GUU	Val	GCU	Ala	GAU	Asp	GGU	Gly	U
	GUC	Val	GCC	Ala	GAC	Asp	GGC	Gly	C
	GUA	Val	GCA	Ala	GAA	Glu	GGA	Gly	A
	GUG	Val	GCG	Ala	GAG	Glu	GGG	Gly	G

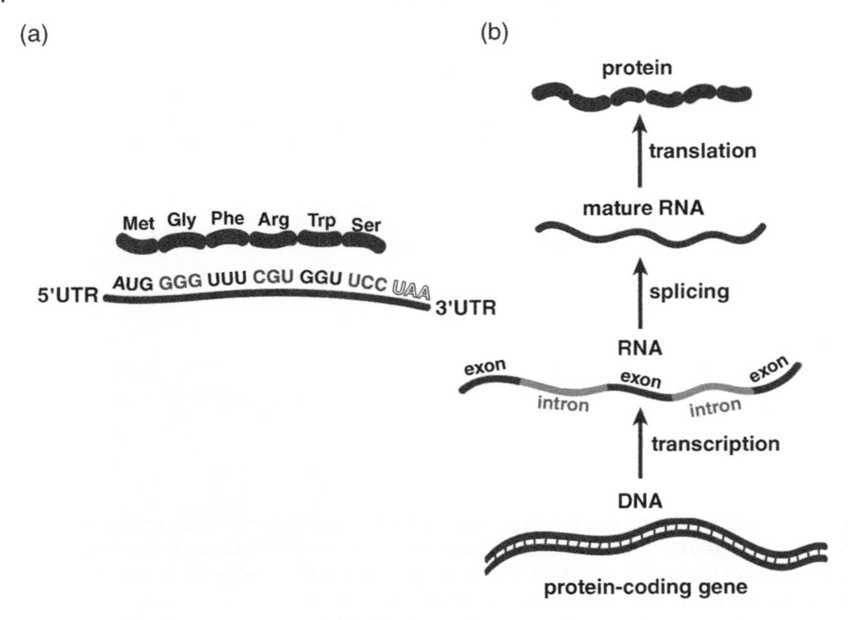

Figure 1.12 (a) Translation of a mature mRNA. (b) Relationship between a protein-coding gene and the encoded protein.

(5'- and 3'- untranslated regions). Still, it is convenient to regard the whole gene which encodes an mRNA as a protein-coding gene, although only a part of it (exons minus UTRs) is really encoding the protein (Figure 1.12).

In eukaryotes, coding genes rarely overlap and, instead, are separated by segments of non-coding DNA. Many of these segments are far from being functionally mute. In particular, non-coding genome segments contain many sites to which DNA-binding proteins can become attached. The most important function of these proteins is to regulate transcription (expression) of coding genes. Sites of attachment of DNA-binding proteins can be regarded as non-coding genes. Regulation of gene expression is particularly complex in multicellular organisms, because cells of, say, brain, heart, liver, and kidney do rather different things and, thus, need to produce different proteins.

Still, a sizeable proportion of a bloated genome consists of segments of junk DNA, untranscribed or transcribed only accidentally, which do essentially nothing, so that any changes of their sequences have (almost) no effect on the cell and the organism. One kind of junk segments, called pseudogenes, represents broken copies of protein-coding genes (see Chapter 7). In plants, sizes of the genomes of closely related species can be different by a factor of 20, due to vastly different proportions of junk DNA. By contrast, genome sizes of mammals are rather uniform. It seems that one factor which can cause a genome to get bloated is inefficient natural selection against mildly detrimental junk DNA (see Chapter 8).

1.4 Cells, Mitosis, and Meiosis

Life consists of cells, small compartments surrounded by membranes. When a cell divides, each of the two daughter cells usually receives exactly the same genotype as that of the mother cell, save new mutations (see Chapter 4). In order to achieve this,

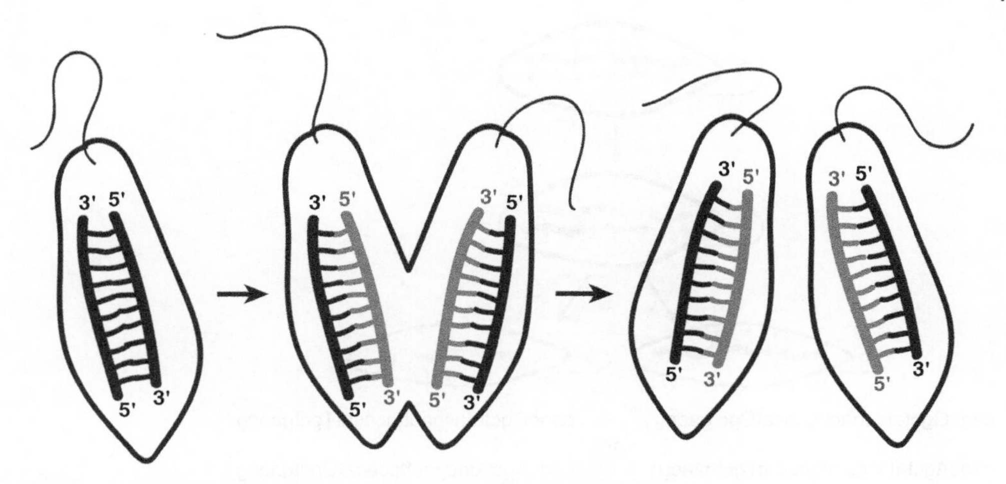

Figure 1.13 Coordination between DNA replication and cell division.

DNA replication must occur before every cell division (Figure 1.13). Although the genetic outcome of a cell division is the same in prokaryotes and eukaryotes, this process is called mitosis only in eukaryotes. Mitosis provides the mechanism both for building of the body of a multicellular organism and for asexual reproduction, in which case one or both daughter cells start a new organism. Fascinating cellular particulars of mitosis – how a membrane that separates the nucleus from the rest of the cell disappears and reappears, how sister DNA molecules, produced by premitotic DNA replication, are pulled into different daughter nuclei, and so on – are not essential for us.

On top of mitosis, another kind of cell division, meiosis, is needed in sexually reproducing eukaryotes. A zygote contains two genotypes, maternal and paternal, which were contributed by the two gametes which produced the zygote through fertilization. Thus, the opposite process leading to two-fold reduction of the amount of DNA in a cell is needed, and this process is meiosis. The easiest way to achieve this reduction would be to simply have a cell division not preceded by DNA replication. There is controversial evidence on whether this process, called one-step meiosis, exists in some unicellular eukaryotes. Still, in a vast majority, if not in all, eukaryotes meiosis begins with pre-cell division DNA replication, followed by two cell divisions. A likely reason for this unnecessary complication is that meiosis evolved from mitosis, and getting rid of pre-cell division DNA replication would require its radical redesign, which evolution could not achieve (see Chapter 10). Still, for us, the two-step mechanism of meiosis is of little importance.

By contrast, it is very important that, on top of the reduction, meiosis also involves recombination between maternal and paternal genotypes (Figure 1.14). Meiotic recombination is called reciprocal, because it consists, as a good approximation, of unbiased reassortment of the corresponding segments of the maternal and paternal genotypes, without any gains or losses. Two processes contribute to meiotic recombination. First, different chromosomes are transmitted independently in meiosis, leading to recombination at the level of chromosomes. Secondly, recombination within a chromosome, called crossing-over, which involves physical reassortment of pieces of maternal and paternal DNA (see Chaper 6) can occur at multiple locations.

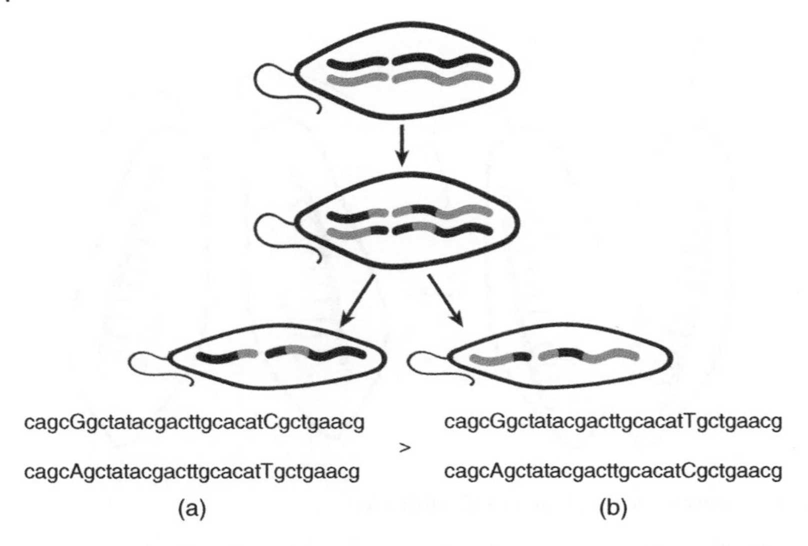

Figure 1.14 (a) Hypothetical one-step meiosis with crossing-over. Maternal and paternal genotypes, each consisting of two chromosomes, each made of double-stranded DNA, are in black and grey, respectively. (b) Segments of maternal and paternal genotypes (left), and segments of two genotypes obtained as a result of their reciprocal recombination (right) (differences between the two genotypes are in upper case).

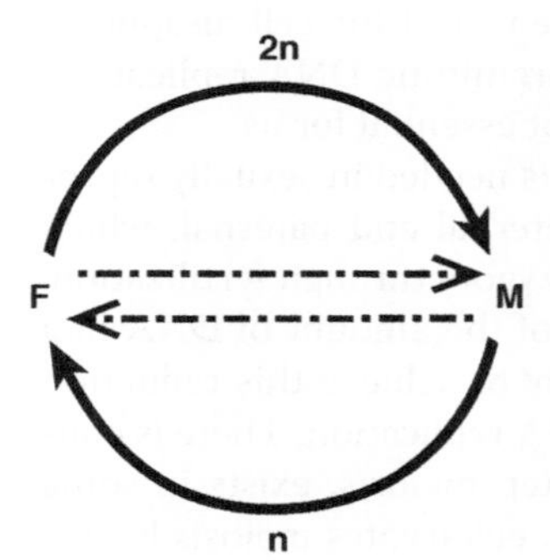

Figure 1.15 Sexual life cycle. Haploid and diploid phases, separated from each other by meiosis (M) and fertilization (F), are shown by n and 2n. Reduction of a phase is shown by a dotted line.

Alternation of fertilization and meiosis constitutes the sexual life cycle, which consists of the haploid (from meiosis to fertilization) and the diploid (from fertilization to meiosis) phases (Figure 1.15). In humans, all mammals, and, more generally, in almost all animals the haploid phase is reduced, in the sense that haploid cells produced by meiosis immediately act as gametes, and there are no haploid mitoses. Thus, we are diploid: an individual possesses two genotypes, maternal and paternal, and two copies of (almost) every gene. The only systematic exception are sex and mitochondrial chromosomes. While a female possesses two large X chromosomes, a male possesses an X chromosome and a much smaller Y chromosome, and, thus, carries only one copy of most genes located on the X chromosome, which are called sex-linked genes. Several mitochondrial chromosomes, sometimes with slightly different sequences, are transmitted to an offspring from the mother.

Diploidy, in a sense, is similar to two-strandness of DNA: it persists, after fertilization, because maternal and paternal genotypes stay together. However, diploidy is not the only option. For example, the sexual life cycle of a multicellular green alga *Chara* is a mirror image of ours: the diploid phase is reduced, because meiosis occurs in a zygote, so that there are no diploid mitoses, and multicellular organisms develop in the haploid phase, through haploid mitoses. Intermediate situations, where neither phase is reduced, are also common in plants and fungi. Still, throughout the book I will assume that the haploid phase is reduced.

Sex is not necessary for reproduction, even for multicellular animals. A group of them, called bdelloid rotifers, has reproduced only asexually for over 100 million of years, producing eggs by mitosis, and is doing very well. Similarly, crossing-over is not needed for meiosis: males of the fruit flies *Drosophila* do not have crossing-over but have no problems with producing sperm. We still do not know why sex and meiotic recombination evolved early in the history of eukaryotes and mostly persist since then. Why reshuffle successful genotypes (if they were not successful, their owners would not be propagating) and produce something new and unpredictable? What makes genotypes which served parents well undesirable for offspring? Over 20 hypotheses aimed at answering these questions have been proposed, and some of them claim that sex and recombination evolved because they make selection against deleterious mutations more efficient. However, neither of these hypotheses is yet universally accepted (see Chapter 11).

1.5 From Genotype to Phenotype

A genotype contains instructions, which need to be implemented. Indeed, an organism is much more than its genotype: we are, first of all, countless molecules of interacting proteins, blood vessels and nerves, hearts, hands, eyes, brains, movements, thoughts, feelings, and so on. All the features of an organism on top of its genotype are collectively called phenotype. It is convenient to think of at least four levels of organization of phenotypes, that of molecules, cells, organs, and multicellular organisms (Figure 1.16).

How the phenotype of an organism develops, on the basis of information written in its genotype, is one of the deepest mysteries of life. Despite the astonishing recent progress of developmental biology, the relationship between genotypes and phenotype is so complex that we are still very far from complete understanding of the development of any facet of any phenotype. Nevertheless, some basic properties of this relationship are already clear.

First, in all living beings there is always some continuity of phenotypes, on top of continuity of their genotypes. Indeed, only viruses can propagate by injecting their genotypes – and nothing else – into the host cell. By contrast, a dividing cell always provides its daughter cells with membranes, RNA molecules, various proteins, ribosomes, and so on. Still, when a multicellular organism reproduces sexually, no features of phenotypes above the cellular level are transmitted to an offspring – instead, they must develop *de novo*, on the basis of the genotype and the phenotype of the egg (Figure 1.17).

Secondly, emergence of spatial structures of phenotypes from essentially linear DNA texts is based on self-organization – indeed, this is the only available option (Figure 1.18).

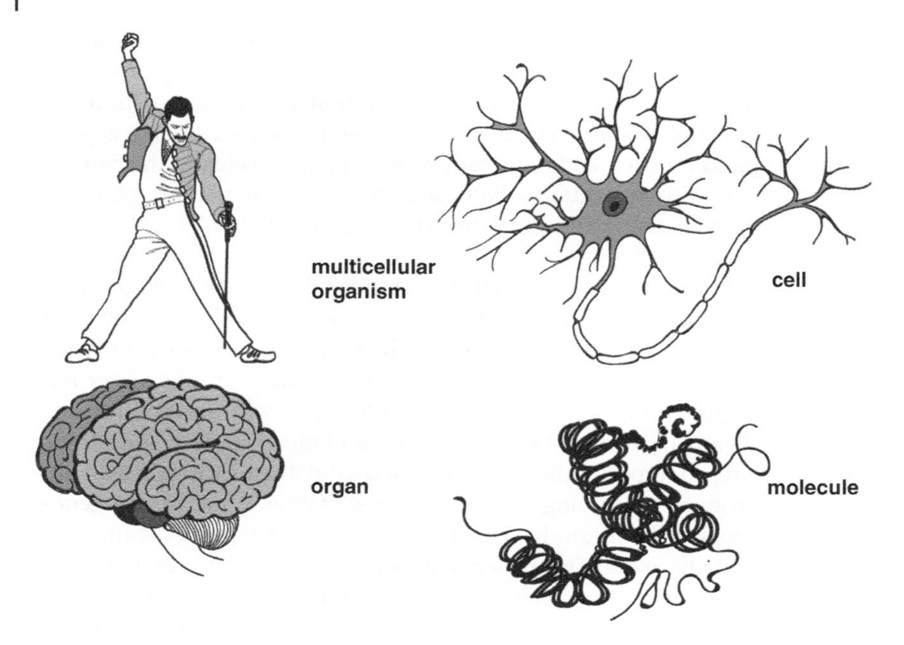

Figure 1.16 Levels of organization.

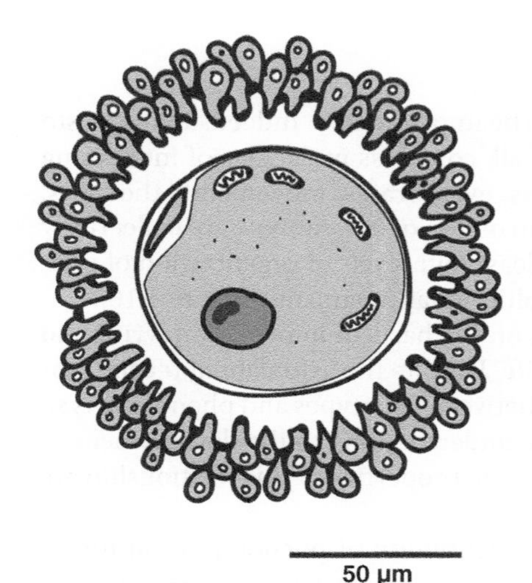

Figure 1.17 A human fertilized egg – this is what is transmitted from parents to offspring, not a naked DNA – and a human, or identical twins, have to develop from this entity.

RNA and protein molecules effectively create themselves, as 3D entities, by folding into the proper conformations, informed by their sequence. Cell membranes are capable of self-assembly from small molecules. Even multicellular organisms perform their own ontogenesis (development), which involves incredibly complex, coordinated processes in all parts of a developing embryo.

Thirdly, there is an essentially one-to-one correspondence between functional parts of the genotype, genes, and parts of the phenotype at its lowest level, RNA and protein

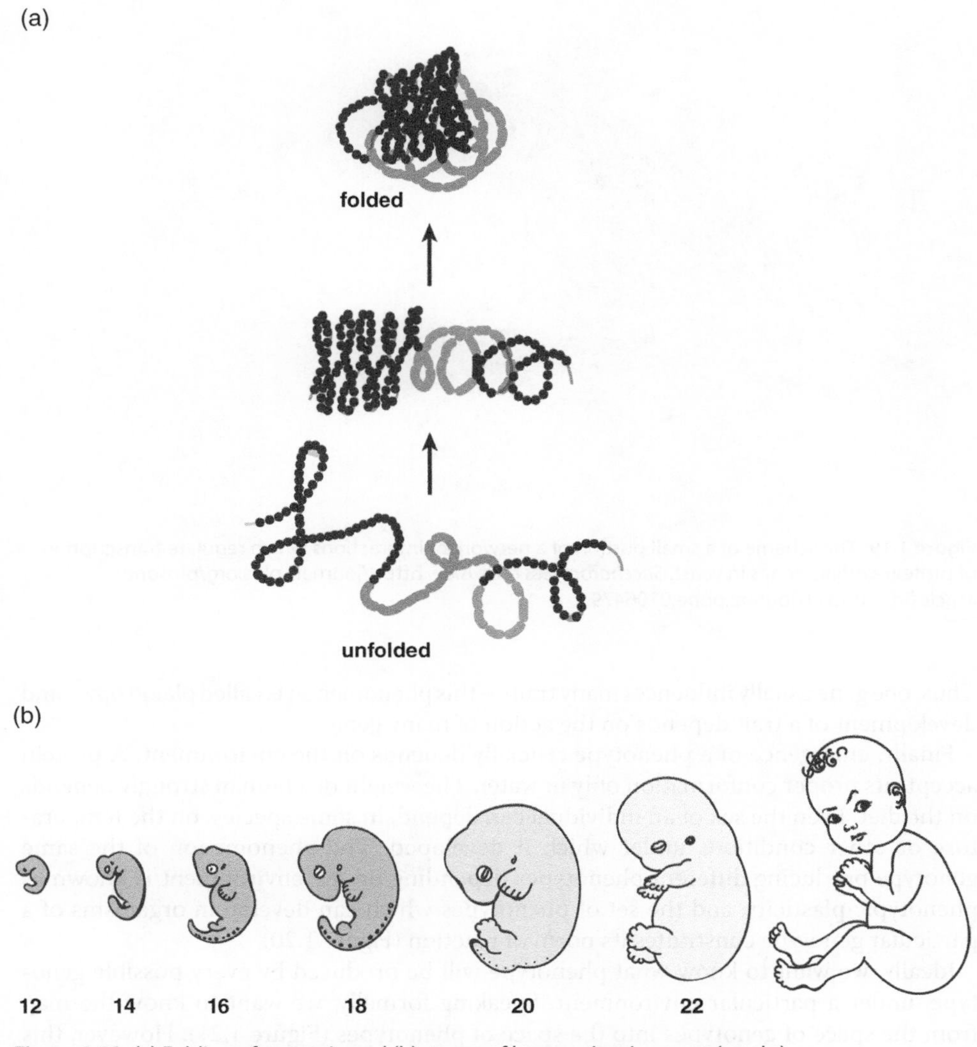

Figure 1.18 (a) Folding of a protein and (b) stages of human development (weeks).

molecules. By contrast, there is no such correspondence between genes and parts of higher level phenotypes. In fact, subdividing a higher level phenotype into parts is even more arbitrary than subdividing a genotype into genes. We may want to subdivide the phenotype into traits, for the purpose of describing it (see Chapter 3), or into adaptations, for the purpose of understanding how it works (see Chapter 7). In both cases, we strive to do this in such a way that traits and adaptations that we define vary and function as independently as possible. Still, no matter how we carve traits and adaptations out of a high-level phenotype, relationships between them and genes are complex and tangled. This is not surprising, because cells, organs, and multicellular organisms are products of astonishingly complex interactions between genes and proteins (Figure 1.19).

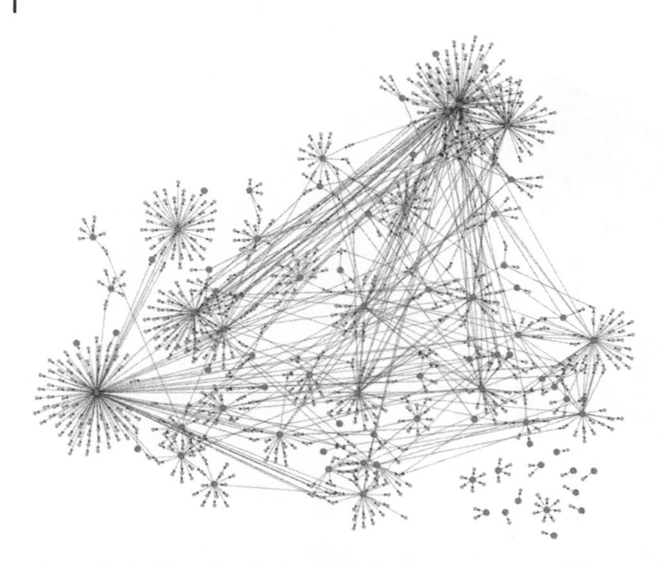

Figure 1.19 The scheme of a small portion of a network of interactions which regulate transcription of protein-coding genes in yeast, *Saccharomyces cerevisiae*. http://journals.plos.org/plosone/article?id=10.1371/journal.pone.0106479.

Thus, one gene usually influences many traits – this phenomenon is called pleiotropy – and development of a trait depends on the action of many genes.

Finally, emergence of a phenotype crucially depends on the environment. A protein accepts its proper conformation only in water. The weight of a human strongly depends on the diet. Even the sex of an individual can depend, in some species, on the temperature or other conditions under which it developed. The phenomenon of the same genotype producing different phenotypes depending on the environment is known as phenotypic plasticity, and the set of phenotypes which can develop in organisms of a particular genotype constitutes its norm of reaction (Figure 1.20).

Ideally, we want to know what phenotype will be produced by every possible genotype, under a particular environment. Speaking formally, we want to know the map from the space of genotypes into the space of phenotypes (Figure 1.21). However, this goal is too ambitious for now: there are way too many possible genotypes to study them all experimentally, and inferring theoretically the phenotype produced by a genotype is currently out of the question. Instead, a less ambitious, although still very difficult, goal – to understand how variation of existing genotypes affect the corresponding phenotypes – makes more sense. Information on this is provided by comparison of genomes and phenotypes of different species, and of genotypes and phenotypes of individuals of the same species.

Two complementary – and contradictory – patterns characterize how the genotype and the environment determine the phenotype of an individual. On the one hand, phenotypes are often quite vulnerable to both genetic and environmental insults. When replacement of a single nucleotide within a human gene *RPSA* renders the ribosomal protein SA, encoded by this gene, non-functional, this results in the absence of spleen at birth and, often, in early death due to severe bacterial infections. About 1 μg of the

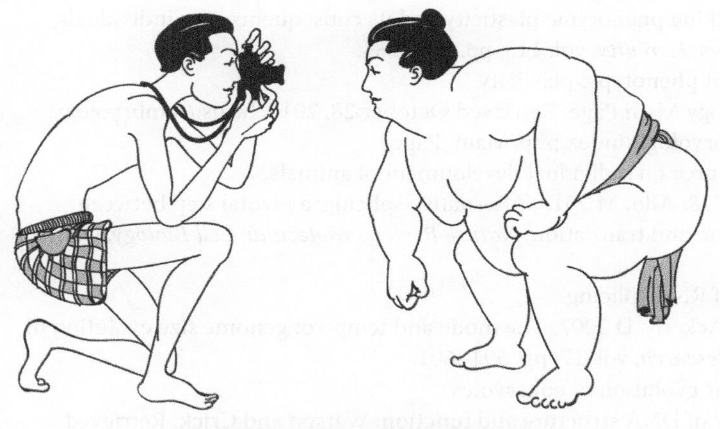

Figure 1.20 Norm of reaction: different phenotypes can develop on the basis of the same genotype.

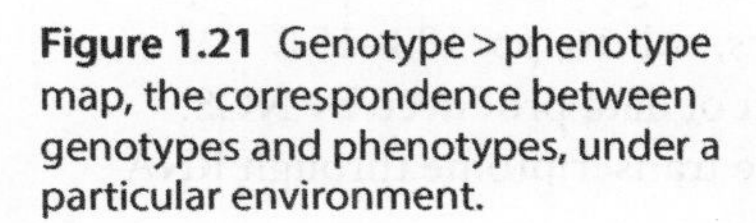

Figure 1.21 Genotype > phenotype map, the correspondence between genotypes and phenotypes, under a particular environment.

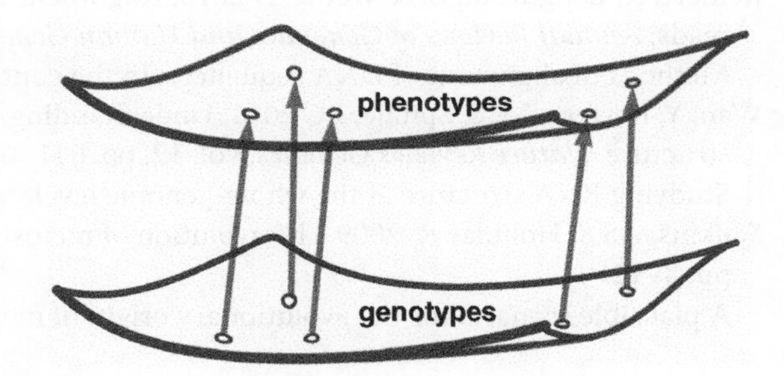

botulinum toxin, which prevents release of neurotransmitter acetyl choline from the presynaptic membrane, is lethal to a human.

On the other hand, phenotypes can also be very resilient. Every "normal" human being carries thousands of mildly deleterious defects in its genotype and, nevertheless, functions at an acceptable level. Every organism can maintain homeostasis – the near constancy of some key parameters, such as body temperature and blood pH – despite wide fluctuations of the environment. In a sense, the Main Concern is about an interplay between vulnerability and resilience of human phenotypes.

Further Reading

Barabási A-L, Zoltán N, & Oltvai, ZN 2004, Network biology: understanding the cell's functional organization, *Nature Reviews Genetics*, vol. 5, pp. 101–113.
On the crucial role of networks in cell functioning.

Fersht, AR 2008, From the first protein structures to our current knowledge of protein folding: delights and scepticisms, *Nature Reviews Molecular Cell Biology*, vol. 9, pp. 650–654.
On protein structure and its self-assembly.

Forsman, A 2015, Rethinking phenotypic plasticity and its consequences for individuals, populations and species, *Heredity*, vol. 115, pp. 276–284.
On the phenomenon of phenotypic plasticity.

Hill, MA 2016, Embryology Main Page, Retrieved October 28, 2016, https://embryology.med.unsw.edu.au/embryology/index.php/Main_Page.
A comprehensive resource on individual development of animals.

Kornblihtt, AR, Schor, IE, & Allò, M 2013, Alternative splicing: a pivotal step between eukaryotic transcription and translation, *Nature Reviews Molecular Cell Biology*, vol. 14, pp. 153–165.
On the complexities of RNA splicing.

Oliver, MJ, Petrov, D, & Ackerly, D 2007, The mode and tempo of genome size evolution in eukaryotes, *Genome Research*, vol. 17, pp. 594–601.
Genome sizes and their evolution in eukaryotes.

Pray, LA 2008, Discovery of DNA structure and function: Watson and Crick, Retrieved January 6, 2017, http://www.nature.com/scitable/topicpage/discovery-of-dna-structure-and-function-watson-397.
Basics of DNA and of the history of its discovery.

Reinert, K, Langmead, B, & Weese, D 2015, Alignment of next-generation sequencing reads, *Annual Reviews of Genomics and Human Genetics*, vol. 16, pp. 133–151.
Methods of alignment of DNA sequences, in the context of data produced by NGS.

Wan, Y, Kertesz, M, & Spitale, RC 2011, Understanding the transcriptome through RNA structure, *Nature Reviews Genetics*, vol. 12, pp. 641–655.
Studying RNA structure at the whole-genome level.

Wilkins, AS & Holliday R, 2009 The evolution of meiosis from mitosis, *Genetics*, vol. 181, pp. 3–12.
A plausible scenario for the evolutionary origin of meiosis.

2

Mendelian Inheritance and Population Genetics

To the students of heredity the inborn errors of metabolism offer a promising field of investigation.

Archibald Garrod, 1908.

Inheritance of simple phenotypic traits demonstrates that genetic variation consists of discrete units, loci, which exist in discrete states, alleles. A diploid individual carries two alleles at each locus, but transmits only one of them, chosen randomly, to each gamete. Before reproduction, alleles at different loci recombine, making genotypes of gametes produced by an individual different from genotypes of the two gametes from which it originated. All populations are genetically variable, mostly due to the presence of diallelic polymorphic loci, where new rare alleles are often deleterious. Polymorphic loci can affect genes and their functions in a wide variety of ways. Effects of individual alleles on higher level phenotypes can be drastic or only mild, and many alleles are phenotypically mute. In humans, there are many hundreds of Mendelian diseases, each caused by drastic alleles that affect a particular gene.

2.1 Inheritance is Discrete

Texts consist of letters. Thus, any text, including a DNA sequence, is a discrete thing: it cannot change continuously, like a plasticine sculpture, but only by small leaps. Discreteness of heredity was discovered in 1866, almost a century before its molecular nature, by the Catholic monk Gregor Mendel. At that time, there were no tools to study how inherited information is processed within an organism. Thus, Mendel, who was interested in the phenomenon of heredity and experimented on garden peas, studied how states of discrete traits are transmitted from generation to generation. Back then, the prevailing hypothesis was that heredity is continuous and that hereditary materials from both parents blend forever within an offspring. On the basis of very simple data, Mendel refuted this hypothesis. Today, the depth of his insight may be hard to appreciate, because Mendelian thinking permeates all biology and is often taken for granted.

Let us consider a cross between parental plants (generation P, in Mendel's terminology) with red and white flowers, in which all offspring of the first generation (F_1) have pink flowers. This is consistent with continuous (blending) inheritance. If so, one would expect all offspring of the second generation (F_2), produced in crosses between F_1 plants (or by their self-fertilization, which is possible in many plants), to also be uniformly pink. However, this is not the case and, instead, if the difference between phenotypes of parents has a simple genetic basis, F_2 offspring are a mixture of individuals with white, pink, and red flowers. Within a large enough sample, these phenotypes appear in approximately 1:2:1 ratio: out of 1000 F_2 plants, there are about 250 reds, 500 pinks, and 250 whites (Figure 2.1).

This result can be easily explained if we make the following three assumptions:

1) Different states of the trait are controlled by one heritable entity (locus) A, which can exist in two states (alleles); in our case, there is a red-color allele (A) and a white-color allele (a).
2) An individual carries two alleles at a locus (flowering plants are diploid, see Chapter 1), but transmits only one of them, chosen randomly, to each offspring (gametes are haploid).
3) An individual which carries two red-color alleles, one red-color and one white-color allele, or two white-color alleles has red, pink, or white flowers, respectively.

Indeed, with these assumptions everything makes sense (Figure 2.2). The two parental plants, each of which came from a genetically uniform strain, possess allele combinations (referred to as genotypes, even if we do not know their DNA sequences) AA and aa. Thus, all F_1 offspring must have genotype Aa, because each of them obtained allele A from one parent and allele a from the other parent. When two F_1 Aa offspring are crossed, each transmits either A or a to their F_2 offspring with probability 0.5. As long as gametes fuse randomly, the probability that an F_2 offspring gets two A's, or two a's, is $0.5 \times 0.5 = 0.25$ (think or tossing a coin twice and getting two heads or two tails). An Aa F_2 offspring can be obtained in two ways: when A comes from the mother and a from the father, or vice versa, with the overall probability $0.25 + 0.25 = 0.5$ (heads followed by tails or tails followed by heads). This process of random reassortment of alleles at one locus, in the course of their transmission from parents to offspring, is now called Mendelian segregation. Mendel called individuals possessing identical and different alleles at a locus homozygous and heterozygous, respectively. Thus, we can say that in F_2 about 1/4 of all offspring are homozygotes of each of the two possible kinds, and 1/2 are heterozygotes.

In fact, Mendel had to deal with a complication, the phenomenon of dominance, which we have so far ignored. The 1:2:1 ratio of phenotypes in the F_2 offspring appears if heterozygotes Aa have a phenotype different from those of both homozygotes, AA and aa. When this is the case, we say that dominance is absent (or intermediate, or incomplete). However, sometimes Aa individuals are phenotypically indistinguishable from AA (or aa) individuals, in which case we say that allele A (or allele a) is (completely) dominant and the other allele is (completely) recessive. In Mendel's experiment, A was dominant, so that in the first generation all offspring were red, and in the second generation a 3:1 red to white ratio was observed (Figure 2.3). Obviously, dominance at the level of phenotypes obscures Mendelian segregation at the level of genotypes but, nevertheless, Mendel was able to see through it. In fact, dominance is only rarely complete, and a careful examination often makes it possible to distinguish a heterozygote from both homozygotes on the basis of their phenotypes. Of course, these days a heterozygote at a particular locus can always be recognized by sequencing its genotype.

Mendel also performed crosses between parents that differed from each other by two discrete traits (e.g., color of flowers and shape of seeds), and observed their

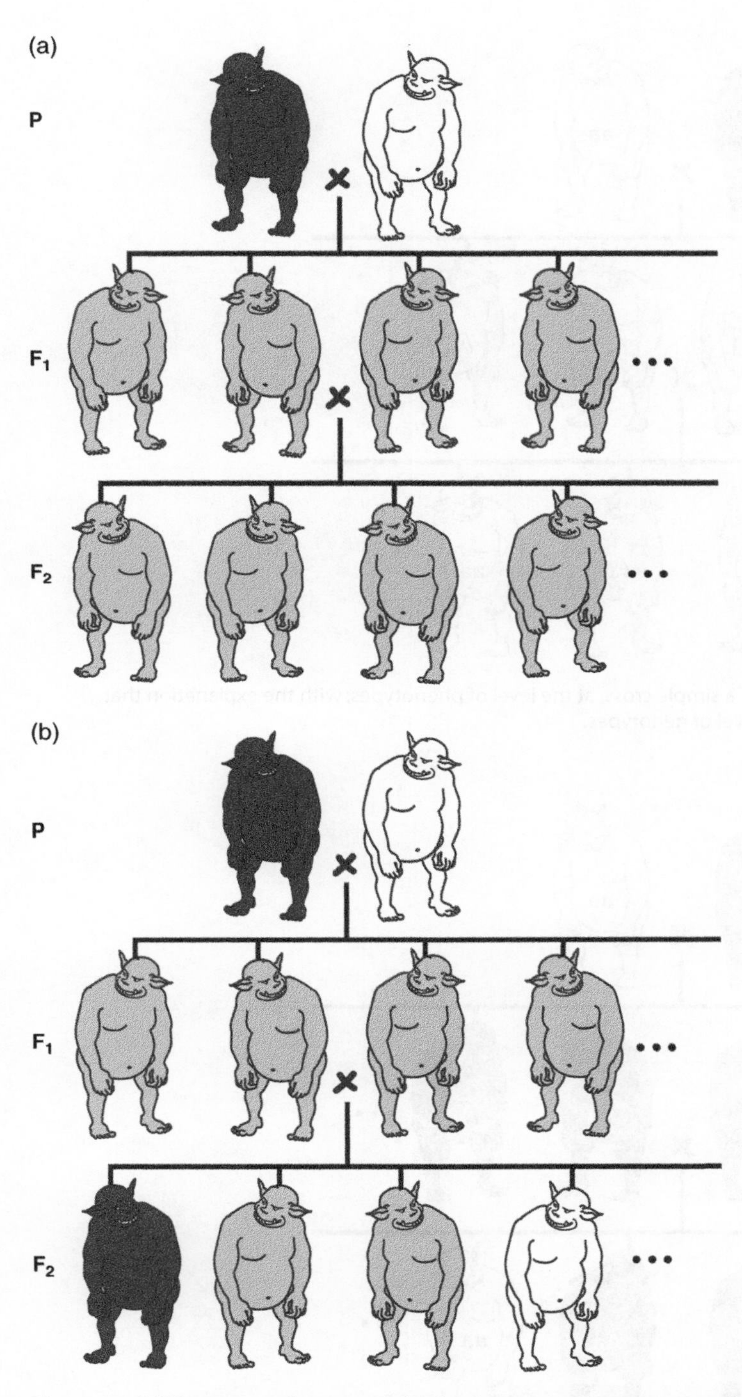

Figure 2.1 (a) The expected outcome of a simple cross between parents with two contrasting phenotypes, if inheritance were continuous. (b) The actual outcome of such a cross in the simplest case.

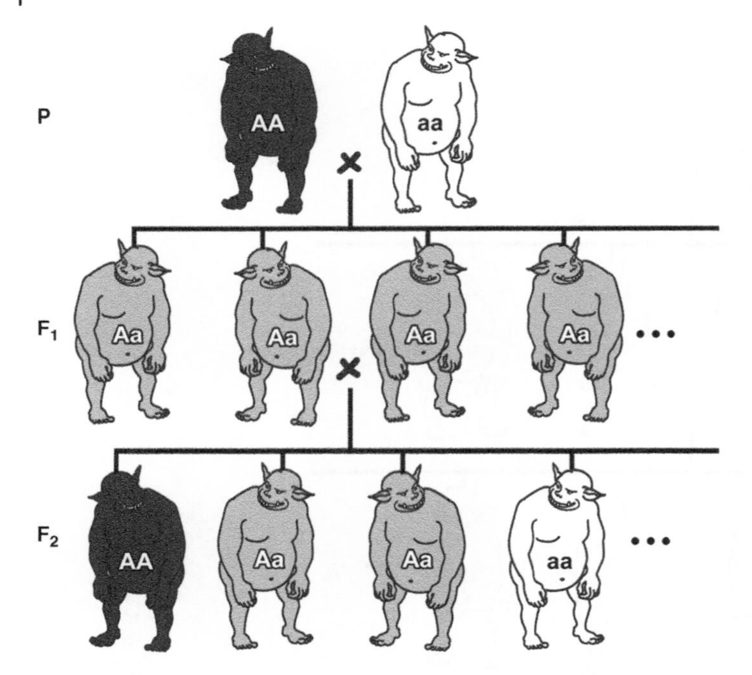

Figure 2.2 The outcome of a simple cross, at the level of phenotypes, with the explanation that Mendel proposed, at the level of genotypes.

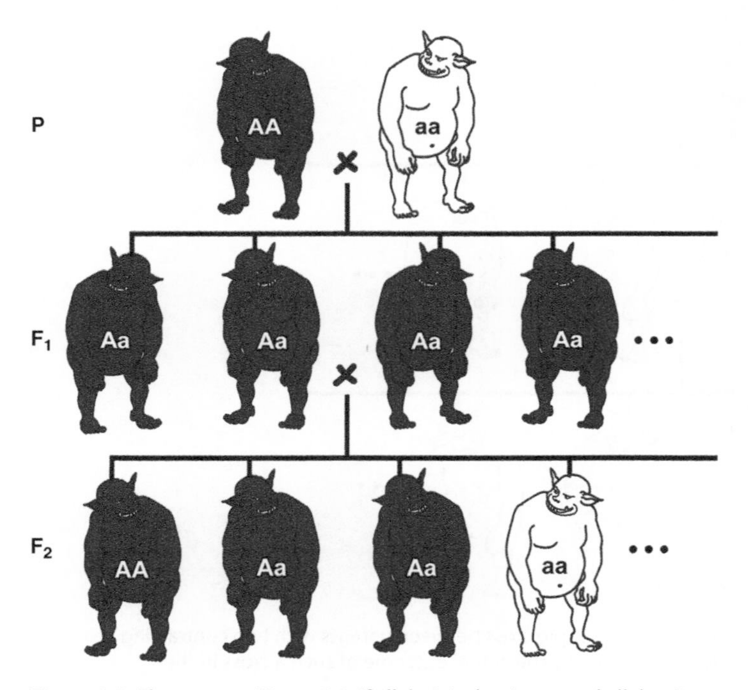

Figure 2.3 The same as Figure 2.2, if allele A is dominant and allele a is recessive.

independent transmission. This phenomenon results from free meiotic recombination between two loci, each of which is responsible for variation of one trait. Recombination is free when the two loci are situated either on different chromosomes or far enough from each other on the same chromosome, so that crossing-over between them is very likely (see Chapter 1). When an individual produces gametes, there is no tendency to transmit alleles at such loci, called unlinked, in the same combinations in which the individual received them from its parents. Thus, an F_1 AaBb offspring of an AABB × aabb cross (locus B, with alleles B and b, is responsible for variation of the shape of seeds), produced AB, Ab, aB, and ab gametes in equal proportions, 0.25 of each. Random assortment of these gametes leads to the following proportions of different genotypes in the F_2 offspring: 1/16 for AABB, AAbb, aaBB, and aabb, 1/8 for AABb, aaBb, AaBB, and Aabb (each of which can be obtained in two ways), and 1/4 for AaBb (which can be obtained in four ways). In the absence of dominance, the same proportions are observed among phenotypes (Figure 2.4).

The phenomenon of linkage is easier to understand if we consider analyzing, in Mendel's terms, cross of AaBb double heterozygote with the aabb double homozygote (Figure 2.5). In this case, the gamete which an offspring receives from the second parent always has genotype ab, and, if loci A and B are unlinked, all of the four possible genotypes of offspring appear at the same frequency, 0.25. By contrast, if these loci are linked, that is, do not recombine freely, there will be an excess of non-recombinant gametes (AB and ab, if the AaBb parent was produced in an AABB × aabb cross, or Ab and aB, if the AaBb parent was produced in an AAbb × aaBB cross), and of the corresponding offspring. The proportion of recombinant offspring, equal to the probability of recombination of loci A and B, is called the coefficient of recombination between them.

Mendel's work remained neglected for 35 years, and discrete alleles of discrete loci were rediscovered only in the early 20th century. By that time direct microscopic observations of meiosis demonstrated that chromosomes are transmitted from parents to offspring exactly as hypothetical Mendelian alleles. In 1909, Thomas Morgan and his students began their epic work on the fruit fly *Drosophila melanogaster*. Soon they discovered linkage, and showed that all loci in *D. melanogaster* can be subdivided into three linkage groups, such that only loci that belong to the same group can be linked to each other. Because the haploid genome of *D. melanogaster* consists of three chromosomes (not counting the tiny fourth nuclear chromosome and mitochondrial chromosomes), this result firmly established the connection between chromosomes and Mendelian loci.

2.2 Populations are Genetically Variable

Within a sexually reproducing species, every individual belongs to a group of many interbreeding individuals, a population. Such a group can consist of any number of individuals, from ~1000 (smaller populations soon go extinct) to many billions. Crucially, no two individuals are genetically identical to each other. The only exception is identical, or monozygotic, twins (Figure 2.6), as long as we discount postzygotic (somatic) mutations (see Chapter 4). When founders of genetics studied inheritance of simple phenotypic differences, they put to good use variation among genotypes of

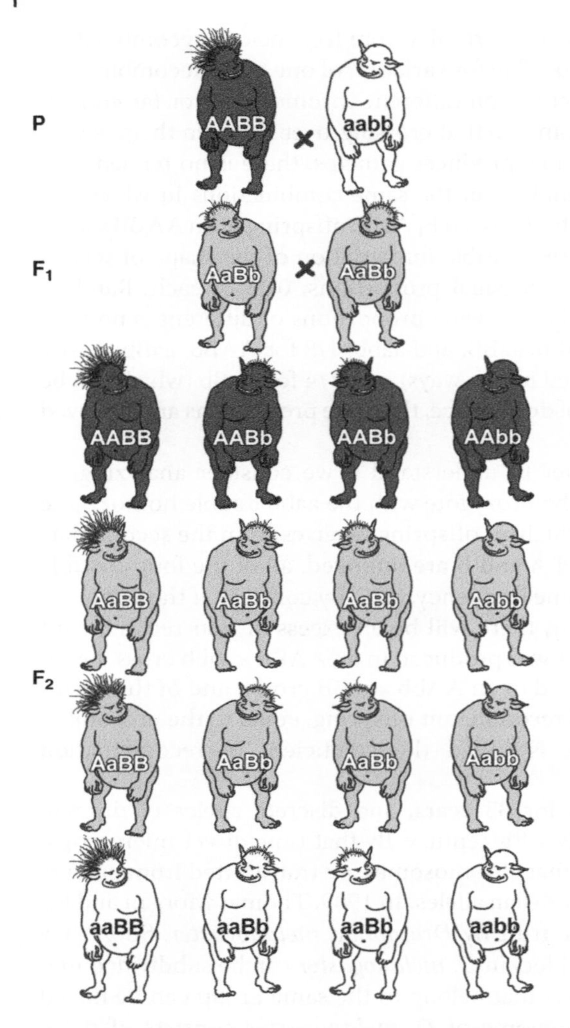

Figure 2.4 A cross between parents which differ from each other by two Mendelian traits, controlled by two unlinked loci. All 16 possible combinations of maternal and paternal gametes, each occurring with probability 1/16, are shown in F_2, although they represent only nine different genotypes, as long as we do not care if an allele came from the mother or from the father.

individuals of the same population or species. This variation is what makes evolution, both adaptive and undesirable, possible and unavoidable.

For over a century after Mendel's work, studies of genetic variation had to rely on indirect inferences from the data on variation of overt phenotypic traits. In 1926, Elena A. and Nikolay V. Timofeev-Ressovsky pioneered quantitative studies of recessive alleles with drastic effects in wild populations, using *D. melanogaster* as a model. In 1938, Theodosius G. Dobzhansky and Alfred H. Sturtevant discovered large-scale polymorphic loci in populations of another fruit fly, *D. pseudoobscura*, by taking advantage of the so-called polythene chromosomes, which are present in some tissues and are

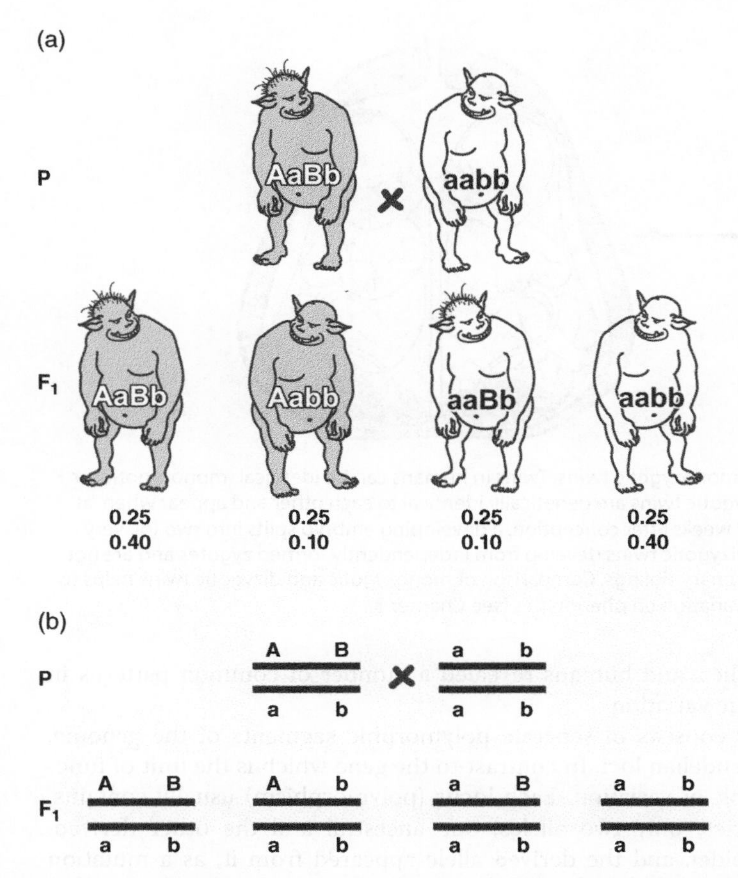

Figure 2.5 (a) Analyzing (or test) cross between AaBb (obtained from a AABB × aabb cross, so that alleles A and B, and a and b, were transmitted in the same gamete) and aabb parents produces offspring of four genotypes. The top numbers are proportions of offspring if loci A and B are unlinked, and the bottom numbers are proportions of offspring if loci A and B are linked, with recombination coefficient 0.2. (b) A diagram showing what happens if loci A and B are on the same chromosome, in which case recombinant offspring appear only if these loci recombined, due to an odd number (1, 3, 5, …) of cross-overs between them.

so large that many individual genes can be observed, as bands, with a light microscope. In 1966, Richard C. Lewontin, Jack L. Hubby, and Harry Harris studied mobility in an electric field of proteins obtained from multiple individuals from *D. pseudoobscura* and human populations and demonstrated that proteins are highly polymorphic.

Invention of efficient DNA sequencing methods in 1977 made it possible to study within-population genetic variation directly. The first such study was performed in 1983 by Martin Kreitman, who sequenced a gene which encodes a protein called alcohol dehydrogenase in 11 genotypes from a wild *D. melanogaster* population and observed 43 polymorphic loci within this gene. However, until the advent of next-generation sequencing (NGS), we could only dream of obtaining data on more-or-less complete genotypes of hundreds and thousands of individuals from the same population. After these dreams became a reality, studies of species as different as fungi,

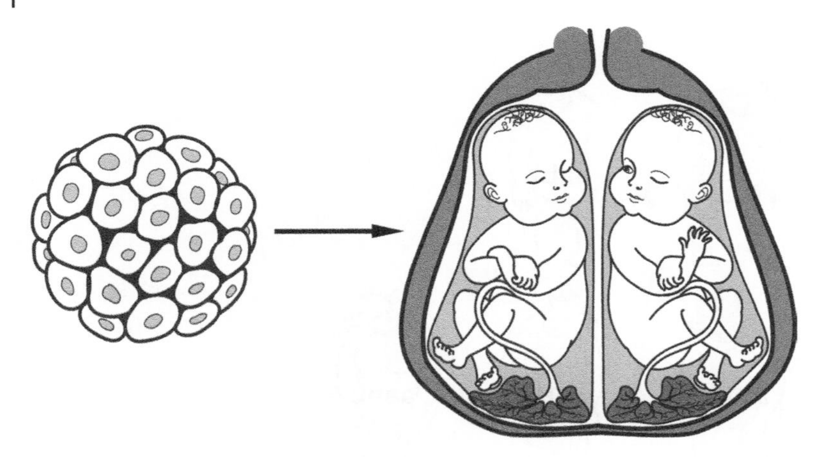

Figure 2.6 Development of monozygotic twins. Twins in humans can be identical (monozygotic) or fraternal (dizygotic). Monozygotic twins are genetically identical to each other and appear when, at some point during the first 2 weeks after conception, a developing embryo splits into two (or, very rarely, up to five) embryos. Dizygotic twins develop from independently formed zygotes and are not genetically different from ordinary siblings. Comparison of monozygotic and dizygotic twins helps to study the effects of genetic variation on phenotypes (see Chapter 3).

flowering plants, fruit flies, and humans revealed a number of common patterns in within-population genetic variation.

This variation mostly consists of separate polymorphic segments of the genome, which correspond to Mendelian loci. In contrast to the gene which is the unit of function, the locus is the unit of variation. Each locus (polymorphism) usually contains, at appreciable frequencies, only two alleles, one ancestral and the other derived. The ancestral allele is older, and the derived allele appeared from it, as a mutation (see Chapter 4), at some moment in the past. Thus, we can use data on the genome of another species, called an outgroup, which diverged from the species we study before the present variation within its populations emerged, in order to determine which allele is ancestral (Figure 2.7). Of course, an outgroup must be tightly related to the species we study, so that the ancestral allele is preserved in the outgroup with a high probability. For humans, the common chimpanzee is a suitable outgroup. On the one hand, human and chimpanzee genomes are 98.7% identical. On the other hand, almost all genetic variation, present in modern human (as well as in modern chimpanzee) populations, emerged after the lineages of these species split ~8 million years ago.

A polymorphic locus with two alleles needs to be described in several ways. First, the difference between the DNA sequences of the ancestral and the derived alleles is essential. Loci can be, somehow arbitrarily, subdivided into small-scale, in which the difference between the two alleles is confined to a DNA segment shorter than, say, 50 nucleotides, and large-scale. A large-scale polymorphism can encompass thousands or even millions of nucleotides, including a full chromosome. Still, >99% of polymorphic loci are small-scale, and can be subdivided into the following four kinds (Figure 2.8):

1) single-nucleotide substitutions (~90% of all loci);
2) short deletions (~6%);
3) short insertions (~2%);
4) more complex short polymorphisms (~1%).

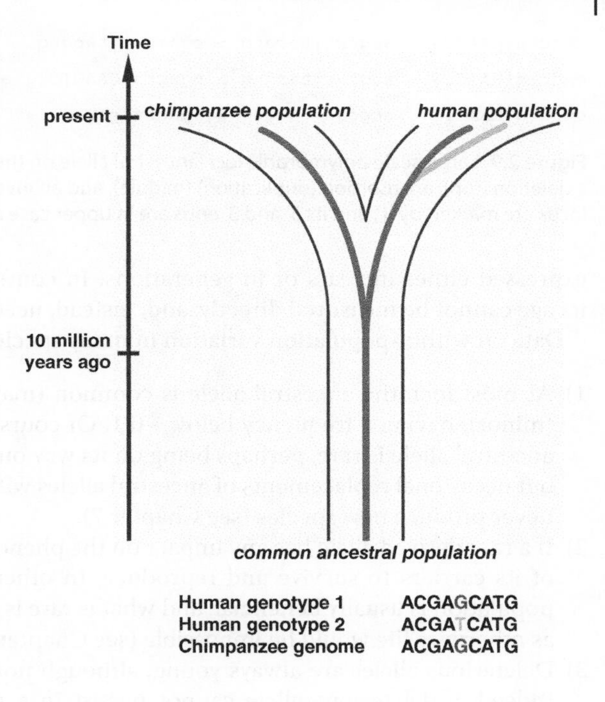

Figure 2.7 Determining which allele is ancestral and which is derived by using an outgroup. At a human polymorphic locus, the allele that is also present in the chimpanzee genome (dark grey) is likely to be ancestral and the one unique to humans (light grey) is probably derived.

```
acgtgactGcagcatctagata-----gctcagctacgactagctagaCacgtagaccgaA-tggatcag
acgtgactAcagcatctagataTATCCgctcagctacgactagctaga-acgtagaccgaCGtggatcag
```

Figure 2.8 Small-scale polymorphic loci (in upper case), revealed by comparison of two aligned genotypes (the one carrying derived alleles is on top): an A > G single-nucleotide substitution, a deletion of five nucleotides, and insertion of one nucleotide, and a complex polymorphism (CG > A).

At large-scale polymorphic loci, the most common differences between the derived and the ancestral alleles are a deletion, an insertion, and an inversion of a continuous DNA segment (Figure 2.9). A large-scale inserted sequence does not appear out of nowhere, but represents a copy, exact or imprecise, of a pre-existing sequence. In the simplest case of a tandem duplication, the inserted sequence is a copy of an adjacent DNA segment. Because the two DNA strands are antiparallel, inversion involves not only rotating the sequence, but also a strand switch, so that atgc after an inversion becomes gcat, and not cgta.

Another characteristic of a polymorphic locus is the frequency of its derived allele within the population. If a large enough sample of genotypes from the population is studied, this frequency can be determined with good precision. Of course, it takes a much smaller sample to detect an allele with frequency 0.1 than with frequency 0.001. In the first case, a sample of 100 genotypes is enough, and in the second case, we will likely miss the derived allele altogether with such a small sample, and at least ~10 000 genotypes are needed to measure such a low frequency.

Finally, a polymorphic locus can be characterized by its age, the time which lapsed from the moment when the derived allele appeared in the population, as a result of mutation (see Chapter 4) or, perhaps, of immigration. The age of an allele can be

cgacta|CAACTG...*acgtac*|tacgtg > cgacta||tacgtg

cgacta|CAACTG...*acgtac*|tacgtg > cgacta|CAACTG...*acgtac*CAACTG...acgtac|tacgtg

cgacta|CAACTG...*acgtac*|tacgtg > cgacta|*gtacgt*...CAGTTG|tacgtg

Figure 2.9 Large-scale polymorphic loci (ancestral allele on the left and derived allele on the right): a deletion (top), an insertion (duplication) (middle), and an inversion (bottom). The boundaries of a locus are marked by "|", and its 5' and 3' ends are in upper case and in italics, respectively.

expressed either in years or in generations. In contrast to the frequency of an allele, its age cannot be measured directly, and, instead, needs to be inferred (see Chapter 11).

Data on within-population variation in many species led to several generalizations:

1) At most loci, the ancestral allele is common (major) and the derived allele is rare (minor), having a frequency below ~0.1. Of course, there are exceptions, when the ancestral allele is rare, perhaps being on its way out of the population. Indeed, without occasional replacements of ancestral alleles with derived alleles, evolution would never produce new species (see Chapter 7).
2) If a rare derived allele has any impact on the phenotype, it usually impairs the ability of its carriers to survive and reproduce. In other words, what is common in the population is usually beneficial, and what is rare is deleterious. This is not surprising as otherwise life would be impossible (see Chapter 10).
3) Deleterious alleles are always young, although not all young alleles are deleterious. Indeed, a deleterious allele cannot persist in a population for a very long time, because its carriers, by definition, tend to leave fewer offspring (see Chapter 9).
4) In diploid individuals, alleles at a locus are usually distributed independently of each other. This pattern is known as the Hardy–Weinberg law and holds as long as individuals mate, and gametes fuse, independently of their genotypes at the locus. The probability of a joint occurrence of several independent events is equal to the product of their individual probabilities. Thus, if the frequency of allele A is x (so that the frequency of allele a is $1-x$), frequencies of individuals with genotypes AA, Aa and aa are x^2, $2x(1-x)$, and $(1-x)^2$, respectively. The factor of 2 in the frequency of heterozygotes appears because there are two ways of being heterozygous: one can have maternal A and paternal a, or vice versa. For us, the most important implication of the Hardy–Weinberg law is that a rare allele is almost exclusively present in heterozygous genotypes. For example, the chances that an individual receives an allele of frequency 0.01 from both mother and father is only $0.01 \times 0.01 = 0.0001$. Thus, only ~1% of copies of such an allele reside in homozygous genotypes aa, and ~99% reside in Aa heterozygotes.
5) Alleles at different loci are also usually distributed more or less independently of each other within the population, unless they are situated very close to one another in the genome and thus are tightly linked to each other (see Chapter 11).

All species harbor an enormous amount of genetic variation. In particular, the total number of human polymorphic loci at which the frequency of the derived allele is above 0.5% is over 20 million (see Chapter 11). Quantitatively, diversity of genotypes within the population can be characterized in two ways.

On the one hand, we can define locus diversity of a population H as a number of loci which are occupied by different alleles in two randomly picked genotypes, divided over the length of their alignment. For example, H estimated from the two genotypes shown

Table 2.1 Genetic diversity within populations of several species.

Species	Description	Locus diversity H	Comment
Lynx lynx	Eurasian lynx	0.0001	The lowest genetic diversity known. Mammals and other vertebrates generally possess low values of H
Homo sapiens	Human	0.001	Relative to the most of multicellular eukaryotes, we are a rather genetically uniform species
Arabidopsis thaliana	An annual flowering plant	0.005	There are flowering plants with both lower and higher values of H
Drosophila melanogaster	A fruit fly	0.02	This is a rather typical value of H for an insect
Caenorhabditis brenneri	A tiny nematode worm	0.16	In a closely related model species *C. elegans*, genetic diversity is ~50 times lower
Schizophyllum commune	A wood-decaying fungus	0.20	This species possesses the highest known value of H among eukaryotes

in Figure 2.8 is 4/70. On the other hand, we can define sequence diversity of a population H' as a number of nucleotide sites which are occupied by different nucleotides in two randomly picked genotypes, divided over the length of their alignment. For example, H' estimated from the two genotypes shown in Figure 2.8 is 9/70.

Large-scale polymorphic loci, due to their small number, make only a minor contribution to H. By contrast, because of their length, and despite their rarity, the contribution of large-scale loci to H' may exceed that of small-scale loci. For example, two human genotypes, including the maternal and the paternal genotypes of a person whose parents are not close relatives, differ from each other at ~3 000 000 individual loci, but the total number of different nucleotides in their alignment is ~12 000 000, mostly due to long gaps caused by deletions and insertions. The same is true for between-species differences: genomes of humans at chimpanzees are different at ~30 000 000 loci but at over 100 000 000 nucleotide sites.

Although general patterns in genetic variation are rather uniform in different populations and species, their quantitative features can vary widely. Table 2.1 presents data on H in a number of species. In the most diverse of them, polymorphic loci at which three or even four alleles have non-trivial frequencies are much more common than in humans. We will discuss factors that determine the level of genetic diversity of a population in Chapters 8 and 9.

2.3 Loci and Genes

As we have seen in Chapter 1, genomes are functionally heterogeneous – their different segments perform different jobs. In order to understand how within-population genetic variation affects phenotypes of individuals, we need to relate it to functional heterogeneity of the genome of the species. Here, three patterns are salient.

First, there is a contrast between evolutionary conservatism of genetic function and flexibility of genetic variation. Indeed, it makes sense to talk about functional annotation of a species genome, instead of individual genotypes, because genotypes of different individuals mostly possess the same sets of protein-coding genes and other functional segments. Between-genotype variation in the number and arrangement of genes does exist, mostly due to large-scale polymorphic loci where one allele corresponds to the presence of a gene and the other to its absence, but such variation is relatively uncommon. Moreover, even genomes of closely related species mostly possess the same sets of genes, arranged in the same order. Humans and chimpanzees share the vast majority of their ~20 000 genes: there are only ~100 human-specific (not found in chimpanzees) or chimpanzee-specific (not found in humans) genes. By contrast, even different populations within a species often possess rather different sets of polymorphic loci: what is polymorphic in one population may not be so in another. For example, African and non-African human populations possess mostly non-overlapping sets of substantially deleterious alleles (see Chapter 11), and only a very small proportion of human polymorphisms are shared with chimpanzees.

Secondly, the more important the function of a genome segment, the fewer polymorphic loci are present within this segment, and the rarer are the derived alleles at these loci. In particular, protein-coding genes are generally less variable than intergenic regions of the genome. Within a gene, polymorphisms that affect its function strongly are rarer than polymorphisms with minimal or no functional impact. This pattern is due to negative natural selection, which removes deleterious derived alleles from the population (see Chapter 7).

Thirdly, most of polymorphic loci are much shorter than the key class of functional genome segments, protein-coding genes. Indeed, a typical human gene encompasses thousands (if we discount introns), or tens of thousands (including introns) nucleotides. By contrast, single-nucleotide polymorphisms (SNPs) are the most common class of loci. Within a large enough population, there are usually many polymorphic loci which affect a gene, and most of them fit entirely within it. Still, there are also rare large-scale polymorphic loci which can affect two or even more adjacent genes (Figure 2.10).

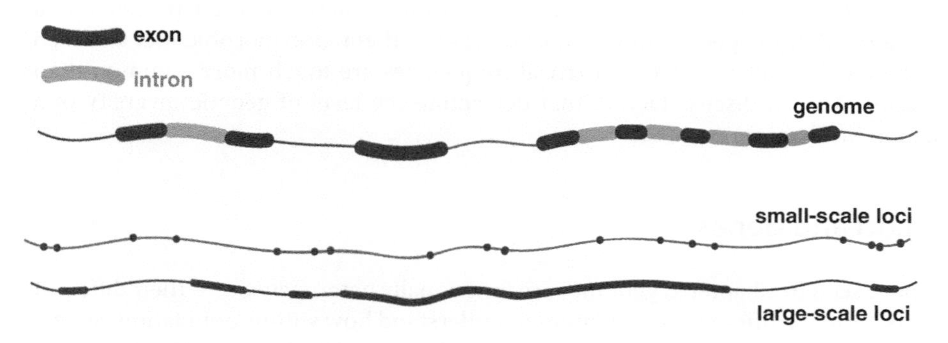

Figure 2.10 The relationship between protein-coding genes and polymorphic loci. Small-scale and large-scale polymorphic loci are shown on separate lines below the genome.

Derived alleles at small-scale loci which reside within a protein-coding gene can affect the encoded protein in a variety of ways (Figure 2.11):

1) A start-abolishing allele disturbs the ATG codon from which translation begins (see Chapter 1), which usually prevents synthesis of the protein.
2) A nonsense nucleotide substitution converts an amino-acid-encoding codon into a stop codon, thus leading to a truncated, and usually non-functional, protein.
3) A missense substitution converts a codon that encodes one amino acid into a codon that encodes another amino acid, thus leading to an amino acid replacement within the protein.

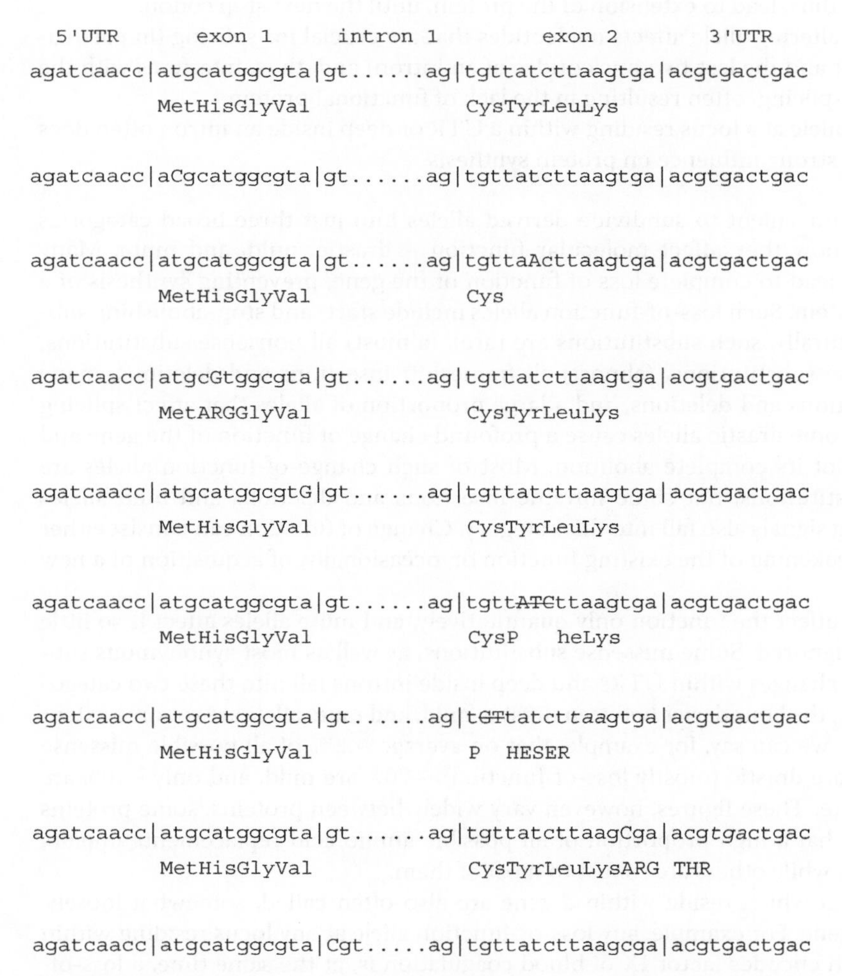

Figure 2.11 A hypothetical short protein-coding gene ("|" symbols within the sequence separate its functional parts), which encodes a protein consisting of eight amino acids, and derived alleles at polymorphic loci within it. From top to bottom (deviations are in upper case): a start-abolishing substitution, a nonsense substitution, a missense substitution, a synonymous substitution, an inframe deletion of three nucleotides, a frameshift deletion of two nucleotides (new stop is in italics), a stop-abolishing substitution (new stop is in italics), and a splicing-disturbing insertion.

4) A synonymous substitution converts a codon into another codon that encodes the same amino acid and, thus, does not affect the amino acid sequence of the protein.
5) An inframe deletion or insertion (or a complex allele) changes the length of a gene by 3, 6, 9, ... nucleotides and, thus, leads to insertion or deletion of one or several amino acids within the protein.
6) A frameshift deletion or insertion (or a complex allele) changes the length of the gene by a number that is not a multiple of three and, thus, leads to an entirely new protein sequence, usually soon terminated by a stop codon that was out of frame in the ancestral gene.
7) A stop-abolishing allele disturbs the stop codon which terminates the protein synthesis and, thus, lead to extension of the protein, until the next stop codon.
8) A splicing-altering allele affects nucleotides that are crucial for splicing (in particular, the first and the last two nucleotides of an intron) and, thus, interferes with the process of splicing, often resulting in the lack of functional protein.
9) A derived allele at a locus residing within a UTR or deep inside an intron often does not exert a strong influence on protein synthesis.

It is often convenient to subdivide derived alleles into just three broad categories according to how they affect molecular function – drastic, mild, and mute. Many drastic alleles lead to complete loss of function of the gene, preventing synthesis of a functional protein. Such loss-of-function alleles include start- and stop-abolishing substitutions (naturally, such substitutions are rare), (almost) all nonsense substitutions, some missense substitutions, (almost) all frameshift insertions and deletions, many inframe insertions and deletions, and a large proportion of alleles that affect splicing signals. Still, some drastic alleles cause a profound change of function of the gene and protein, but not its complete abolition. Most of such change-of-function alleles are missense substitutions, but some inframe insertions and deletions and mild alleles within splicing signals also fall into this category. Change of function can consist either of a simple weakening of the existing function or, occasionally, of acquisition of a new function.

Mild alleles affect the function only quantitatively, and mute alleles affect it so little that it can be ignored. Some missense substitutions, as well as most synonymous substitutions and changes within UTRs and deep inside introns fall into these two categories. Of course, the boundaries between drastic, mild, and mute allele are to some extent arbitrary. Still, we can say, for example, that on average ~20% of all possible missense substitutions are drastic (mostly loss-of-function), ~70% are mild, and only ~10% are effectively mute. These figures, however, vary widely between proteins: some proteins are fragile so that a high proportion of all possible amino acid replacements impairs their function, while others are robust to most of them.

Alleles at loci which reside within a gene are also often called, somewhat loosely, alleles of the gene. For example, any loss-of-function allele at any locus residing within the gene which encodes factor IX of blood coagulation is, at the same time, a loss-of-function allele of this gene. Complete loss of function of the factor IX gene leads (in boys, who carry only one copy of this X-chromosome gene) to severe hemophilia B, and a partial but substantial loss leads to moderate or mild forms of this disease. So far, over 1000 different alleles of the factor IX gene have been described. Table 2.2 lists several of them.

Table 2.2 A sample of derived alleles affecting the factor IX-encoding gene.

Position of a polymorphic locus within the gene	Sequences of the ancestral and the derived allele	Part of the gene affected	Sequence class	Effect on function	Severity of hemophilia B
10392	a>g	Exon 4	Missense substitution Asp93Gly	Partial loss	Mild
10396–10397	ag>–	Exon 4	Frameshift deletion	Complete loss	Severe
10406	g>t	Exon 4	Nonsense substitution	Complete loss	Severe
10428	g>t	Exon 4	Missense substitution Gly105Val	Partial loss	Moderate
10506	g>t	Intron 4	Splicing signal	Complete loss	Severe
20422	g>a	Exon 6	Missense substitution Ala194Thr	None	No disease
30053	t>c	Exon 7	Synonymous substitution	None	No disease
Between 30802 and 30803	–>a	Intron 7	Insertion	None	No disease
31071–31077	ggagatc>a	Exon 8	Frameshift complex	Complete loss	Severe

Functioning of intergenic DNA segments is understood much less than functioning of protein-coding genes. Still, a derived allele of a small-scale locus within such a segment can also be drastic, quantitative, or mute. In particular, such an allele can either abolish a binding site of a DNA-binding protein, create a new one, change the strength of an existing one, or have no functional effect.

Not surprisingly, alleles at large-scale polymorphic loci usually affect function drastically. A long deletion may simply remove a gene, or even two or more adjacent genes, from the genotype. If a boundary of a long inversion is located within a gene, the function of the gene is usually destroyed (Figure 2.12). By contrast, a long duplication can lead to emergence of a new functional gene, and an extra chromosome (trisomy) leads to an extra copy of every gene located on the chromosome.

2.4 Effects of Alleles on Phenotypes

From the point of view of their effects on higher level phenotypes, derived alleles of polymorphic loci can also be subdivided into drastic, mild, and mute classes. Although much less common than mild and mute alleles, drastic alleles are still very important due to their drastic effects. Thus, let us consider them first.

A drastic allele can be lethal and/or visible. A lethal allele can cause death, and a visible allele can lead to easily detectable changes of the phenotype, such as those studied by Mendel. The state of a Mendelian trait that is caused by drastic, derived alleles is often pathological and, in humans, constitutes a Mendelian disease. Often, the line between lethal and visible alleles is blurred. From the perspective of population

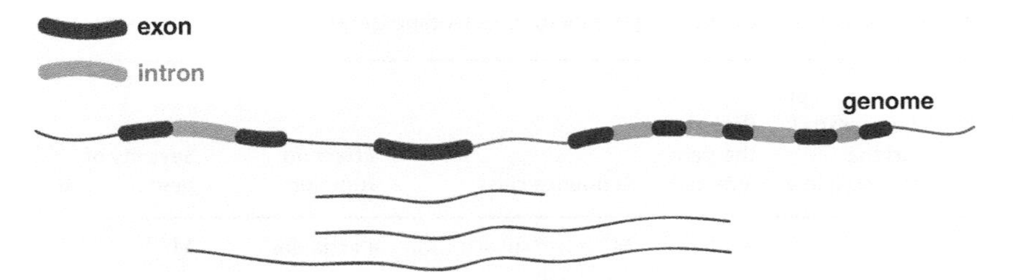

Figure 2.12 Three inversions within a genome segment, shown in Figure 2.10. The top inversion is unlikely to affect the function of either of the flanking genes, the second one destroys the function of the right gene, and the bottom one destroys the function of both the left and the right gene. Function of the central gene may be unaffected in all three cases.

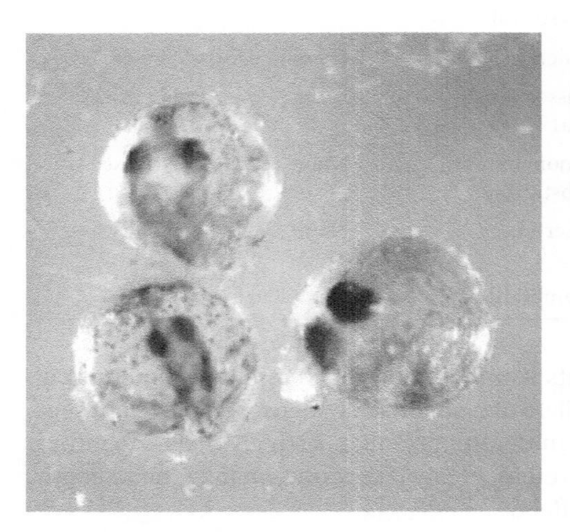

Figure 2.13 Recessive lethals that are present in natural populations of fish usually cause death before or soon after hatching, and produce drastic morphological changes. The embryo on the right is normal, and the other two carry different homozygous recessive lethal alleles (McCune *et al.*, *Science*, vol. 296, p. 2398, 2002).

genetics, an allele that leads to a phenotype that is totally incapable of reproducing is a lethal. Still, even those lethals that cause death early in the course of individual development may also produce easily visible morphological changes (Figure 2.13). From the point of view of a doctor, an allele which leads to the birth of a sick child is, first of all, a visible, even if the Mendelian disease is inconsistent with reproduction. Even if a drastic allele does not prevent reproduction, it may still cause death later in life. For example, people who carry a loss-of-function allele of gene APC (adenomatous polyposis coli) can reproduce, but, unless treated, develop colorectal cancer with a probability of over 90% by the age of 50.

All derived drastic alleles are pleiotropic. Indeed, a lethal is necessarily pleiotropic, as death affects all traits. A visible allele also affects many phenotypic traits at the same time. In particular, a visible allele, even if not lethal, is usually deleterious, that is, reduces fitness, the ability of an individual to survive and reproduce (see Chapter 7). However, there are exceptions, and a drastic visible allele may be beneficial under a suitable environment.

Because derived drastic alleles are usually rare, in population and medical genetics the term dominance is often understood differently from its original usage. Mendel called allele A dominant if the phenotype of the heterozygote Aa cannot be easily distinguished from that of a homozygote AA. In contrast, in population and medical genetics a rare derived allele A is called dominant if the phenotype of heterozygote Aa is different from that of the common homozygote aa. Almost all derived visible alleles that doctors call dominant are lethal in the homozygous state, so that Aa and AA have very different phenotypes (Figure 2.14). However, this usually does not matter, because homozygotes AA are very rare. I will follow this population usage of the term dominant throughout the book.

By contrast to dominance, recessivity of derived drastic alleles in population and medical genetics is understood traditionally. An allele a is called recessive if it affects the phenotype only when homozygous, so that AA and Aa have one phenotype, and aa has a different phenotype. Obviously, a dominant lethal is immediately removed from the population, but a rare recessive lethal can persist for a while, being present almost exclusively in heterozygotes.

Almost all phenotypically drastic alleles affect molecular function drastically, mostly by altering protein-coding genes, but occasionally by altering functional genome segments located outside genes. For example, a DNA-binding protein PTF1A is essential for development of the pancreas. People who carry homozygous loss-of-function alleles of the PTF1A-encoding gene lack the pancreas from birth, and the same is true for people who carry homozygous alleles that strongly disrupt a short (~400 nucleotides) segment of intergenic DNA, located ~25 000 nucleotides away from this gene, which regulates its transcription.

A drastic allele can be dominant in two ways. A loss-of-function allele is dominant only when possessing just one functional allele is not enough for the development of a normal (healthy, wild-type) phenotype (haploinsufficiency). By contrast, a change-of-function allele can also be dominant if its new function is disruptive in the heterozygous state (negative dominance). There are genes where loss-of-function alleles are recessive but some change-of-function alleles are dominant, being, therefore, more disruptive. Still, haploinsufficiency is the most common cause of a drastic phenotype of a heterozygote.

Almost all loss-of-function alleles reduce the expected fitness of their carriers, at least under a harsh environment. Indeed, if the function of a gene were unimportant, its loss-of-function-alleles would accumulate in the population, eventually destroying the gene (see Chapter 7). Still, not all loss-of-function alleles are drastic. In mammals, only ~6000 (30%) of protein-coding genes are essential, in the sense that a complete loss of their function in loss-of-function homozygotes is lethal. Thus, in many genes, even homozygous loss-of-function alleles produce only a quantitative phenotypic effect. Even a long deletion that removes several adjacent genes may cause no drastic phenotypic effects, even when homozygous.

Naturally, loss-of-function alleles of essential genes are recessive lethals. In a smaller fraction of genes, even heterozygous loss of function is lethal, due to haploinsufficiency, so that loss-of-function alleles of such genes are dominant lethals. Homozygous and heterozygous loss of function are known to cause visible (drastic, but viable after birth) phenotypes for ~1200 and ~2000 of human genes, respectively (see Chapter 12).

Dominant drastic alleles are easy to detect, with the only major exception of early-acting lethals (see Chapter 11). However, such alleles are very rare in populations, because negative selection rapidly removes them. Drastic alleles that are recessive, at least as far as their drastic effects are concerned, are more common. Still, even recessive

Figure 2.14 (a) In classical genetics, we say that allele A that causes purple flower color is dominant because Aa has the same phenotype as AA. (b) In population and medical genetics, allele A that causes achondroplasia is called dominant because Aa causes a phenotype different from wild-type aa. However, homozygotes AA are lethal, and, thus, are also phenotypically different from Aa, so that neither of these alleles is dominant in the classical, Mendelian sense.

drastic alleles tend to remain rare because they usually cause some reduction of fitness even when heterozygous. As long as mating within the population is more or less random, rare recessive drastic alleles are mostly shielded from observation and strong negative selection, due to the Hardy–Weinberg law.

However, such alleles can be revealed by inbreeding, mating between close relatives. When parents share some (or even all) of their not-too-distant ancestors, the mother and the father can transmit to their offspring a pair of alleles at a locus that originated from the same ancestral allele. Such alleles are called identical by descent. The probability that at a locus within the genotype of an individual two alleles are identical by descent is called the coefficient of inbreeding F of this individual. If both parents of an individual have the same genotype (being either the same individual or monozygotic twins), $F = 1/2$; if the parents are full siblings (including dizygotic twins), $F = 1/4$; and if the parents are first cousins, $F = 1/16$ (Figure 2.15).

The closest form of inbreeding is impossible in mammals, because their monozygotic twins are of the same sex and, thus, cannot mate. However, many species of animals and plants are hermaphrodites, and self-fertilization is often possible, in which case the same individual acts as both the mother and the father. In fact, Mendel used self-fertilization to produce F_2 in his experiments: garden peas are mostly selfers, and it takes special efforts to outcross them. Self-fertilization of a heterozygote Aa produces 25% of aa offspring, so that a heterozygous drastic recessive allele becomes exposed with probability 1/4 (Figure 2.3).

When self-fertilization is impossible, brother–sister matings can be used to expose recessive alleles. If one of the parents that produced a sibship carries a heterozygous recessive allele a at locus A, 1/4 of crosses between siblings will be Aa × Aa, and 1/4 of offspring from each such cross will have genotype aa (Figure 2.16). Thus, if at least ~20 brother–sister crosses were made within an F_1 sibship, and at least ~20 F_2 offspring were produced in each such cross, (almost) every drastic recessive allele hidden in the diploid genotype of either parent will be observed. Such an analysis of genotypes from the wild population was first performed by Elena A. and Nikolay V. Timofeev-Ressovsky in 1926. Detecting drastic alleles in *D. melanogaster* can also be aided by genetic tricks, specific to this model organism (see Chapter 6). In humans, effects of drastic recessive alleles can be studied in children from marriages between first cousins, which are common in many countries.

Naturally, an individual cannot carry any dominant lethals, because such most drastic alleles are inconsistent with life. By contrast, heterozygous recessive lethals are quite common in natural populations. Starting in the 1920s, a number of studies showed that in several *Drosophila* species the typical per genotype number of recessive lethals is ~2. In 2002, similar estimates were obtained for zebrafish, *Danio rerio* and bluefin killifish, *Lucania goodei*.

Studying recessive lethals in mammals is difficult due to their internal development. Still, some estimates for humans are close to those for fishes and flies. Indeed, if an average human carried, say, 30 or more heterozygous recessive lethals, fertility of brother–sister marriages would be very low, because most of zygotes produced would by homozygous by at least one recessive lethal. However, this is not the case: although incest and other forms of close inbreeding are definitely deleterious to the offspring (inbreeding depression, see Chapter 9), in those few cultures where they were acceptable, brother–sister marriages proved to be fertile. In fact, in some periods and places in ancient Egypt as many as 20% of legal marriages were between full siblings.

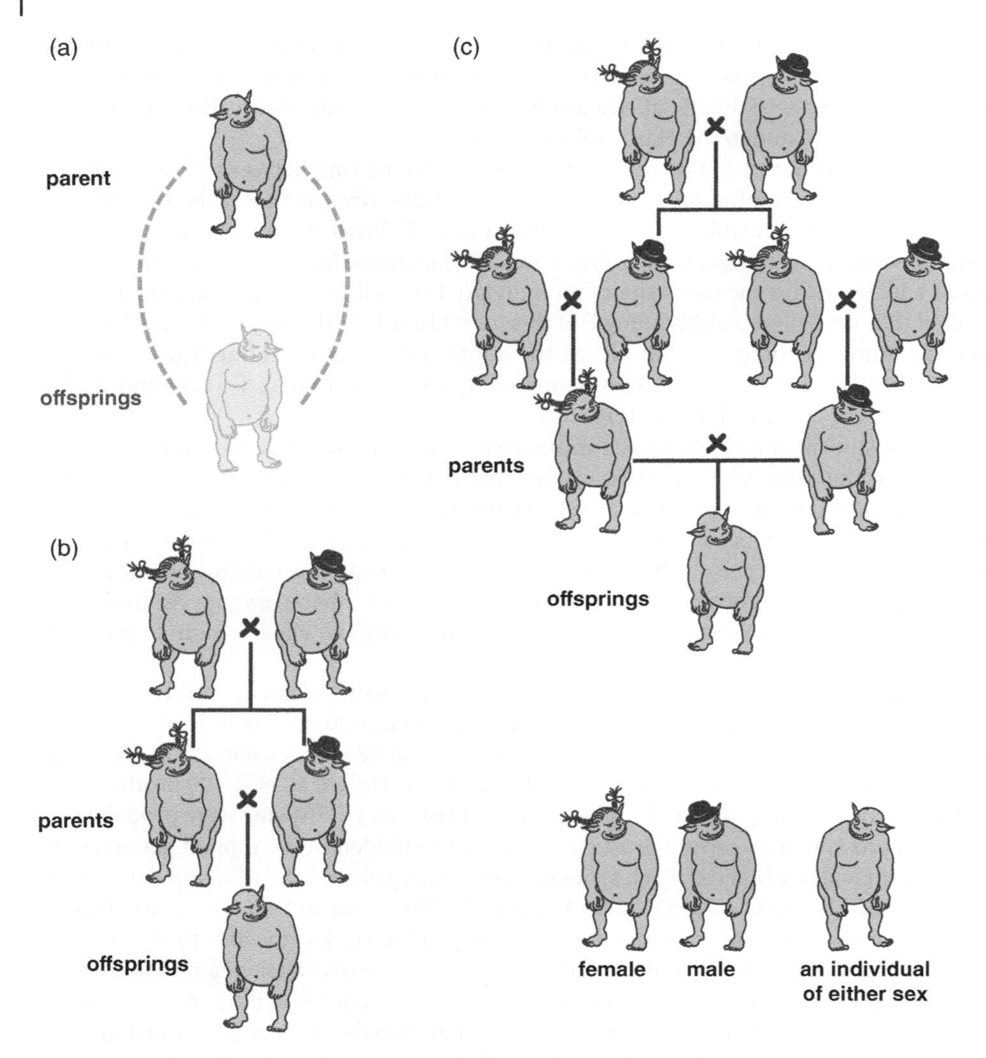

Figure 2.15 Pedigrees showing offspring of different kinds of related parents: (a) one individual acting as both the mother and the father, (b) siblings, and (c) first cousins.

In *Drosophila*, the average per-genotype number of dominant visible alleles is only ~0.002. In humans, the per-genotype number of dominant alleles causing Mendelian diseases is ~0.02 (see Chapters 12 and 13). The same pattern is observed for recessive alleles. The average fly carries ~0.1 heterozygous recessive visible alleles, while in humans the number of heterozygous alleles that cause Mendelian diseases when homozygous is ~1 per genotype. Perhaps, the difference is at least partially due to our much better ability to score drastic phenotypes in humans: many Mendelian diseases would likely be overlooked if we were flies.

There is no sharp boundary between drastic and mild alleles: some "borderline" quantitative alleles exert such a large influence on the phenotype that they may be also

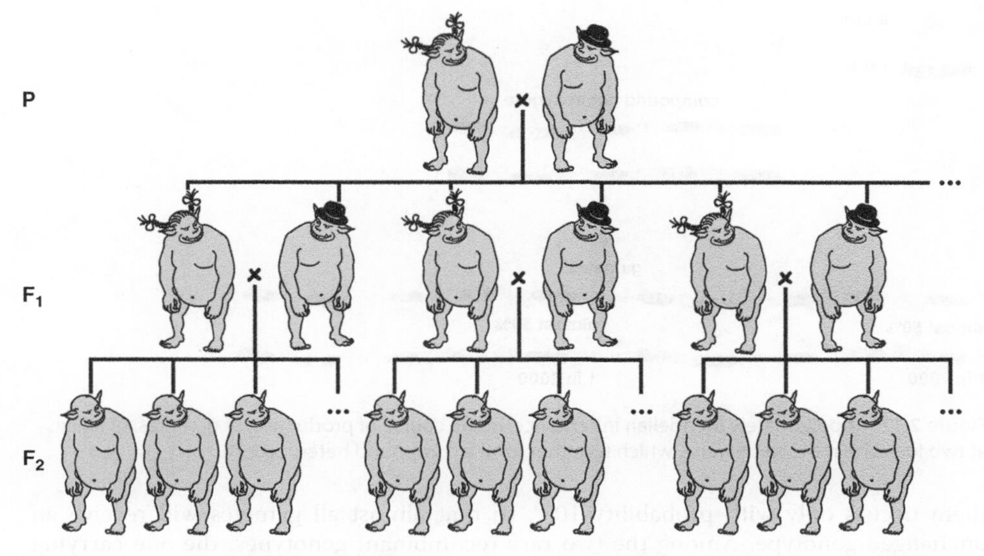

Figure 2.16 The simplest way of exposing recessive drastic alleles, by multiple matings within a sibship.

regarded as drastic. Still, a typical mild allele always affects the phenotype only slightly and, thus, cannot be detected individually without data on genotypes. However, only those alleles that are truly mute at the molecular level can be truly mute at the higher phenotypic level. Again, there is also no sharp boundary between mild and mute alleles, because many mild alleles have only very slight effects. Due to their ubiquity, mild alleles are even more important than drastic alleles (see Chapter 3). Mute alleles will be considered in Chapter 8.

2.5 Mendelian Traits and Diseases

Let us now consider the same issue, the relationship between simple variation of genotypes and of higher level phenotypes, from the opposite, phenotypic, perspective. Here we deal only with simple traits, whose variation mostly depends on variation at individual loci. All such traits, regardless of the length of the underlying polymorphic loci, can be called Mendelian. However, it is inconvenient to associate a separate Mendelian trait with each polymorphic locus, because alleles at different loci that affect the same gene often have identical impacts on the molecular function and phenotype (Table 2.2). Thus, it is better to treat phenotypes caused by all loci that affect a particular gene as states of the same Mendelian trait. For example, all alleles of the factor IX gene that result in a complete loss of function of the encoded protein lead, in boys, to severe hemophilia B (Table 2.2), the same state of the same Mendelian trait.

Indeed, because recombination within a gene is very rare, even if we lump changes caused by different polymorphic loci within a gene into the same trait, its inheritance still remains essentially Mendelian (Figure 2.17). In humans, the probability of crossing-over is ~10^{-8} per nucleotide. Thus, even if two loss-of-function alleles affect the opposite sides of a relatively long protein-coding gene of 10^5 nucleotides, crossing-over between

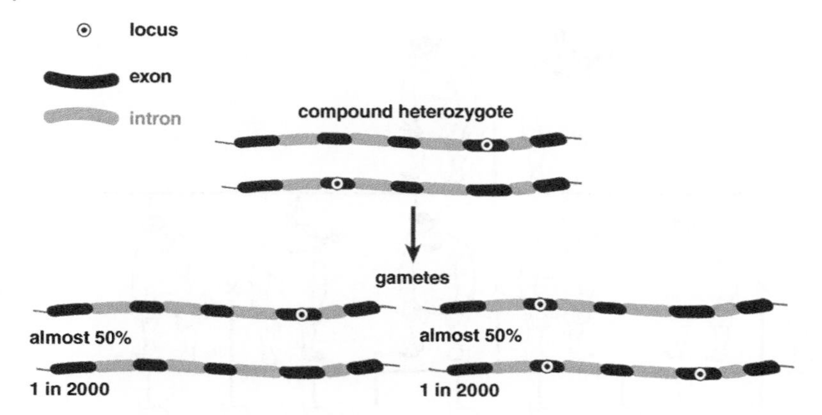

Figure 2.17 Approximately Mendelian inheritance, in the course of production of gametes, of alleles at two loci within the same gene, which together form a compound heterozygote.

them occurs only with probability 10^{-3}, so that almost all gametes will receive an unchanged genotype. Among the two rare recombinant genotypes, the one carrying two loss-of-function alleles would still cause the same loss of function, and the other one would have its function restored. A genotype carrying two alleles with similar effects on the molecular function at different loci within the same gene is called a compound heterozygote, to distinguish it from a homozygote which carries two alleles at the same locus. Still, in terms of inheritance, a compound heterozygote is very close to a homozygote.

Sometimes, impairments of function of two or more genes have very similar phenotypic manifestations. For example, loss-of-function alleles of a gene encoding factor VIII of blood coagulation lead to hemophilia A, which is clinically very similar to hemophilia B. This is not surprising, because factor VIII and factor IX proteins work is succession in the blood coagulation functional cascade, and it does not matter much at which point the cascade malfunctions.

Still, it is convenient to consider hemophilia A and hemophilia B as states of different Mendelian traits, because there is a sharp difference between individuals who carry two recessive loss-of-function alleles at different loci versus at the same locus (Figure 2.18). A woman carrying a wild-type ancestral and a loss-of-function derived allele in both factor VIII and factor IX genes would not have hemophilia – she would have only 50% of functional factors VIII and IX, but this is enough to be mostly free of clinical symptoms. Genetically speaking, in this case the wild-type allele "complements" the loss-of-function allele in both genes. By contrast, a woman carrying two different loss-of-function alleles of either gene, and two normal alleles of the other gene, would suffer from hemophilia A or B (hemophilia in women is very rare, due to rarity of loss-of-function alleles of factor VIII and factor IX genes). This pattern provides a useful definition of gene, as a unit of function: as long as maternal and paternal loss-of-function alleles at two different loci have the same phenotypic effect as two copies of either allele, the two loci affect the same gene.

To be recognized as states of a distinct trait, phenotypes produced by different alleles at a locus must differ drastically. Thus, each Mendelian trait has states corresponding to lethal or visible drastic alleles, although mild alleles may also contribute to variation of such traits. Mendelian diseases of humans were discovered in 1909 by the British

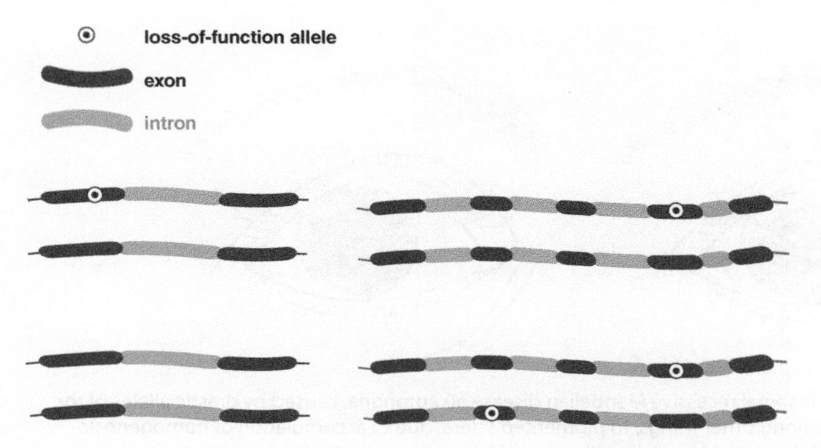

Figure 2.18 Two loss-of-function alleles of different genes (top) and of the same gene (bottom), where they form a compound heterozygote. If both alleles are recessive, in the first case neither of them affects the phenotype, because each allele is complemented by a wild-type allele of the same gene. By contrast, in the second case the compound heterozygote affects the phenotype.

physician Archibald Garrod, who recognized, in his terminology, four "inborn errors of metabolism". One of them was alkaptonuria, a condition caused by loss of function of enzyme homogentisate 1,2-dioxygenase, which leads to pathological accumulation of compounds called homogentisic acid and alkapton (Figure 2.19). Garrod noticed that affected children are often born to related parents, and correctly concluded that alkaptonuria is an autosomal recessive disease.

Genes and alleles behind Mendelian diseases form biased samples from all human genes and alleles. One the one hand, even complete loss of function of many mammalian genes does not produce a clear-cut phenotype. On the other hand, even heterozygous loss of function of some genes leads to death before birth. Thus, genes that cause Mendelian diseases must be important but not too important. By contrast, alleles that cause Mendelian diseases usually affect the molecular function drastically, often leading to its complete loss. Still, some diseases, such as X-linked glucose-6-phosphate dehydrogenase deficiency that provides protection against malaria, are caused only by functionally mild alleles, because a complete loss of function of the affected gene is prenatally lethal and, thus, is not observed in children.

Mendelian diseases vary widely in their severity. Some of them are fatal, in early childhood or later, and only palliative care can be offered. Examples are Tay–Sachs disease and Duchenne muscular dystrophy. Others are grave or even fatal without specific care but eminently treatable. Hemophilia A and B can be treated by infusions of factors VIII and IX, respectively, and severe intellectual disability in people affected by phenylketonuria can be prevented by a special diet, low in the amino acid phenylalanine. Still other Mendelian diseases, such as several forms of color blindness, may result in only a mild inconvenience, at least under an Industrialized environment (see Chapter 10). Heterozygous carriers of a number of "recessive" diseases experience mild symptoms, because their recessivity is not complete.

A Mendelian trait may be not a Mendelian disease, if all states of the trait are not pathological. However, in humans, there is only a small number of benign Mendelian

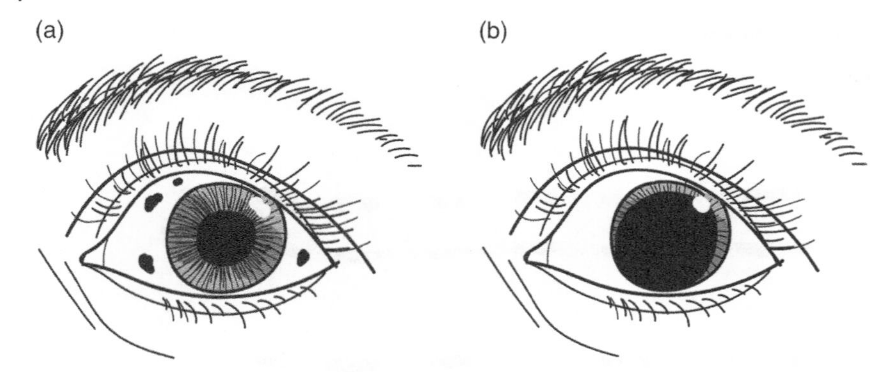

Figure 2.19 (a) Autosomal recessive Mendelian disease alkaptonuria, caused by drastic alleles of the HGD gene, leads, among other things, to pigmented sclera, due to accumulation of homogentisic acid. (b) Autosomal dominant Mendelian disease aniridia, whose most salient manifestation is the absence of the iris, is caused by drastic alleles of the PAX6 gene.

within-population polymorphisms. One example is wet versus dry earwax: dry earwax is due to a derived, recessive allele. Another example is (in)ability to sense a bitter taste of a substance called phenylthiocarbamide: inability is due to a derived, loss-of-function allele of the corresponding receptor protein. A well-known example of benign Mendelian variation, although not directly manifested in higher level phenotype, are the ABO blood groups, which are due to three alleles at one locus, A, B, and O. Although six diploid genotypes can be formed by three alleles, only four blood groups are recognized (A: AA or AO, B: BB or BO, AB: AB, and O: OO). By contrast, genetic variation underlying some other two-state human traits, such as (in)ability to roll the tongue, is more complex (see Chapter 3).

Further Reading

Chien, YH, Huang, HP, & Hwu, WL 2009, Eye anomalies and neurological manifestations in patients with PAX6 mutations, *Molecular Vision*, vol. 15, pp. 2139–2145.
Abnormal alleles of the gene PAX6 can cause aniridia, congenital cataracts, and other eye and developmental anomalies.

Kreitman, M 1983, Nucleotide polymorphism at the alcohol dehydrogenase locus of *Drosophila melanogaster*, *Nature*, vol. 304, pp. 412–417.
The first direct study of within-population variation of genotypes.

Leffler, EM, Bullaughey, K, & Matute, DR 2012, Revisiting an old riddle: what determines genetic diversity levels within species?, *PLoS Biology*, vol. 10: e1001388.
A comprehensive review of genetic diversity across species.

Li, T, Miller, CH, & Payne, AB 2013, The CDC Hemophilia B mutation project mutation list: a new online resource, *Molecular Genetics and Genomic Medicine*, vol. 1, pp. 238–245.
Currently, 1083 different alleles of the gene F9 that cause hemophilia B are known.

McCune, AR, Fuller, RC, & Aquilina AA 2002, A low genomic number of recessive lethals in natural populations of bluefin killifish and zebrafish, *Science*, vol. 296, pp. 2398–2401.
A direct measurement of the per genotype number of recessive lethals in two fish species.

Savage, DB, Tan, GD, & Acerini, CL 2003, Human metabolic syndrome resulting from dominant-negative mutations in the nuclear receptor peroxisome proliferator-activated receptor-γ, *Diabetes*, vol. 52, pp. 910–917.

An example of a human Mendelian disease caused by dominant-negative alleles.

Shaw, BD 1992, Explaining incest: brother-sister marriage in Graeco-Roman Egypt, *Man*, vol. 27, pp. 267–299.

In Graeco-Roman Egypt, a significant proportion of marriages was between full brothers and sisters.

Varki, A & Altheide, TK 2005, Comparing the human and chimpanzee genomes: searching for needles in a haystack, *Genome Research*, vol. 15, pp. 1746–1758.

Comparison of human and common chimpanzee genomes.

Wang, T, Birsoy, K, & Hughes, NW 2015, Identification and characterization of essential genes in the human genome, *Science*, vol. 350, pp. 1096–1101.

A study of human essential genes.

Weedon, MN, Cebola, I, & Patch, A-M 2013, Recessive mutations in a distal PTF1A enhancer cause isolated pancreatic agenesis, *Nature Genetics*, vol. 46, pp. 61–64.

A drastic phenotype, lack of pancreas, is produced by regulatory alleles of gene PTF1A.

3

Complex Traits and Their Inheritance

Which of you by taking thought can add one cubit unto his stature?

Jesus, Matthew's Gospel 6:27, circa 0080.

Most phenotypic variation is complex, and the state of a complex trait in an organism depends on alleles of many unlinked genes in its genotype, as well as on its environment. Two opposite extreme kinds of complex traits, two-state and quantitative, are of particular importance. An individual can be characterized by the actual, inherited, and transmittable value of a complex trait. Distribution of actual values of the trait within the population has the highest variance, and distribution of transmittable values has the lowest variance. Variances of inherited and transmittable values of a trait among individuals characterize the contribution of genetic variation to variation of the trait. A number of methods can be used to estimate this contribution, which is usually quite large. Most populations harbor many polymorphic loci which affect variation of every complex trait. Next-generation sequencing helps to identify these loci, which, however, remains a difficult task. Often, the effect of an allele on a trait depends on the rest of the genotype. This phenomenon, called epistasis, is of central importance for genetics and evolutionary biology.

3.1 Complex Inheritance of Phenotypes

In the previous chapter, we considered simple Mendelian phenotypic traits whose states correspond to different alleles of one gene, or to different alleles at a large-scale polymorphic locus that overlaps many genes. Although Mendelian traits are many, in a vast majority of them the states that correspond to derived alleles are rare and pathological. Thus, most phenotypic variation present within a population is more complex, in the sense that each trait is affected by alleles at many polymorphic loci that alter multiple unlinked genes. Pervasive complexity of the connection between variations of genotypes and of phenotypes (Figure 1.21) is evident from observations of two kinds.

First, for most variable traits, outcomes of matings and marriages do not obey Mendel's laws and, instead, these traits are inherited in more complex ways. This is the case even when a trait can accept only two states. For example, when one parent suffers retinal detachment and the other does not, most of their children do not have this condition, which is consistent with retinal detachment being the recessive state of

Crumbling Genome: The Impact of Deleterious Mutations on Humans, First Edition. Alexey S. Kondrashov.

a Mendelian trait. However, among grandchildren whose both grandmothers suffered retinal detachment and both grandfathers did not, only a small proportion will have this condition, instead of the 25% expected if the trait were Mendelian and grandparents were homozygotes (Figure 2.3).

Violation of Mendel's laws at the level of phenotypes is even more obvious when a trait varies quantitatively. If white Finnish women marry black Maasai men, their children are more or less uniformly brown, which is consistent with Mendelian uniformity of F_1, in the absence of dominance (Figure 2.1). However, if such F_1 biracial children intermarry, there will be nothing like 1:2:1 Mendelian ratio of white, brown, and black among F_2 grandchildren. Instead, almost none of them will be as white as their Finnish grandmothers, or as black as their Maasai grandfathers, and they all will be of different shades of brown.

The same is true for inheritance of within-population quantitative variation. When tall and short Maasai (or Finns) intermarry, there is nothing like 1:2:1 tall–intermediate–short (or, perhaps, 3:1 or 1:3 tall–short) Mendelian ratio among their grandchildren. Instead, variation is again essentially continuous, with an excess of intermediate-stature grandchildren (Figure 3.1). In fact, in the case of stature, even F_1 children are far from being uniform. Nevertheless, such complex variation must

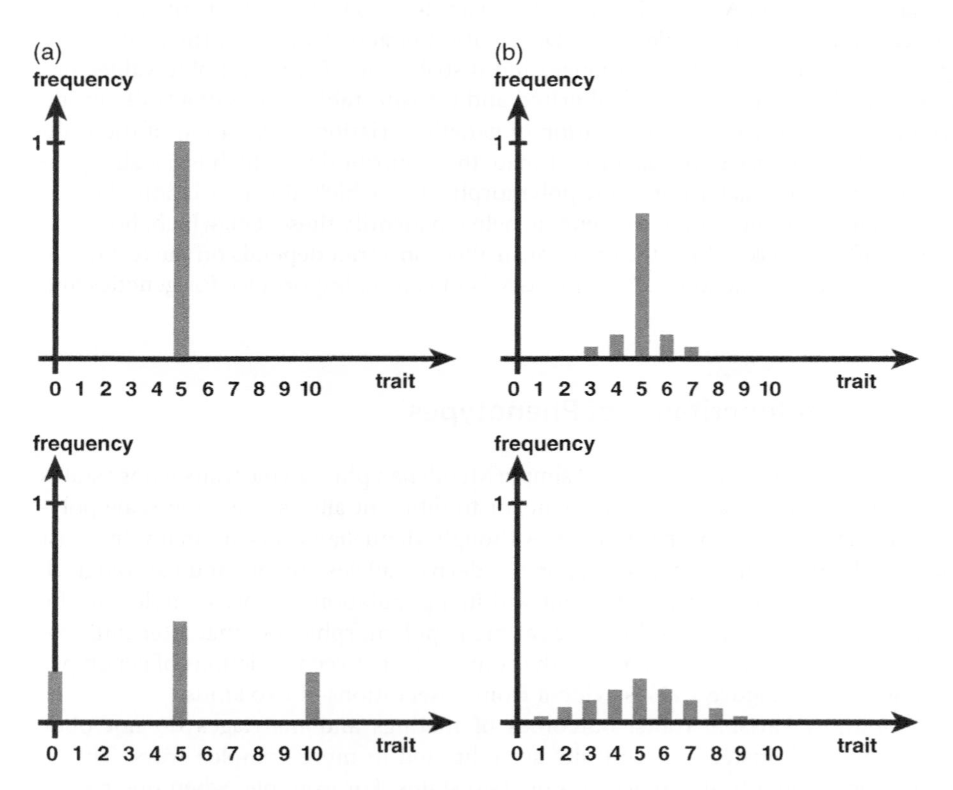

Figure 3.1 Mendelian (a) and complex (b) inheritance of a quantitative trait which can accept states 0, 1, …, 10. One parent had phenotype 0, and the other parent had phenotype 10. Frequencies of the trait states in F_1 (top) and F_2 (bottom) are shown.

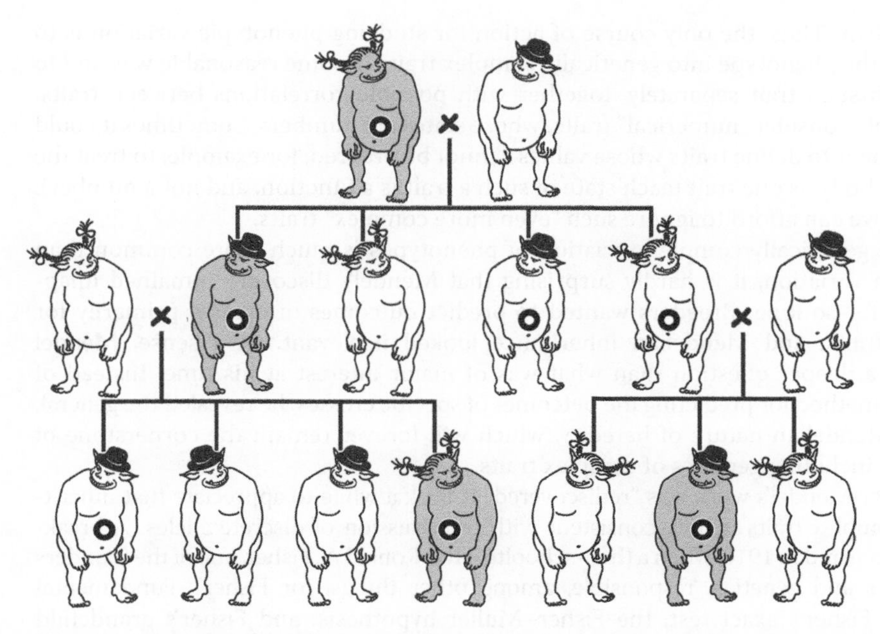

Figure 3.2 A hypothetical pedigree illustrating incomplete penetrance of an autosomal allele. Individuals who carry this allele in a heterozygous state are marked by circles, and the phenotype which appears only in some of them is shown by gray.

be at least partially heritable, because relatives tend to be more similar to each other, in all facets of their phenotypes, than unrelated individuals.

Secondly, phenotypic manifestation of a particular allele, which can be directly detected by DNA sequencing, often varies from individual to individual. We have already encountered this phenomenon at the level of one gene: a recessive allele is manifested only in homozygotes (see Chapter 2). However, in many cases, varying manifestation of an allele cannot be explained by what happens inside the gene. For example, some people heterozygous for a particular missense allele of Connexin 46 gene have congenital cataracts, but others do not. Almost every person heterozygous for a loss-of-function allele of gene PAX6 has aniridia (Figure 2.19), but its severity varies widely. We say that the Connexin 46 allele is incompletely penetrant (in some individuals its effect does not penetrate all the way to the higher level phenotype, Figure 3.2) and PAX6 loss-of-function alleles, although completely penetrant, have variable expressivity.

Factors that determine whether the effect of an allele penetrates to the level of phenotypes in a particular individual and, if so, to what extent, are usually obscure, although it is safe to say that genotype, environment, and chance may all play a role. As it is often the case in biology, there is no sharp boundary between Mendelian and complex traits. One can say, purely arbitrarily, that if an allele causes an obvious change of the phenotype in over 50% of individuals, this phenotype is Mendelian, even if this allele is not completely penetrant. By contrast, if such a change appears with a lower probability, the allele only causes a "susceptibility", and the trait is complex.

Observing phenotypes is enough to recognize individual alleles only at a small minority of polymorphic loci. In this sense, most phenotypic variation is complex, instead of Mendelian. Thus, the only course of action for studying phenotypic variation is to subdivide the phenotype into genetically complex traits in some reasonable way and to study each such trait separately, together with possible correlations between traits. We will only consider "numerical" traits, whose states are numbers. Sometimes it could be convenient to define traits whose values cannot be ordered; for example, to treat the shape of a body as one trait (each state of such a trait is a function, and not a number). However, we can afford to ignore such "even more complex" traits.

Because genetically complex variation of phenotypes is much more common than Mendelian variation, it is hardly surprising that Mendel's discovery remained unappreciated for so long. Breeders wanted to predict outcomes of crosses primarily for complex traits, and Mendelian inheritance looked irrelevant. In a sense, Mendel answered a deeper question than what was of major interest at his time. Instead of devising a method for predicting the outcomes of specific crosses, he revealed the general, discrete, Mendelian nature of heredity, which will forever remain the cornerstone of genetics – including genetics of complex traits.

Even after Mendel's work was "rediscovered", it took a while to appreciate that inheritance of complex traits is fully consistent with transmission of discrete alleles. A breakthrough occurred in 1918, when a then-schoolteacher Ronald A. Fisher, one of the founders of statistics and genetics, responsible, among other things, for Fisher's Fundamental Theorem, Fisher's exact test, the Fisher–Muller hypothesis, and Fisher's grandchild argument, published a paper titled "The correlation between relatives on the supposition of Mendelian inheritance". Still, even today identifying Mendelian alleles behind complex variation remains a daunting task.

3.2 Properties of a Complex Trait

Complex traits come with a wide variety of properties, three of which are of particular importance. The first is the number of genes whose variation can affect the trait strongly. The set of such genes is called the target of this trait. This term originated from consideration of mutation (see Chapter 4): in order to affect a trait, a mutation must hit its target.

Fitness, the fundamentally important trait which characterizes the ability of organisms to survive and reproduce, has the largest target, which normally includes all protein-coding genes in the genome (see Chapter 7). Targets of other, less inclusive, traits are smaller. The boundaries of the target of a trait may be fuzzy, because variation of a gene can affect a particular trait only slightly, making it unclear whether the gene should be included in the target. Because variation of a pleiotropic gene affects many traits, targets of different traits often overlap.

The second property of a trait is the number of states it can accept. One extreme is two-state (all-or-none, binary, dichotomous) traits. For example, some people suffer detached retina but most do not. The other extreme is quantitative traits that can vary continuously and, thus, mathematically speaking, can accept an infinite number of states. Examples of such quantitative traits are skin color, height, body mass index (BMI, weight over the square of height), blood pressure, and so on. However, some quantitative traits, such as the number of children of an individual, can accept only a finite number of states.

To some extent, the number of states we recognize may be a matter of convention. On the one hand, cases of retinal detachment vary in their severity. On the other hand, even if a trait is inherently continuous, it is often convenient, as a rough approximation, to subdivide it into only a small number of states. For example, one can regard any value of intelligence quotient (IQ) above 70 as normal, and subdivide lower values into three degrees of intellectual disability – mild (51–70), moderate (31–50), and severe (30 and below). Still, some traits are more naturally described as two-state and others as quantitative.

One state of many two-state complex traits is pathological, such as retinal detachment or schizophrenia. However, in some of such traits both states are benign. For example, about 70% of people can roll their tongue, but 30% cannot, and variation of this trait is due to a number of genetic and environmental factors. Still, most benign variation is inherently quantitative, although extreme values of a quantitative trait, such as very low or very high BMI, or very low or very high blood pressure, are clearly pathological.

A radical difference between a quantitative trait and a Mendelian trait is obvious. Indeed, within a simple pedigree a Mendelian trait can accept only two or three values, and a quantitative trait can always accept a continuum of values. By contrast, Mendelian and two-state complex traits can be rather similar phenotypically. For example, breast cancer may represent a Mendelian disease (caused by mutations at genes BRCA1 and BRCA2) or, in a majority of cases, have multiple genetic and environmental causes. Still, two-state complex traits are also fundamentally different from Mendelian traits. In contrast to a two-state Mendelian disease, loss or change of function of any one gene may not be enough to lead to the pathological state of a two-state complex trait, as long as other relevant genes are functioning well. Instead, a complex disease may appear only when several genes are compromised (see Chapter 12). In other words, incomplete penetrance of individual alleles is rampant when inheritance of a two-state trait is complex.

There is no simple correspondence between the number of states of a trait and the size of its target. On the one hand, even a Mendelian trait can accept many states, due to action of alleles with intermediate properties (cases of hemophilia are of different severity, see Chapter 2), although only two, with dominance, or three, without dominance, states of a Mendelian trait are observed within F_2 offspring from a mating between two homozygous individuals. On the other hand, the target of even a two-state complex trait can be large: many genes are known to influence the risk of retinal detachment.

The third property of a trait is the role of the environment in its variation. This role is substantial for (almost) all complex traits. Even monozygotic twins, although their genotypes are essentially identical, often possess different values of even two-state traits. Only rarely both twins have retinal detachment, and if one twin has schizophrenia, the other one also has it with probability ~60%, which is much higher that the population average (~1%), but still well below 100%. Similarly, monozygotic twins, even when raised together, may have substantially different heights, weights, or blood pressures, although, on average, these differences are much smaller than those between unrelated individuals or even dizygotic twins.

Still, some complex traits depend on the environment more than others. Jesus used height, and not weight, as an illustration of futility of human effort (see epigraph). Indeed, it is not at all that difficult to add to your weight, especially if you have some money and there are fast food restaurants nearby. The norm of reaction (see Chapter 1) is wider for weight than for height.

It is important to distinguish systematic (controllable) and random (chance) components of the environment. Individuals who live together may be affected by some components of the environment in more or less the same way, receiving the same amount of sunlight, experiencing the same temperature, consuming the same foods, and so on. However, there are also unpredictable environmental influences, which cannot be standardized and are thus unique for each individual, such as traffic accidents. Moreover, there may be random accidents in the course of development of an individual (developmental noise), which cannot be attributed to any external influences, resulting, in particular, in fluctuating asymmetry between the left- and the right-hand sides of a human or other bilateral animal. Such random environmental influences, external and internal, prevent even monozygotic twins from having exactly the same phenotypes, even if every care is taken to make their environments identical.

To study complex traits and their inheritance, we need to introduce three fundamental concepts – actual, inherited, and transmittable value of a trait in an individual. Let us do this for a quantitative trait. My height is 189 cm – this is my actual trait value. Now suppose that I have 1000 identical twins (or clones), which all grew up independently under the same controllable environment as every child in the population. The average height of these twins (say, 190 cm) is what my genotype is expected to produce under this environment, or my inherited value of the trait. Of course, there will always be variation among actual heights of these twins, due to chance differences between their environments, so that the actual height of an individual can be both above and below their inherited height and cannot be exactly inferred from their genotype alone. One can say that the actual value of a trait in an individual is equal to their inherited value plus random environmental scattering.

By definition, environmental scattering has zero mean. Its shape is usually Gaussian (or normal, Figure 3.3), due to a fundamental law of nature, known as Central Limit Theorem. Informally, it states that when something (a trait value) is affected by many mostly independent influences of comparable magnitudes, the value of this something has Gaussian distribution. Because random environmental influences that affect the path from genotype to phenotype are many, this theorem is applicable here.

Like inherited value of a trait of an individual, transmittable value is also a property of its genotype, but in the context of reproduction, within a particular population. In contrast to inherited value, which describes the relationship between genotypes and phenotypes within a generation, transmittable value describes connection between successive generations, without explicitly taking into account Mendelian inheritance. Suppose that each of my 1000 clones marries a woman, picked randomly from the population, and produces a son (or, perhaps, I produce 1000 sons by 1000 random women). Suppose that the average height of these sons is, say, 181 cm, and the average height of all males in the population is 175 cm. Then, my transmittable value of height is $175 + [(181-175)\times 2] = 187$ cm: it deviates from the mean by twice the average deviation of a child. The factor of 2 appears because a parent is responsible for only a half of genes of each child (Figure 3.4).

Inherited and transmittable values of a two-state trait (with states 0 and 1) are defined analogously. In this case, inherited value of a trait of an individual is the probability that its genotype will lead, under a particular environment, to state 1. Transmittable value of

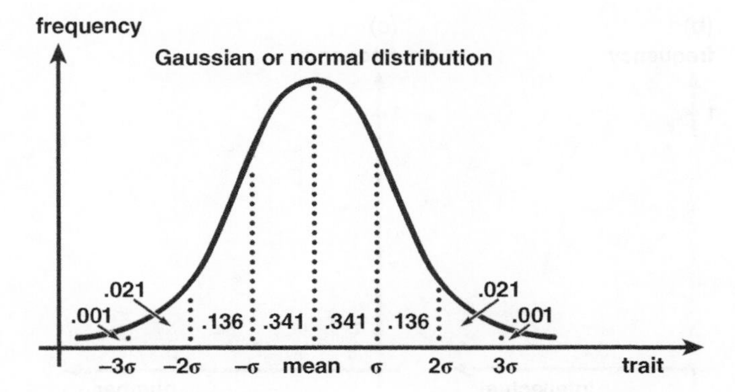

Figure 3.3 Gaussian distribution. σ stands for standard deviation. The exact properties of this fundamentally important distribution are not important for us. It is enough to remember that Gaussian distribution is symmetric about its mean and that the probability of a particular deviation from the mean declines very rapidly with the size of the deviation.

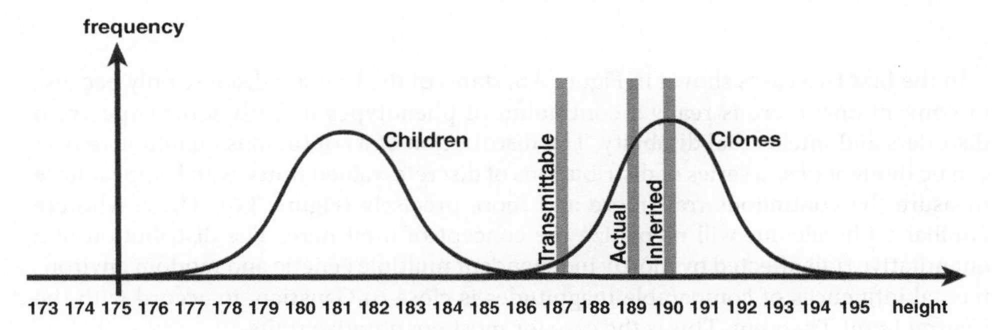

Figure 3.4 An example of actual, inherited, and transmittable values of a trait of an individual (vertical bars). Curves show distributions of actual values of the trait in multiple individuals having this genotype and in multiple offspring of individuals having this genotype.

a trait of an individual is twice the deviation of the probability that its offspring has state 1 from the frequency of this state in the population. Thus, both inherited and transmittable values of a two-state trait can vary continuously.

3.3 Complex Traits in Populations

So far, we have considered values of complex traits of individuals. However, individuals that constitute a population are not uniform and, instead, possess different phenotypes. How do we deal with a complex trait at the population level?

A (numerical) complex trait in the population is described by its distribution. When the number of states of a trait is finite, its distribution is an array of frequencies of all possible states (Figure 3.5). Because the sum of frequencies of all states is 1, the frequency of any one state can be inferred from that of the remaining states. In particular, the distribution of a two-state trait is fully described by the frequency of either state.

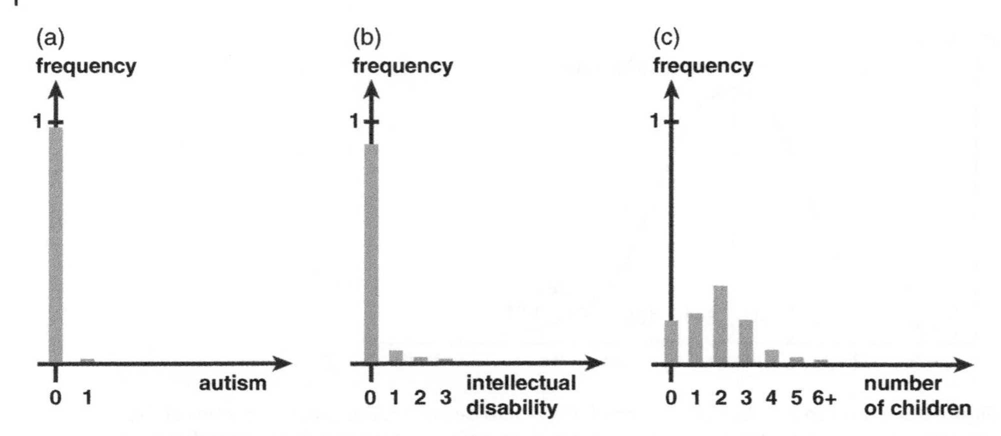

Figure 3.5 Distributions of two-, four-, and seven-state traits. (a) Autistic spectrum disorder: absent (0) or present (1) (at frequencies 0.99 and 0.01, respectively). (b) Intellectual disability: absent (0), mild (1), moderate (2), or severe (3) (0.98, 0.012, 0.005, 0.003). (c) Lifetime number of children of a woman in the USA (0.178, 0.184, 0.362, 0.178, 0.061, 0.025, 0.012).

In the first two cases shown in Figure 3.5, states of the trait are discrete only because of conventions: there is really a continuum of phenotypes in both autism spectrum disorders and intellectual disability. The distribution of a continuous quantitative trait can be thought of as a series of distributions of discrete-valued traits, which appear if we measure the continuous trait more and more precisely (Figure 3.6). Those who are familiar with calculus will recognize the concept of limit here. The distribution of a quantitative trait affected by mostly independent multiple genetic and random environmental influences of comparable magnitudes is close to Gaussian, in accord with the Central Limit Theorem. This is the case for most quantitative traits.

Mathematically speaking, the distribution of a continuous trait is a function and, thus, a rather complex entity. Often, it is desirable to describe a distribution in a simplified way, by a number. This can be done in a variety of ways, but the two most important numerical characteristics of a distribution are its mean and variance (together, they define a Gaussian distribution completely).

Let us introduce mean and variance using the distribution of human height rounded to 10 cm. In Canada, proportions of men having heights 145–155, 156–165, 166–175, 176–185, 186–195, and 196–205 cm are 0.004, 0.085, 0.408, 0.404, 0.097, and 0.002, respectively (Figure 3.6a). Then, the mean man height in this population is (150×0.004) + (160×0.085) + (170×0.408) + (180×0.404) + (190×0.097) + (200×0.002) = 175 cm.

Variance is a sum of squared deviations from the mean, weighted by frequencies (if we simply sum all weighted deviations from the mean, without squaring them first, the result would always be zero because positive and negative deviations would cancel each other). Thus, in our case the variance is $[(150-175)^2 \times 0.004] + [(160-175)^2 \times 0.085] + \ldots = 65\ \text{cm}^2$. Variance characterizes the spread of distribution and is equal to 0 only when there is no variation, so that everybody is of the same (mean) height.

Often, it is more intuitive to consider square root from variance, called standard deviation, because it is measured in the same units as the trait and its mean. For example, human height and its mean and standard deviation are measured in centimeters

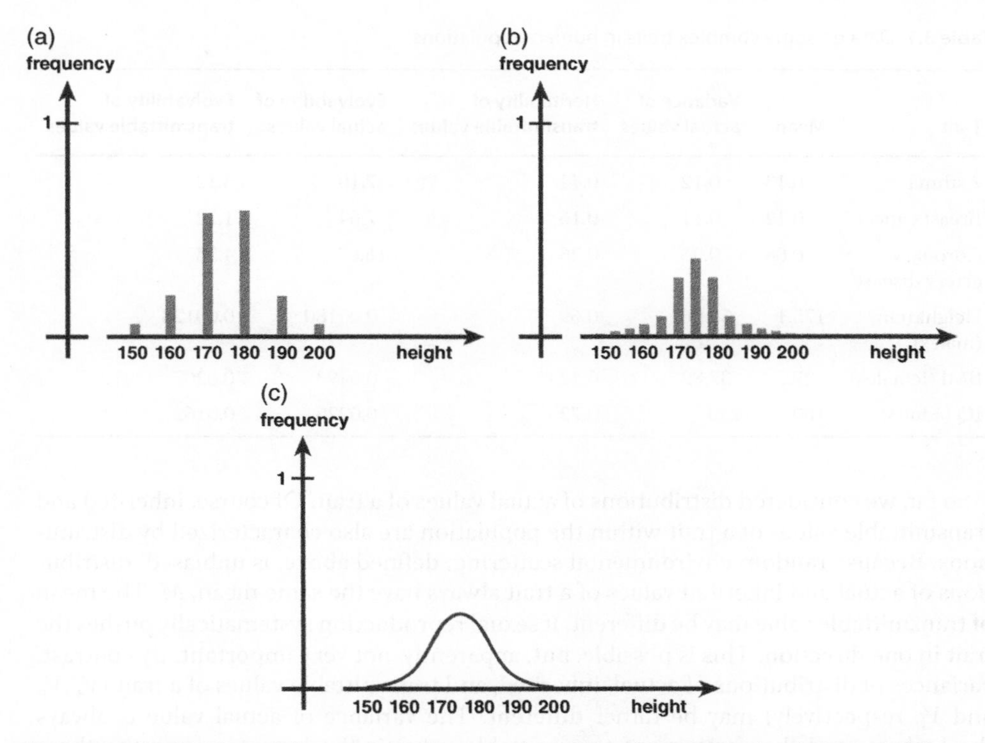

Figure 3.6 Distributions of discrete-valued quantitative traits which appear when height of adult males in Canada is measured with low (a) and intermediate (b) precision which approximate the essentially Gaussian distribution (c) of this continuous quantitative trait. Data from https://tall.life/height-percentile-calculator-age-country/.

(or any other unit of length), but its variance is measured in squared centimeters. Thus, if we switch to measuring height in meters, the mean and standard deviation become 100 times smaller, but variance becomes 10 000 times smaller, which may be inconvenient.

To ascertain the distribution of a trait within the population, one needs a good sample, which is representative (unbiased) and consists of enough individuals. A trait that can accept only a small number of states may be invariant within a population – every human has two eyes and seven neck vertebrae – but variation is present for all continuous traits. Naturally, distributions of a trait can be very different in different populations. Moreover, for many human traits, such as height, it makes sense to consider males and females separately, even within the same population.

Table 3.1 presents data on 3 two-state and 3 quantitative human traits. The exact figures presented in this table should not be taken too seriously, because they may be rather different in different populations. Moreover, different studies sometimes produce rather diverse estimates of heritability and evolvability (these concepts will be introduced shortly) even for the same population. Nevertheless, the figures presented appear to be typical.

Often, values of different traits are correlated. For example, those who are taller also tend to be heavier. Pleiotropy of alleles at variable loci is one of the reasons for such correlations, which should be taken into account when traits are defined. For example, weight *per se* is not a sensible characteristic of obesity, and BMI is a better one.

Table 3.1 Data on some complex traits in human populations.

Trait	Mean	Variance of actual values	Heritability of transmittable values	Evolvability of actual values	Evolvability of transmittable values
Asthma	0.13	0.12	0.44	7.10	3.12
Breast cancer	0.12	0.11	0.16	7.64	1.22
Coronary artery disease	0.06	0.06	0.26	16.67	4.33
Height (cm) (males)	175.1	55.06	0.69	0.00180	0.00123
BMI (females)	27.7	37.82	0.42	0.0493	0.0207
IQ (adults)	100	225	0.72	0.0225	0.0162

So far, we considered distributions of actual values of a trait. Of course, inherited and transmittable values of a trait within the population are also characterized by distributions. Because random environmental scattering, defined above, is unbiased, distributions of actual and inherited values of a trait always have the same mean, M. The mean of transmittable value may be different, if sexual reproduction systematically pushes the trait in one direction. This is possible, but, apparently, not very important. By contrast, variances of distributions of actual, inherited, and transmittable values of a trait (V_a, V_i, and V_t, respectively) may be rather different. The variance of actual value is always the highest, and the variance of transmittable value is the lowest among the three (Figure 3.7).

Why is this so? $V_a > V_i$ because the actual value of a trait in an individual deviates from the population mean due to both genetic and random environmental scattering. Even in a genetically uniform population, where all individuals have the same inherited value of a trait and, thus, $V_i = 0$, there still would be environmental scattering of actual values of a continuous trait, leading to $V_a > 0$.

The reason why $V_i > V_t$ is subtler. Let us consider an example. Suppose that a variable quantitative trait is unaffected by the environment and that genetic variation behind it is due to a single locus A with two equally frequent alleles A and a, so that, in a population with random mating, genotypes AA, Aa, and aa have frequencies 0.25, 0.5, and 0.25, respectively. Let us further assume that actual (as well as inherited) values of the trait of individuals of genotypes AA, Aa, and aa are always 1, 2, and 1, respectively. The situation when the heterozygote produces a higher value of the trait than both heterozygotes is called overdominance (see Chapter 10). In the course of reproduction, each individual mates with a partner with genotypes AA, Aa, and aa with probabilities 0.25, 0.5, and 0.25, respectively. Taking into account Mendelian segregation, we can see that, for individuals of every genotype, half of their offspring is homozygous and another half heterozygous, so that transmittable value of the trait is always the same, and equal to 1.5 (Table 3.2). Thus, although the value of the trait of an individual is exactly determined by its genotype, it is not at all transmitted to its offspring, and $V_t = 0$!

How could this be possible? Because in our example, the phenotype of an individual depends only on the combination of alleles in its genotype. An isolated allele determines nothing: neither A nor a *per se* leads to a higher value of a trait. In such a situation,

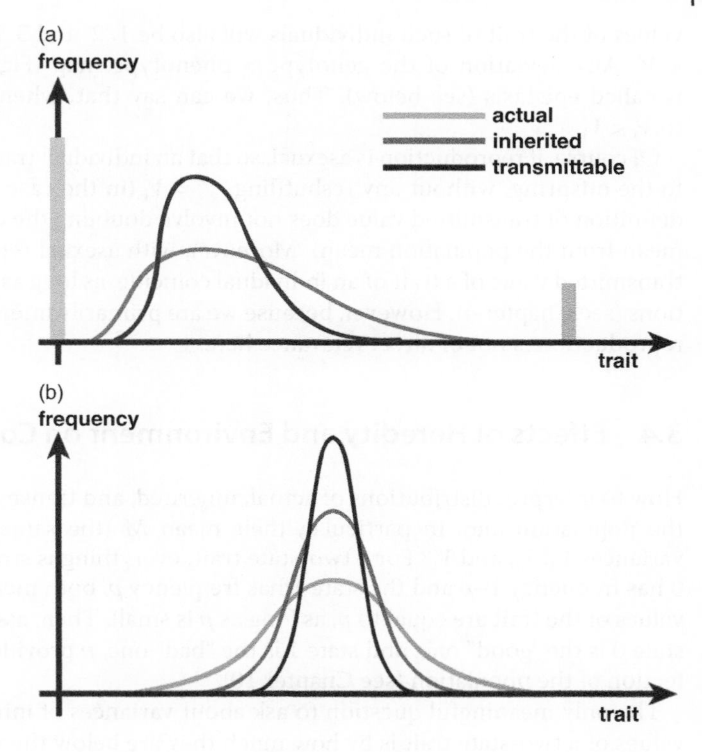

Figure 3.7 Distributions of actual, inherited, and transmittable value of a two-state (a) and quantitative (b) complex traits within a population.

Table 3.2 The case of no inheritance of genetic influences (see text).

	Offspring from matings with			
Genotype	**AA**	**Aa**	**aa**	**Total offspring**
AA	1.0 AA	0.5 AA, 0.5 Aa	1.0 Aa	0.5 AA, 0.5 Aa
Aa	0.5 AA, 0.5 Aa	0.25 AA, 0.5 Aa, 0.25 aa	0.5 Aa, 0.5 aa	0.25 AA, 0.5 Aa, 0.25 aa
aa	1.0 Aa	0.5 Aa, 0.5 aa	1.0 aa	0.5 Aa, 0.5 aa

reshuffling of alleles in the course of sexual reproduction, due to Mendelian segregation within loci and to reciprocal recombination of loci, erases resemblance between parents and offspring. If there is only one polymorphic locus, so that only Mendelian segregation is relevant, genotypes are fully randomized, and parent–offspring resemblance destroyed, in just one generation.

Thus, $V_t = V_i$ only when the inherited value of a trait is simply a sum of contributions of alleles that constitute the genotype of an individual, in which case sexual reshuffling is of no consequence. In scientific parlance, we can say that in this case different alleles affect the trait additively. For example, if contributions of A and a to the trait are 0.5 and 1.5, actual and inherited values of the trait in individuals of genotypes AA, Aa, and aa would be 1, 2, and 3, respectively. Then, it is easy to verify that transmittable values of the trait of such individuals will also be 1, 2, and 3. In all other situations, $V_t < V_i$. Any deviation of the genotype > phenotype map (Figure 1.21) from linearity is called epistasis (see below). Thus, we can say that, when present, epistasis leads to $V_t < V_i$.

Of course, if reproduction is asexual, so that an individual transmits its intact genotype to the offspring, without any reshuffling, $V_t = V_i$ (in the case of asexual reproduction, definition of transmitted value does not involve doubling the deviation of the offspring mean from the population mean). Moreover, with asexual reproduction inherited and transmitted value of a trait of an individual coincide, as long as we ignore *de novo* mutations (see Chapter 4). However, because we are primarily interested in humans, asexual reproduction is not of direct relevance here.

3.4 Effects of Heredity and Environment on Complex Traits

How to interpret distributions of actual, inherited, and transmittable values of a trait in the population and, in particular, their mean M (the same in all three cases) and variances V_a, V_i, and V_t? For a two-state trait, everything is straightforward. If the state 0 has frequency $1-p$ and the state 1 has frequency p, both mean and variance of actual values of the trait are equal to p, as long as p is small. Then, assuming that the common state 0 is the "good" one, and state 1 is the "bad" one, p provides the measure of imperfection of the population (see Chapter 10).

The only meaningful question to ask about variances of inherited and transmittable values of a two-state trait is by how much they are below the variance of its actual values. The answer to these question is provided by ratios $h^2_i = V_i/V_a$ and $h^2_t = V_t/V_a$, which are called heritabilities of inherited and transmittable values, respectively (or broad-sense and narrow-sense heritability). Theoretically, a heritability can vary from 0 to 1.

The situation is more complex for a quantitative trait. Suppose that the mean height within a population is 175 cm and its variance is 55 cm^2 – are they "small" or "large"? In order to answer such questions, we need to relate means and variances to some standards. Mean can be related to the optimal, or perfect, value of the trait, but, by contrast to a two-state trait, this value is not known *a priori* and cannot be simply inferred from within-population variation. The question "is this mean of a quantitative trait large?" will be addressed in Chapter 10.

Here we will concentrate on interpreting V_a, V_i, and V_t of a quantitative trait, which characterize the whole scattering of the actual trait value, the part of scattering due to variation of genotypes of individuals, and the part of scattering due to variation of genotypes of parents of individuals, respectively. By contrast to the case of a two-state trait, there is no natural unit of measurement for a quantitative trait. Of course, a parameter which tells us whether a variance is large or small must be dimensionless, because the answer must not depend on the choice of the unit of measurement.

Heritabilities of inherited and transmittable values of a trait obviously satisfy this condition. However, there is another way of obtaining dimensionless parameters which characterize variances: they can be normalized not over the variance of actual values of the trait, but over the square of its mean (values of traits that characterize a living being usually cannot be negative, so that their mean values are not arbitrary). In 1992, David Houle proposed the term evolvability for such variances of mean-normalized traits, and

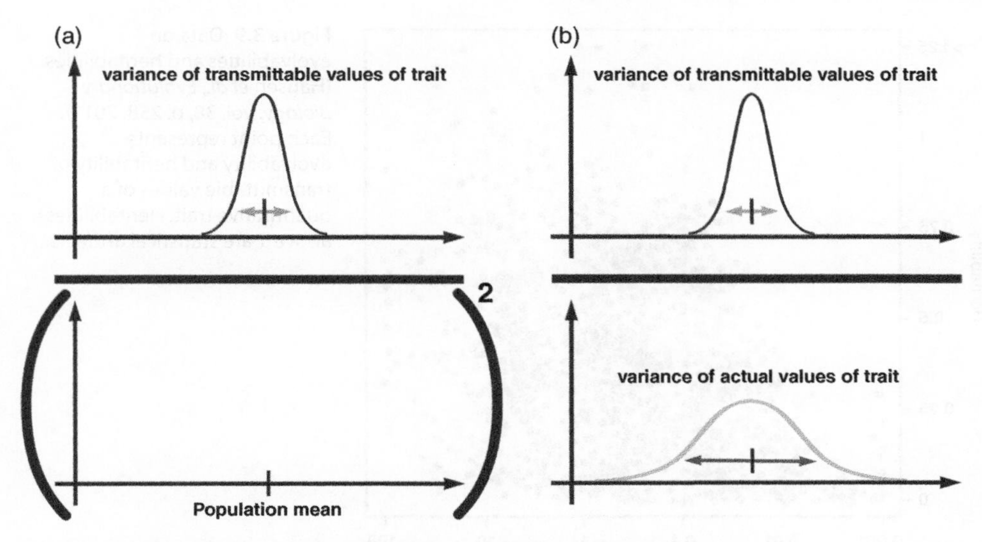

Figure 3.8 Evolvability (a) and heritability (b) of transmittable values of a trait.

we can consider evolvability of actual, inherited, or transmittable value of a trait, $E_a = V_a/M^2$, $E_i = V_i/M^2$, or $E_t = V_t/M^2$, respectively (Figure 3.8) Theoretically, evolvability can vary from 0 to infinity. The square root of an evolvability, or the standard deviation of a trait over its mean, is known in statistics as coefficient of variation (CV). For example, if $E_t = 0.01$, the corresponding CV_t is 0.1, and the standard deviation of the trait is 10% of its mean. As long as we know both M and V_a (these parameters can be measured directly), knowing an evolvability is enough to calculate the corresponding heritability, and vice versa.

Both evolvabilities and heritabilities are useful characteristics of a quantitative trait, because they address different questions. Evolvability tells us how large scattering of the actual, inherited, or transmittable value of the trait is relative to its mean. Heritability tells us what proportion of total scattering is due to inherited or transmittable genetic influences.

Thus, high evolvability does not necessarily imply high heritability, and vice versa. Indeed, even if all scattering of a trait is transmittable, so that heritability is 1.0, evolvability can still be low if this scattering is narrow. Figure 3.9 presents a compilation of data on evolvabilities and heritabilities of transmittable values of many quantitative traits in many species. We can see that median evolvability is ~0.005 (which corresponds to standard deviation of transmittable values constituting 7% of the mean), and median heritability is ~0.4 (implying that 40% of actual scattering is due to transmittable genetic influences). Still, both evolvabilities and heritabilities vary widely, and are not tightly correlated. Data on humans will be considered in Chapter 12.

How can heritabilities and evolvabilities be measured? This is a rather technical issue, so only some key ideas will be outlined here. The easiest part is to study actual values of the trait in the population. Indeed, this only requires determining these values for a large enough unbiased sample of individuals. As a result, the whole distribution of actual values of the trait can be ascertained. Of course, doing this is much

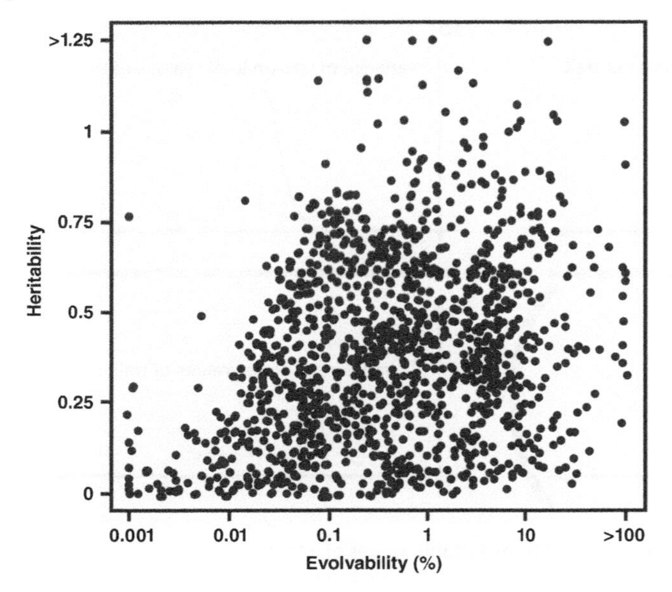

Figure 3.9 Data on evolvabilities and heritabilities (Hansen *et al.*, *Evolutionary Biology*, vol. 38, p. 258, 2011). Each point represents evolvability and heritability of transmittable values of a quantitative trait. Heritabilities above 1 are statistical artifacts.

simpler for directly measurable traits, such as body mass, and not for traits such as life-time reproductive success in the wild (see Chapter 7).

By contrast, one cannot determine inherited or transmittable value of the trait of an individual, unless the trait can be measured in many identical twins (or clones) or children of this individual, which usually is not the case. Thus, it is impossible to directly ascertain distributions of inherited or transmittable values of traits. Fortunately, several indirect methods make it possible to estimate overall characteristics of these distributions, in particular, their variances, which is sufficient for many purposes. Scattering of both inherited (see Chapter 7) and transmittable (see Chapter 12) values of traits can be of primary importance, depending on a problem considered.

Traditional methods of estimating V_i or V_t are based on comparison between similarities of actual values of a trait in related versus unrelated individuals. Excessive phenotypic similarity between relatives is a ubiquitous phenomenon, and the closer the relatives the more similar they are. V_i can be estimated from data on pairs of identical twins, because differences between actual values of a trait within such pairs provide information on the variance of these values due to random environmental scattering. Then, we can estimate V_i by subtracting this environmental variance from the directly measurable V_a. An important caveat is that identical twins, but not other kinds of relatives, share all their *de novo* prezygotic mutations (most mutations are singletons, so that siblings do not share them, see Chapter 4). Thus, high estimates of V_i obtained from data on identical twins likely reflect the contribution of *de novo* mutations (see Chapter 13).

Genotypes of parents and offspring differ from each other due to Mendelian segregation and meiotic recombination, and excessive similarity between their phenotypes provides information only on V_t (Figure 3.10). For example, excessive similarity between a parent and its offspring can be characterized by the slope of the straight line, called the regression coefficient, which provides the best fit for the data on actual values of the

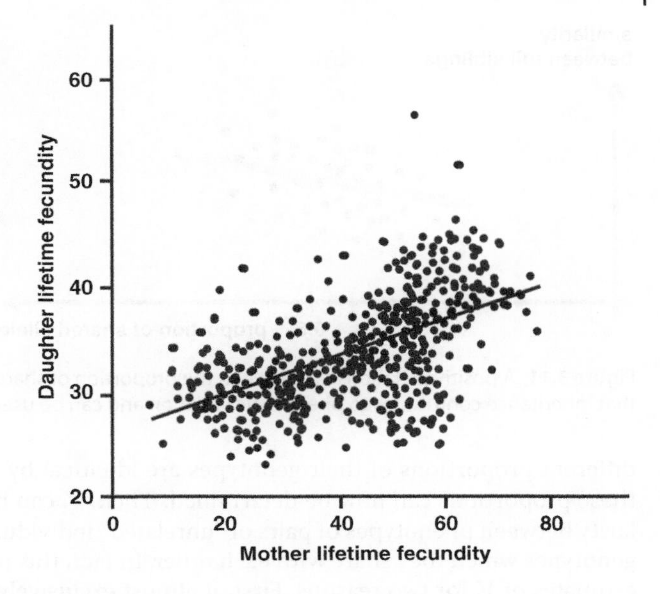

Figure 3.10 The line which shows linear dependence on the life-time fecundity of *Drosophila* individuals on that of their mothers (Long *et al.*, *Journal of Evolutionary Biology*, vol. 22, p. 637, 2009). The slope of this line, the parent–offspring regression coefficient, provides the simplest way to estimate V_t.

trait in (parent, offspring) pairs. The regression coefficient multiplied by V_a is called the parent–offspring covariance, and V_t is equal to twice the covariance (and regression coefficient *per se* is equal to $h^2{}_t/2$). Data on excessive similarities between other kinds of relatives, such as siblings or cousins, as well as on complex pedigrees, may shed light on interactions between alleles (epistasis, see below), which makes it possible to estimate both V_t and V_i.

However, such estimates of V_i or V_t may be too high because relatives not only possess excessively similar genotypes, but often also experience excessively similar controllable environments. Data on twins separated at birth, or on parents and their children adopted immediately after birth, should mostly be free from this bias, although twins always share the same environment before birth.

Recently, genotype-level data made it possible to develop two new methods of estimating V_t. One of them takes advantage of the fact that different pairs of relatives of the same kind may, nevertheless, share different proportions of their genotypes. For example, a pair of full siblings is expected to share 50% of alleles, but the actual proportion varies substantially, due to randomness of meiotic recombination. While ~10% of pairs of full siblings in humans share less than 45% of their alleles, ~10% share more than 55% (of course, a parent and a child always share exactly 50% of their alleles, located on autosomes). By comparing genotypes of a pair of siblings, it is now easy to determine what proportion of them is actually shared. Siblings that are more genetically similar are also more similar phenotypically, and the strength of this effect can be used to estimate V_t (Figure 3.11). Here, environmental biases are unlikely: sisters sharing 45% versus 55% of their alleles probably experience equally similar environments.

The other modern method of estimating V_t is based on comparison of genotypes of "unrelated" individuals who do not know each other, live in different locations, and, indeed, are not closely related to each other. Of course, every two individuals within a population are, in fact, relatives, because they share some common ancestors, usually no more than 10–20 generations ago. For different pairs of "unrelated" individuals,

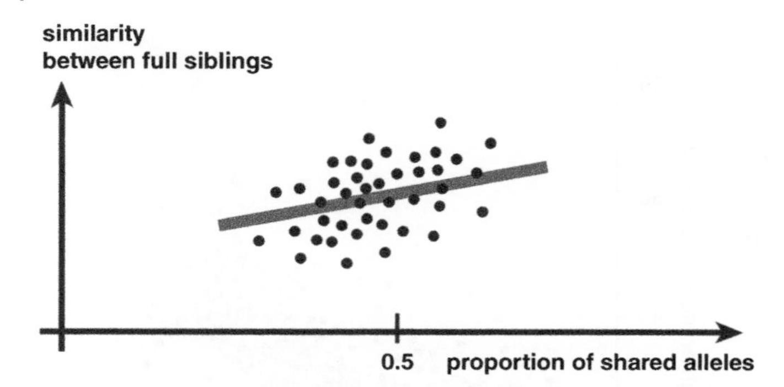

Figure 3.11 A positive relationship between the proportion of shared alleles and similarity indicates that inheritance contributes to phenotypic variation and can be used to estimate V_t.

different proportions of their genotypes are identical by descent (see Chapter 2), and these proportions can now be determined. Then, V_t can be estimated by relating similarity between phenotypes of pairs of "unrelated" individuals to the proportions of their genotypes which they share with each other. In fact, this method provides conservative estimates of V_t for two reasons. First, it almost exclusively measures separate effects of alleles, while epistasis (see below) can also contribute to similarity between parents and offspring and to V_t. Secondly, it ignores contributions of very young alleles, produced by mutation (see Chapter 4) in recent generations.

The above methods can be used to estimate the contribution of genetic variation into variation of both quantitative and two-state complex traits. In the first case, similarity between two individuals is characterized by a number, the difference between their actual values of the trait. In the second case, similarity can accept only two values – two individuals can be concordant (possess the same actual value of the trait) or discordant.

Table 3.1 uses data on variances of transmittable values of six human traits (data on inherited values are scarce for humans). Both heritabilities and evolvabilities are presented for all traits, although for two-state traits evolvabilities do not make much sense. It seems that situations such as that shown in Table 3.2, where transmittable genetic effects on a complex trait are absent or very weak are rare, because usually epistasis, if present, is only moderate. Thus, in most cases V_t is likely to be only slightly below V_i.

Contributions of genetic variation to phenotypic variation of complex traits may be very different in different populations. Heritable within-population differences do not necessarily imply that heredity plays a role in between-population differences, which can be exclusively due to different environments.

3.5 Polymorphic Loci Behind Complex Variation

Knowing the overall contribution of heredity to variation of a complex trait, characterized by V_i and V_t, is important, but not enough. We also want to identify polymorphic loci that are behind it. Indeed, if a population is genetically uniform, the contribution of heredity to variation of every trait must be zero. Here, a locus that affects variation of a complex trait will be referred to as a CTL (complex trait locus) for this trait. The widely used term QTL (quantitative trait locus) is inaccurate, because many important complex traits are two-state. Two characteristics of a CTL are of particular importance. The first is the derived allele frequency, or DAF (at most polymorphic loci only two alleles reach substantial frequencies and the derived allele is rarer, see Chapter 2). The second is the average effect of the CTL on the trait, which can be described by the change of the value of the trait due to replacement of the ancestral allele with the derived one. For example, if individuals carrying homozygous common allele at some CTL have, on average, systolic blood pressure of 120, and those carrying heterozygous derived allele have a value of 121.2, the absolute effect of this allele is 121.2–120 = 1.2 and the relative effect is (121.2–120)/120 = +0.01. We will mostly consider relative effects of alleles, in particular, because such effects, called selection coefficients, are used for describing how polymorphic loci affect the most important complex trait, fitness (see Chapter 7).

Identifying CTLs is difficult. Some of the methods that were developed for this purpose rely on information obtained in the course of many generations. For example, a method known as linkage analysis is based on observing genotypes and phenotypes of individuals in a multigeneration pedigree (genealogy), and detecting alleles whose transmission is correlated with inheritance of a particular phenotype. However, the most straightforward, and, currently, the most widely used method, known as genome-wide association study (GWAS) uses information obtained in the course of only one generation. GWAS directly explores the genotypes > phenotypes map (Figure 1.21) and consists of recording genotypes and phenotypes of very many individuals and identifying polymorphic loci whose different alleles are associated with different values of the trait under study (Figure 3.12).

In contrast to alleles that cause Mendelian diseases, alleles at CTLs can – and often do – have only small effects on the trait. Reliable identification of a CTL is particularly difficult when DAF is low and the effect of the allele is small. Suppose that at a CTL with DAF = 0.01 the heterozygous derived allele increases the risk of schizophrenia by 3%. If we compare genotypes of, say, 10 000 individuals with schizophrenia and of 10 000 healthy controls, there will be 200 copies of the derived allele in the control genotypes versus 206 copies in genotypes of affected people. Of course, this difference is statistically insignificant, and a much larger sample is needed to discover such a CTL.

This problem is exacerbated by a statistical phenomenon known as multiple testing. Because there is almost no correlation between the function of a gene and its location, genes that constitute the target of a particular complex trait are scattered throughout the genome. Moreover, the current level of understanding of the connection between

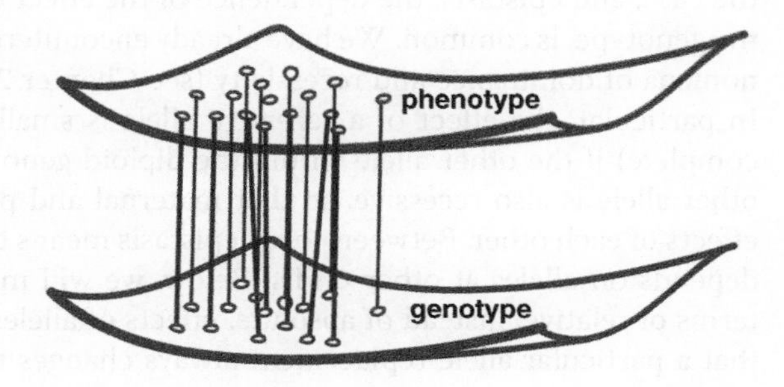

Figure 3.12 Direct detection of CTLs by a GWAS.

genotypes and phenotypes does not make it possible to rule out that any genome segment belongs to the target of a particular complex trait. Thus, when a trait is studied, it is necessary to consider every polymorphic locus, compare average values of the trait in those who carry the two alleles at it, and detect loci for which these values are significantly different. Analyzing very many loci inevitably leads to many random "false positive" associations. The only way to deal with this problem is to further increase the sample size, in order to gain a statistical power to detect rare real associations among multiple spurious ones.

Another problem with GWAS is caused by the structure of genetic variation within populations. Alleles at polymorphic loci that are situated close enough to each other in the genome (in humans, closer than ~100 000 nucleotides, see Chapter 11) are distributed non-independently. Thus, the derived allele at a real CTL may be associated with particular alleles at adjacent polymorphic loci which are phenotypically mute. As a result, it is very difficult to precisely detect a CTL that causally affects a particular phenotypic trait. Instead, one often has to be content with identifying only a long genome segment which harbors one or several CTLs.

On top of these inescapable problems, the power of a GWAS to identify CTLs was limited, until very recently, by incomplete information about genotypes. Instead of actual sequencing of genotypes, which was too expensive, only a sample of loci with high enough DAFs were investigated. This approach, known as genotyping, is much less powerful than sequencing and, in particular, is prone to overlooking CTLs with low DAFs. The first papers that related exact sequences of (almost) complete human genotypes to phenotypes appeared only in 2013. Unfortunately, the sequence-based studies performed so far have involved only thousands of individuals, although identifying CTLs of small effects will likely require much larger samples. However, because revealing the relationship between genotypes and phenotypes is one of a few "Holy Grails" of modern biology, rapid progress can be expected.

Although the target of a complex trait is essentially invariant within a species, the numbers and identities of CTLs may be very different in different populations. Indeed, a locus which is polymorphic in one population may be monomorphic in another. Moreover, between-population CTLs, that is, those loci that are responsible for between-population differences in complex traits, may be monomorphic within populations, and, thus, not be within-population CTLs. This is the case, for example, for genes that are responsible for contrasting skin colors in a pair of human populations with uniformly dark and uniformly light skin.

So far in this section, we implicitly assumed that a CTL always has the same effect on a complex trait, independent of the genetic background. However, this is not necessarily the case, and epistasis, the dependence of the effect of an allele on other alleles within the genotype, is common. We have already encountered within-locus epistasis, the phenomena of dominance and recessivity (see Chapter 2) and overdominance (Table 3.2). In particular, the effect of a recessive allele is small (or even absent, if recessivity is complete) if the other allele within the diploid genotype is dominant, but large if the other allele is also recessive, so that maternal and paternal recessive alleles reinforce effects of each other. Between-locus epistasis means that the effect of an allele at a CTL depends on alleles at other CTLs. Below, we will mostly be interested in epistasis in terms of relative, instead of absolute, effects of alleles. In this case, no epistasis means that a particular allele replacement always changes the value of the trait by the same

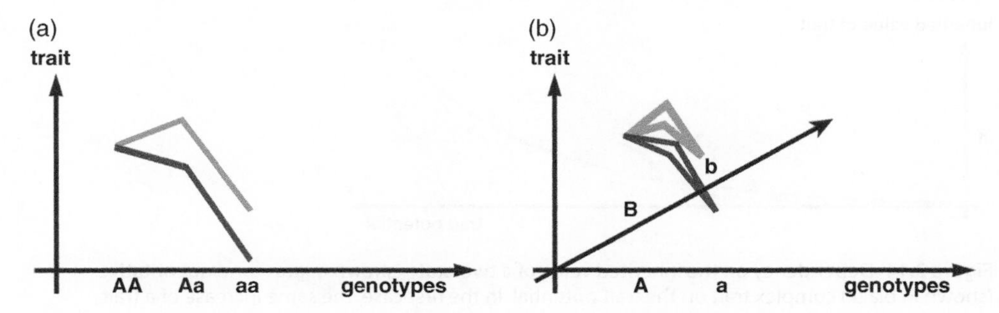

Figure 3.13 Within-locus epistasis in the case of one diploid locus (a) and between-loci epistasis in the case of two haploid loci (b). In the case of monotonic epistasis between derived alleles, shown in lower case, two such alleles, either within the same diploid locus, or at different haploid loci, reinforce effects of each other (shown in black). In the case of sign epistasis, derived alleles present alone (genotype Aa in the case of one diploid locus, and genotypes Ab and aB in the case of two haploid loci) and derived alleles present together (aa and ab) affect the trait in the opposite directions (shown in gray).

proportion and, therefore, makes the same absolute contribution into the logarithm of the value of the trait (see Chapter 7). Then, $V_t = V_i$ on the logarithmic scale.

Epistasis can be monotonic, in the sense that the magnitude of the effect of an allele depends on the genotype, but its direction does not. By contrast, under the extreme form of epistasis, called sign epistasis, the very direction of the effect of an allele depends on the genetic background (Figure 3.13). How important sign epistasis is at the scale of within-species variation is still unknown (see Chapter 15).

When we consider a complex trait, epistasis may involve interactions between alleles at multiple loci, which can be very complex. Then, a useful simplifying assumption is that all variable loci within the target of a trait contribute to just one variable, called a trait potential or genotype score, which somehow determines the inherited value of the trait in an individual. This assumption, called one-dimensional epistasis, is flexible enough to describe a variety of interactions between alleles, but is still much simpler than the general case. It can be traced back to one of the founders of statistics and mathematical biology, Karl Pearson, but became widely used only in the1960s, in particular, after a seminal paper by Douglas C. Falconer on two-state traits. At the same time, the assumption of one-dimensional epistasis also found its way into the theory of natural selection (see Chapter 7).

For the two extreme cases of a two-state and a quantitative trait, the likely dependences of their inherited values on the trait potential are shown in Figure 3.14. In the first case, if we denote the two possible actual values of the trait by 0 and 1, its inherited value is a probability of state 1. Let us assume that low values of the trait potential favor state 0, and high values favor state 1. When the trait potential increases, the increase of probability of state 1 can be rather sharp, although it cannot be perfectly abrupt, due to chance environmental influences on the trait. Still, monotonic epistasis is inherent for a two-state trait: an allele replacement that is unlikely to lead to state 1 in the genotype with a low trait potential can easily do this when the trait potential is high. One can say that penetrance of alleles that can cause state 1 increases with the trait potential. By contrast, in the case of a quantitative trait there is no *a priori* need to reject the simplest assumption of no epistasis. Then, if the relative contribution of

Figure 3.14 Dependency on the inherited value of a two-state (shown in gray) and quantitative (shown in black) complex trait on the trait potential. In the first case, the same increase of a trait potential due to a particular allele replacement substantially increases the probability of state 1 only when the trait potential is already high. In the second case, as long as epistasis is absent (as shown in the figure), the relative effect of a particular allele replacement is always the same.

an allele to the inherited value of the trait is independent of the rest of the genotype, this value depends exponentially on the trait potential.

All kinds of epistasis further complicate finding CTLs. Thus, it should not be surprising that there is a large gap between our knowledge of the overall contributions of heredity to variation of complex traits, which are usually quite large, and rather limited data on individual CTLs, which are responsible for these contributions. Often, a study reveals only a modest number of CTLs which together are capable of explaining only a small proportion of the observed overall effect of heredity on the trait. This situation is known as the missing heritability problem. Moreover, we still do not know for sure whether most alleles that increase susceptibility of individuals to complex diseases are ancestral or derived. Even less is known about epistatic interactions between alleles at CTLs. The available data on complex human traits are considered in Chapters 12 and 15.

Further Reading

Chen R, Shi L, Hakenberg J *et al.* 2016, Analysis of 589,306 genomes identifies individuals resilient to severe Mendelian childhood diseases, *Nature Biotechnology* DOI:10.1038/nbt.3514.

Even alleles that cause classical Mendelian diseases may be not fully penetrant.

Falconer, DS 1966, The inheritance of liability to certain diseases, estimated from the incidence among relatives, *Annals of Human Genetics*, vol. 29, pp. 51–76.

Early development of the concepts of a train potential and of a two-state complex trait.

Fu, W, O'Connor, TD, & Akey, JM 2013, Genetic architecture of quantitative traits and complex diseases, *Current Opinion in Genetics and Development*, vol. 23, pp. 678–683.

A review of the concept of genetic architecture of a phenotypic trait.

Hansen, TF, Pelabon, C, & Houle, D 2011, Heritability is not evolvability, *Evolutionary Biology*, vol. 38, pp. 258–277.

Comparison of different ways of characterizing the genetic effects on variation of a trait.

Lettre, G 2014, Rare and low-frequency variants in human common diseases and other complex traits, *Journal of Medical Genetics*, vol. 51, pp. 705–714.

On the effects of rare alleles on complex phenotypic traits.

Morrison, AC, Voorman, A, & Johnson, AD 2013, Whole-genome sequence-based analysis of high-density lipoprotein cholesterol, *Nature Genetics*, vol. 45, pp. 899–901.
The first study of genetic effects on a complex phenotypic trait based on sequencing of complete genotypes.

Phillips, PC 2008, Epistasis – the essential role of gene interactions in the structure and evolution of genetic systems, *Nature Reviews Genetics*, vol. 9, pp. 855–867.
A comprehensive review of the phenomenon of epistasis.

Polderman, TJC, Benyamin, B, & de Leeuw, CA 2015, Meta-analysis of the heritability of human traits based on fifty years of twin studies, *Nature Genetics*, vol. 47, pp. 702–709.
Review of estimates of genetic effects on human traits based on twin studies.

Robinson, MR, Hemani, G, & Medina-Gomez, C 2015, Population genetic differentiation of height and body mass index across Europe, *Nature Genetics*, vol. 47, pp. 1357–1362.
A study of genetic factors responsible for phenotypic differences between human populations.

Zaitlen, N, Kraft, P, & Patterson, N 2013, Using extended genealogy to estimate components of heritability for 23 quantitative and dichotomous traits, *PLoS Genetics*, vol. 9: e1003520.
Estimates of genetic effects on human traits obtained by comparing phenotypes of distantly related individuals. Data on two-state traits in Table 3.1 are from this paper.

4

Unavoidable Mutation

In short, mutations are accidents, and accidents will happen.

Alfred H. Sturtevant, 1937.

The ultimate cause of all genetic variation is mutation, the appearance of random errors in DNA sequences. Most of individual mutations affect only short segments of genotypes, and single-nucleotide substitutions and short deletions or insertions are particularly common. However, occasionally a mutation affects a very long DNA segment. Mutations can appear as singletons or as clusters. Under normal conditions, most mutations occur spontaneously, due to unavoidable errors in the course of DNA replication, repair, or recombination, and are not caused by any external influences. Apparently, natural selection tries to keep the rate of germline mutation in multicellular organisms as low as possible, without incurring too high a cost of DNA handling. It is not difficult to induce mutations by radiation or, even more efficiently, by chemical mutagens; in contrast, all attempts at germline antimutagenesis have been so far unsuccessful. Tools for active antimutagenesis by means of deliberate genotype modification are now in the early stages of development.

4.1 Phenomenon of Mutation

In the review of genetics, presented in the previous three chapters, it was mentioned in passing that polymorphic loci originate as a result of mutation. This phenomenon is the root cause of the Main Concern, and now it is time to treat it in detail. Most generally, mutation (from the Latin *mutatio* meaning an alteration) is a process of appearance of changes in genotypes. Due to the discrete nature of DNA sequences, mutation consists of separate events, and each such event is also called a mutation. To add to the confusion, a new, altered, mutant DNA sequence, which emerges as the result of a mutation, is also called a mutation (Figure 4.1)! And an individual that carries a *de novo* (fresh) mutation is a mutant.

The length of a DNA segment within which a mutant genotype differs from the original one can vary from just one to millions of nucleotides. In terms of population genetics, a mutation represents a derived allele at a polymorphic locus whose boundaries coincide with that of the affected DNA segment, while the ancestral allele is represented by the sequence that existed before the mutation. A mutation can be novel for the population, in which case it creates either a new polymorphic locus or a new, previously absent,

Crumbling Genome: The Impact of Deleterious Mutations on Humans, First Edition. Alexey S. Kondrashov.

```
atGct > atAct
```

Figure 4.1 In the course of mutation (process), a mutation (event) results in a mutation (new allele). The altered nucleotide is in upper case. Single-nucleotide substitution is by far the most common kind of mutation.

```
           atGct > atAct

atgct          attct        atact
atgct          atgct        atact
atgct          atgct        atgct
atgct          atgct        atgct
atgct          atgct        atgct
```

Figure 4.2 Depending on the genetic composition of the population, the same mutation, shown on top, produces a new polymorphic locus (left column), a new allele at an already-present polymorphic locus (center column), and an already-present allele (right column).

```
cgatgtacagtactgcataCgtg          cgatgtacagtactgcataTgtg
           X                 >
cgatgtacagtactgcataTgtg          cgatgtacagtactgcataCgtg
```

Figure 4.3 Reciprocal recombination between two genotypes that differ from each other at just one locus (in upper case) does not produce anything new.

allele at an already-polymorphic locus. Alternatively, a mutation can be recurrent, producing an extra copy of a derived allele that is already present (Figure 4.2). Most diallelic polymorphic loci (with derived allele frequency >0.01) within modern populations emerged due to a unique ancestral mutation that created the derived allele.

Mutation is the ultimate cause of all genetic variation. Indeed, no other force is capable of producing new polymorphic loci. Reciprocal meiotic recombination (see Chapter 1), which creates new combinations of alleles at already polymorphic loci, would be genetically mute in a monomorphic population, or even in a population polymorphic at just one locus (Figure 4.3).

The word mutation refers to young, rare, and clearly abnormal derived alleles, and the word allele is used when it is older, more common, and not abnormal; however, this convention is rather loose. I reserve the word mutation only to fresh products of the mutation process, and, if a mutation persists for even a short while, it becomes a (mutant) allele.

In most cases, a mutation is soon lost from the population, being removed either by negative selection (see Chapter 7) or by random drift (see Chapter 8). Still, some mutant alleles persist for a long time, reaching substantial frequencies. A very small proportion of mutant, derived alleles eventually become fixed in the population, completely replacing the ancestral alleles. Of course, without such rare allele replacements life would be unable to evolve.

The key premise of evolutionary biology, common ancestry of different species, implies that mutation is the ultimate cause of not only within-population variation, but also of between-species differences. Together with natural selection, mutation is a *sine qua non* of Darwinian evolution. However, mutant alleles responsible for interspecific differences, which passed the test of natural selection, are different from those responsible for within-population variation: many of the former are advantageous, and most of the latter, as long as they have any effect on the phenotype, are deleterious (see Chapters 6, 7, 9, and 11).

4.2 Kinds of Mutations

The mutation process produces a wide variety of changes of DNA sequences. First of all, different mutations involve changes at very different scales. Over 99% of individual mutations are small-scale, affecting DNA sequence only within the span of, say, less than 50 nucleotides. Among those, a vast majority consists of a simple replacement of one continuous DNA segment, of length L_1, with another segment, of length L_2. Among such mutations, single-nucleotide substitutions ($L_1 = L_2 = 1$) constitute ~90% of cases, followed by ~6% of deletions ($L_1 > 0, L_2 = 0$), ~2% of insertions ($L_1 = 0, L_2 > 0$) and ~2% of complex events (both L_1 and L_2 positive, and at least one of them above 1) (Figure 4.4). Exact proportions may be different in different organisms, but the general pattern appears to be universal.

There is a one-to-one correspondence between mutations that affect just one segment of the DNA and polymorphisms. To emphasize this point, mutations shown in Figure 4.4 coincide with polymorphisms shown in Figure 2.8. However, occasionally a mutation results in a mutant allele which is different from the original allele at several distinct segments of the genome (Figure 4.5). One can say that such a complex mutational event creates several new, closely spaced polymorphic loci within the population.

Simple deletions, insertions, or inversions of a continuous DNA segment are the most common kinds of large-scale mutations, and they correspond precisely to long polymorphisms (Figure 2.9). However, a large-scale mutation may also involve rather complex rearrangement of the original genotype. There is a plethora of mishaps in handling of DNA molecules that can result in large-scale mutations (see Chapter 5), and if many things go wrong at once, the mutant allele can be weird at the level of genotype, and drastic at the level of phenotype (Figure 4.6).

In terms of their effects on phenotypes, *de novo* mutations can be, exactly like pre-existing derived alleles, dominant or recessive; drastic, mild, or mute; haploinsufficient

```
actAcag > actGcag
ataTATCCgct > atagct
agaacg > agaCacg
cgaCGtgg > cgaAtgg
```

Figure 4.4 A single-nucleotide substitution, a deletion, an insertion, and a complex event. Three nucleotides are shown on each side of each mutation.

CCAG~~CTGC~~TGGACGCCA~~TGTCCGGGAGGCTGGGG~~CGCGGGGACCTTCCTGGGGAGG
GGA T CT G T

Figure 4.5 A complex mutation (after Chen *et al.*, *Human Mutation*, vol. 30, p. 1435, 2009). Barred sequences denote deleted nucleotides whereas nucleotide substitutions are indicated below the original sequence.

Figure 4.6 Sequencing of the genotype of a boy with severe congenital abnormalities, together with genotypes of his parents, revealed a paternal mutation that involved a complex rearrangement of 16 segments from chromosomes 1, 4, and 10. The figure shows a simplified scheme of three derived, rearranged chromosomes, present in the boy's genotype. Multiple short deletions and insertions at junctions between the rearranged segments are not shown (after Kloosterman *et al.*, *Human Molecular Genetics*, vol. 20, p. 1916, 2011).

or negatively dominant; and completely or incompletely penetrant. Quantitatively, however, *de novo* mutations are, on average, much more drastic than pre-existing alleles at polymorphic loci. Indeed, mild and, especially, mute alleles on average persist in the population for much longer that drastic alleles (see Chapter 7).

A mutation which makes the genotype of an offspring different from that of its parents is an inherited mutation. In multicellular organisms, inherited mutations occur, within a parental organism, in cells that are directly involved in reproduction. In humans, these are gametes and their precursor cells, whose descendant cells will eventually become gametes. Such cells are collectively called germline; thus, inherited mutations in multicellular organisms are germline mutations. Germline mutations which occur late, close to the formation of gametes, affect only a small proportion of them and appear in the offspring as singletons. By contrast, early occurring germline mutations often form clusters, appearing in many gametes and offspring (Figure 4.7).

Mutations that occur in cells outside the germline are called somatic. Somatic mutation obviously cannot affect genotypes of offspring and, instead, causes different cells within an organism to have slightly different genotypes, a phenomenon called genetic mosaicism. Due to a high rate of somatic mutation, we are all mosaics, even in the absence of disease. Somatic mutations that impair regulation of the cells' cycle is the root cause of cancer. In the course of development of the immune system of vertebrates, localized somatic mutations are produced deliberately at a very high rate, to ensure a wide diversity of antibodies (see Chapter 5).

Mutations have been observed since times immemorial because children suffering from severe diseases due to *de novo* mutations are born regularly (see Chapter 13). However, the phenomenon of sudden changes of hereditary information has been recognized, and the very term mutation proposed, only in 1903, by Hugo de Vries, one of the rediscoverers of Mendelian genetics. Soon after this, the key role of germline mutation in genetics and evolution became widely appreciated. In contrast, the key role of somatic mutation in the origin of cancer became apparent only in the last quarter of the 20th century.

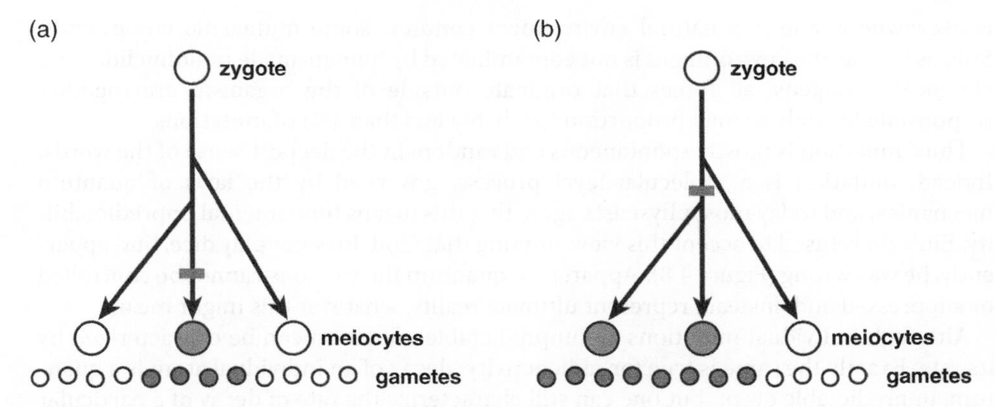

Figure 4.7 Singleton (a) and cluster (b) germline mutation. Cells carrying the mutation are shown in grey. The moments when mutations occurred are shown by horizontal bars.

4.3 Spontaneous Mutation

In contrast to meiotic recombination, which depends on deliberate efforts by the cell (see Chapter 5), mutation mostly occurs spontaneously. Here, analogy with random influences on phenotypes is helpful: monozygotic twins are never exactly the same height (or thumbs of the right and left hand of the same length), no matter how hard we try to standardize the environment (see Chapter 3). Mutation occurs spontaneously in the same sense: it does not require any provocation and, moreover, there is no way to prevent it from happening. Indeed, DNA molecules consist of parts and, thus, are prone to crumbling.

However, the immediate outcome of something going wrong with a DNA molecule is its damage, and not a mutation. A damaged DNA molecule is, strictly speaking, not a DNA anymore, because it no longer conforms to the chemical rules of what it means to be a DNA. A wide variety of damages is possible, and they all must be eventually repaired for the cell to survive (see Chapter 5). In contrast, a mutant DNA molecule is as good a DNA as the original one, although its nucleotide sequence is altered. A mutation emerges in several steps, triggered by an error in one of the three fundamental processes which affect DNA – its replication, repair, or recombination.

DNA replication must occur before every cell division, and its occasional errors, which happen despite multi-tier efforts to prevent them, may result in mutations. In contrast, spontaneous damage to a DNA molecule can happen at any moment, even when the molecule lays idle (see Chapter 5). These damages are primarily caused by normal chemical processes that constitute life, and not by any external insults. The only reason why we are still alive is that almost all damages to our DNA are rapidly and correctly repaired. Still, occasionally errors of DNA repair happen, resulting in mutations. Recombination between different DNA molecules or similar segments of the same molecule is a necessary step in many mechanisms of DNA repair, and reciprocal recombination in meiosis is a key part of sexual reproduction (see Chapter 1). Errors in the course of recombination also lead to mutations.

Besides, mutation can be induced by radiation and a variety of chemical compounds. All organisms are subject to some external mutagenic influences: background radiation is everywhere, and any natural environment contains some mutagenic compounds. Still, as long as the environment is not contaminated by human-made radionuclides and chemical mutagens, all forces that originate outside of the organism are together responsible for only a small proportion (probably less than 1%) of mutations.

Thus, mutation is mostly spontaneous and random in the deepest sense of the words. Indeed, mutation is a molecular-level process, governed by the laws of quantum mechanics, and today most physicists agree that this means fundamental unpredictability. Einstein refused to accept this view, arguing that God does not play dice, but, apparently, he was wrong (Figure 4.8). Apparently, quantum fluctuations cannot be controlled or suppressed and, instead, represent ultimate reality, whatever this might mean.

Although individual mutations are unpredictable, mutation can be characterized by its rate. Exactly the same is true for radioactivity: decay of an individual atom is a quantum, unpredictable event, but one can still characterize the rate of decay of a particular isotope by what proportion of atoms in a large sample decays during a unit of time. Analogously, mutation rate is the expected number of mutations that affect some DNA segment in the course of some time interval. Unpredictability of individual mutations

Figure 4.8 God does play dice.

does not mean that all mutations occur with equal probabilities. Instead, mutation rates can be very different both for mutations of different kinds and for different positions in the genome (see Chapter 6), exactly as different radioactive isotopes decay at widely different rates.

Both per genome and per nucleotide mutation rates are useful. The per genome mutation rate is simply the expected number of mutations in the whole genome, and the per nucleotide mutation rate is this number divided over the genome size in nucleotides. Sometimes it is also convenient to consider mutation rate per gene.

Per cell division and per generation mutation rates are both widely used. In unicellular organisms, these two rates are the same thing, but in multicellular organisms one generation includes many cell divisions. Naturally, per generation mutation rate is K times per cell division rate, where K is the average number of germline cell divisions per generation, from zygote to zygote. In some species, this number is different in the two sexes. In particular, in humans K ~30 in females, but K ~60–500 in males (the older a male, the higher is K, see Chapter 6). In such cases it may be necessary to consider two per generation mutation rates – female and male – even if the per cell division mutation rate is about the same in both sexes.

4.4 Evolution of Mutation Rates

In biology, every phenomenon could – and should – be studied from two complementary perspectives, mechanismal and teleological. The mechanismal question "How does this happen?" is common for all natural sciences. In contrast, the teleological question "For what purpose does this happen?" can be applied only to life. Indeed, it does not make scientific sense to ask "For what purpose is the sky blue?". In contrast, the question "For what purpose is the bluebird blue?" makes perfect sense, as long as we understand "purpose" evolutionarily. Because living beings are products of adaptive evolution (see Chapter 7), we can ask, for every facet of every phenotype, why it was favored or, at least, not strongly opposed by natural selection.

One may wonder if this reasoning applies to mutation. Due to fundamental unpredictability of nature, it is impossible to reduce an error rate in any process to zero. Indeed, it can be shown that any attempt to do this would increase the cost of the process, in terms of both required energy and time, to infinity. Thus, spontaneous mutation is unavoidable because natural selection cannot lead to evolution of perfect fidelity of DNA replication, repair, or recombination.

Nevertheless, the teleological question "For what purpose?" is not irrelevant to mutation. Natural selection cannot abolish mutation altogether, but must be capable of altering the mutation rates. Cells employ a large array of exquisite antimutation tools (see Chapter 5), which clearly shows that evolution does not favor too much mutation. Thus, it makes perfect sense to ask the teleological question quantitatively: "For what purpose are spontaneous mutations as common as they are – and not less or more?".

Natural selection will be treated in some detail in Chapter 7. Here it is sufficient to consider the following thought experiment. Suppose that the laws of physics changed radically in such a way that it became possible to abolish mutation completely, without incurring any cost. Would then an allele which leads to no mutation in individuals that carry it spread in the population? If this were possible, would zero mutation rate evolve?

Obviously, there are forces that can both promote and prevent the spread of a no-mutation allele. On the one hand, if new mutations are beneficial, in the sense that they aid propagation of their carriers, a no-mutation allele will be soon outcompeted and eliminated from the population. On the other hand, if new mutations are deleterious and, thus, impede propagation of their carriers, a no-mutation allele will be eventually fixed in the population.

Both beneficial and deleterious mutations happen, but deleterious mutations are much more common (see Chapter 7). However, an individual beneficial mutation, which can eventually take over the population, can make a larger impact than an individual deleterious mutation, which is doomed to eventual elimination. Thus, even a small proportion of beneficial mutations may be enough to prevent the spread of a no-mutation allele, especially if reproduction is asexual. Even if an original mutation-causing allele leads to a *de novo* deleterious mutation in the genotype with probability 10%, it may eventually oust a no-mutation allele from the population, after producing a beneficial mutation which confers a selective advantage of over 0.1. Indeed, the genotype that carried this beneficial mutation together with the mutation-causing allele will reproduce faster that a no-mutation genotype, despite the need to sacrifice 10% of unlucky offspring every generation. Thus, if an offspring usually receives less than one *de novo* mutation, natural selection may favor an allele that causes some mutation, over a no-mutation allele.

In many, if not all, unicellular organisms the per genome per generation mutation rate is well below one. There are well-described cases when natural selection favors mutator alleles that increase mutation rates in such organisms. Apparently, this happens most often when the population experiences a novel environment. It is likely that a larger proportion of mutations are beneficial under a novel environment than under the old environment to which the population has already been adapted. However, after the population spends some time under the novel environment, mutation rate goes down, due to spread of antimutator alleles, presumably because the proportion of beneficial alleles among *de novo* mutations declines.

In contrast, in many sexual multicellular organisms mutation rates are much higher. In particular, ~100 *de novo* mutations, ~10% of which are deleterious, hit a human genotype every generation (see Chapter 8). Thus, a beneficial mutation never has the luxury of appearing alone. Would natural selection favor a non-zero mutation rate even in such a situation, if it were possible to abolish mutation painlessly? I do not think so, because only an unrealistically high proportion of very strongly beneficial mutations can outweigh the harm caused by deleterious mutations in this case.

Moreover, in multicellular organisms, mutations are both germline and somatic, and somatic mutations, which cause cancer and contribute to aging, are almost never beneficial. The pathological selective advantage of some somatic mutations for cells which carry them (see Chapter 5) only contribute to the harm they inflict at the level of organism. Thus, an allele that abolishes, at no cost, somatic mutation would definitely be beneficial, simply by making its carriers healthier. Indeed, alleles that increase the rate of somatic mutation are deleterious, leading, in particular, to cancer-predisposition syndromes (see Chapter 5).

Apparently, the rates of germline and somatic mutation could evolve independently to some extent. Thus, as long as beneficial germline mutations are of any importance for the evolution of mutation rate, the rate of germline mutation must be higher than of somatic mutation. In reality, however, per cell division mutation rates are ~10 times lower in the germline than in the rest of the organism in humans and mice. This difference can be easily explained if natural selection favors reduction of the rate of germline mutation and would likely abolish it altogether, if it were possible.

Comparative evidence also suggests that evolution of multicellular organisms favors no mutation. Mutation rates are apparently much lower in species which reproduce by self-fertilization than in related species which reproduce by outcrossing. This is what we expect to see if selection tries to reduce mutation rate: such selection must be much stronger in asexual or self-fertilizing species. Indeed, without outcrossing a mutator allele is forever tied to the mostly deleterious consequences of its action. In contrast, in the case of sexual reproduction with outcrossing, a mutator allele can become separated from the deleterious alleles it induced.

Unfortunately, little is known about antimutator alleles. A population of *Drosophila melanogaster* that lived for almost 1000 generations near a source of mutagenic gamma radiation evolved a two times lower rate of spontaneous mutation. However, the genetic basis of this adaptation has not been investigated, and it is not known how costly it was to the individuals. There is a substantial inherited variation of spontaneous mutation rate within human populations, indicating that natural selection likely operates on the naturally occurring mutator and antimutator alleles. However, nothing is known about the identities or dynamics of these alleles.

Nevertheless, it seems that, as far as multicellular organisms are concerned, Alfred Sturtevant was right (see epigraph) and mutation rates are as low as they could be, without incurring too much cost. If so, we encounter a fascinating paradox: evolution keeps going only because the laws of physics prevent natural selection from abolishing mutation altogether, which, of course, would prevent any further evolution. This is not impossible: natural selection is near-sighted, and does not care about long-term consequences of its action (see Chapter 7).

4.5 Artificial Mutagenesis and Antimutagenesis

Although normally they make only a minor contribution to mutation, UV and ionizing radiation and, especially, some chemical compounds have a potential to substantially increase mutation rates. It is convenient to characterize a mutagen by its doubling dose – a dose which causes as many mutations as occur spontaneously under the normal, mostly mutagen-free, environment. Radiation is not a very efficient mutagen for mammals, because its doubling dose is close to its LD-50, a dose which kills 50% of the exposed organisms. Thus, one cannot increase the mutation rate in mice by a factor of 5 by irradiating them – the mice will die.

In contrast, doubling doses of the most powerful chemical mutagens are much lower than their LD-50 (Figure 4.9). As a result, a single application of a mutagen can lead to as much as a 100-fold increase of the frequency of *de novo* mutations in the offspring, without killing the parents. Such powerful mutagens are often used in experiments, when there is a need to obtain many mutants.

When a new chemical compound is created, its potential mutagenicity for humans is a major concern. Indeed, a mutagen, on top of its harmful effect on further generations, is also likely to be a carcinogen (see Chapter 5). Unfortunately, we do not understand

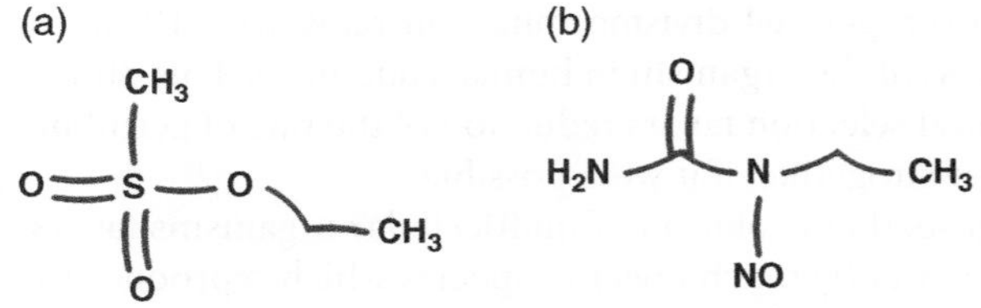

Figure 4.9 Structural formulae of two of the most powerful, and widely used, chemical mutagens: (a) ethyl methanesulfonate (EMS); and (b) *N*-ethyl-*N*-nitrosourea (ENU).

the mechanisms of mutagenesis well enough to simply infer mutagenicity of a compound from its chemical structure. A widely used Ames test assays the effect of a compound on the mutation rate in a bacterium, *Salmonella typhimurium*. Doing this on a bacterium is much simpler and cheaper than on a mammal. The utility of this test demonstrates that many basic mechanisms of mutation are universal.

If mutations can be easily induced, why not try to prevent them? Indeed, some chemical compounds, including polyphenols from green tea, were reported to reduce the rate of spontaneous mutation in bacteria. In contrast, so far all attempts at germline antimutagenesis in multicellular organisms have failed. This should not be surprising: to impair the fidelity of a process must be much simpler than to enhance it. Still, an efficient germline antimutagen would be enormously beneficial to humans (see Chapter 15), and it may be worth it to keep trying.

Preventing mutations from happening can be regarded as passive antimutagenesis. Until recently, there were no efficient means for deliberately reverting mutations that had already happened to the original, wild-type alleles. However, in 2012 a molecular machine which can be used to perform active antimutagenesis, at least in some cases, was discovered. The native function of this machine, known as Cas9 (CRISPR-associate protein 9), is to protect prokaryotes from their viruses, by recognizing short segments of foreign DNA and cutting them. However, it can also be used for modifying all kinds of genotypes.

If we ignore a lot of important and fascinating molecular details, antimutagenesis by means of CRISPR/Cas9 is quite simple (Figure 4.10). First, Cas9 binds a single-stranded RNA molecule of ~20 nucleotides, known as guideRNA. Secondly, the resulting complex finds a segment of double-stranded DNA in which sequence one of the strands is complementary to that of the guideRNA (thus, the other strand has the same sequence). Thirdly, Cas9 cuts both DNA strands within this segment, introducing a double-stranded break (see Chapter 5). Finally, the cell repairs this break through recombination with a "donor" DNA molecule which has segments identical to those at both sides of the break. As a result, the middle part of the donor DNA is incorporated into the genotype.

The key strength of the Cas9 machine is its flexibility. Any guideRNA can be supplied, and its length is often sufficient to precisely identify the unique target for editing within the whole human genotype (there are $4^{20} \sim 1000$ billion different sequences of length 20, and only ~3 billion nucleotides in the human genome). If this is the case, the genotype sequence that emerges after its modification can be precisely controlled by that of the donor DNA. If the distance between segments that promote recombination is the same within the donor DNA and the original genotype, the incorporated sequence is of the same length than the one it replaces, so that genotype editing leads to one or several nucleotide substitutions. If, however, these segments are more (less) distant from each other in the donor DNA than in the genotype, insertion (deletion) also occurs – and they can be quite long.

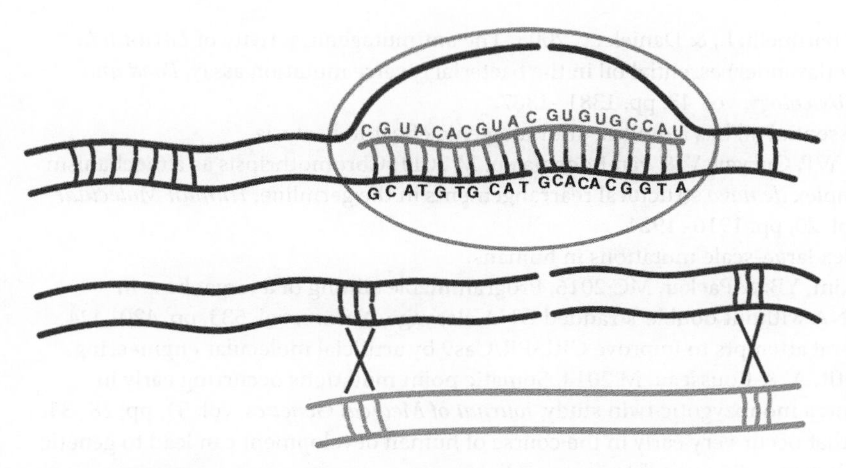

Figure 4.10 Genotype editing by means of CRISPR/Cas9 (see text).

Still, there are many technical difficulties. First, Cas9, guideRNA, and donor DNA must be somehow delivered into the cell. This is currently accomplished by feeding the cell DNA molecules that encode all these necessary components. Secondly, occasionally Cas9 cuts DNA molecules off-target. Thirdly, a double-stranded break may be repaired, instead of recombination with the donor DNA, by an imprecise mechanism called non-homologous ends joining (see Chapter 5), producing an undesired sequence. Currently CRISPR/Cas9, as well as all other available techniques of genotype editing, cannot guarantee that all the intended changes – and nothing else – will be introduced, which makes it unsuitable for modifying human germline genotypes (see Chapter 15). In contrast, CRISPR/Cas9 is already good enough for modifying genotypes of human somatic cells in some situations.

Obviously, deliberate genotype modification is still at the very early stages of its development. However, Cas9 has already been improved to make it more selective, which all but eliminated off-target cutting, as well as to introduce two single-stranded breaks close to each other, instead of one double-stranded break, which prevents undesirable non-homologous end joining. There is little doubt that much better tools for active antimutagenesis, based on CRISPR/Cas9 or on something totally different, will be developed in the not-too-distant future.

Further Reading

Chen, JM, Férec, C & Cooper, DN 2009 Closely spaced multiple mutations as potential signatures of transient hypermutability in human genes, *Human Mutation*, vol. 30, pp. 1435–1448.

Rare complex small-scale mutations in humans.

Collins, GP 2007, The many interpretations of quantum mechanics, *Scientific American*, http://www.scientificamerican.com/article/the-many-interpretations-of-quantum-mechanics/.

On the fundamental randomness of nature.

Evandri, MG, Battinelli, L, & Daniele, C 2005, The antimutagenic activity of *Lavandula angustifolia* (lavender) essential oil in the bacterial reverse mutation assay, *Food and Chemical Toxicology*, vol. 43, pp. 1381–1387.
Lavender essential oil has a modest antimutagenic effect in bacteria.

Kloosterman, WP, Guryev, V, & van Roosmalen, M 2011, Chromothripsis as a mechanism driving complex *de novo* structural rearrangements in the germline, *Human Molecular Genetics*, vol. 20, pp. 1916–1924.
Rare complex large-scale mutations in humans.

Komor, AC, Kim, YB, & Packer, MC 2016, Programmable editing of a target base in genomic DNA without double-stranded DNA cleavage, *Nature*, vol. 533, pp. 420–424.
One of several attempts to improve CRISPR/Cas9 by artificial molecular engineering.

Li, R, Montpetit, A, & Rousseau, M 2014, Somatic point mutations occurring early in development: a monozygotic twin study, *Journal of Medical Genetics*, vol. 51, pp. 28–34.
Mutations that occur very early in the course of human development can lead to genetic differences between "identical" twins.

Lynch, M 2010, Evolution of the mutation rate, *Trends in Genetics*, vol. 26, pp. 345–352.
Somatic mutation rate in mammals is much higher than germline mutation rate.

Neel, JV 1999, Changing perspectives on the genetic doubling dose of ionizing radiation for humans, mice, and *Drosophila, Teratology*, vol. 59, pp. 216–221.
A thorough discussion of the concept of doubling dose.

Wielgoss, S, Barrick, JE, & Tenaillon, O 2013, Mutation rate dynamics in a bacterial population reflect tension between adaptation and genetic load, *Proceedings of the National Academy of Sciences of the USA*, vol. 110, pp. 222–227.
Natural selection favors increased mutation rates in bacteria under a novel, but not under a stable, environment.

Yang, HP, Tanikawa, AY, & Kondrashov, AS 2001, Molecular nature of 11 spontaneous *de novo* mutations in *Drosophila melanogaster, Genetics*, vol. 157, pp. 1285–1292.
Clusters of mutations in siblings.

5
Struggle for Fidelity

DNA is so precious that ... many distinct repair mechanisms ... exist.
Francis Crick, 1974.

At the level of individual nucleotides, spontaneous mutations occur astonishingly rarely. Cells use a variety of ingenious mechanisms in order to achieve a high fidelity of DNA replication, repair, and recombination. A DNA polymerase does not rely on Watson–Crick complementarity and, instead, directs the choice of nucleotides that it attaches, and tries to remove nucleotides that were attached incorrectly. A mechanism called mismatch repair is involved in postreplication detection and correction of errors made by DNA polymerases. DNA damages can occur spontaneously, due to unavoidable insults to vulnerable DNA molecules, can be induced artificially by physical and chemical means, or can be deliberately introduced by the cell in the course of some processes, such as recombination. A DNA damage can affect one strand only, consist of an interstrand covalent cross-link, or involve breakage of both strands. DNA repair machinery consists of an array of interacting mechanisms, responsible for repairing damages of different kinds, with many overlaps. If a cell cannot repair all damages to its genotype, it often commits suicide, by a mechanism known as apoptosis. Mutations that impair different components of the DNA handling machinery lead to many Mendelian diseases, often characterized by neurodevelopmental abnormalities, premature senescence, and cancer predisposition. Spontaneous mutations occur as a result of errors in replication, repair, or recombination, by a wide variety of mechanisms.

5.1 Fidelity of DNA Replication

Although we do not know for sure whether a complete abolition of mutation would evolve if the laws of physics permit this, too rapid a mutation is obviously not an option. In particular, in our species the mutation rate of $\sim 10^{-10}$ per nucleotide per cell division (or $\sim 10^{-8}$ per generation, because there are ~ 100 cell divisions and rounds of DNA replication per human generation, see Chapter 6) leads to $\sim 1\%$ of all newborns being affected by a serious disease due to a *de novo* mutation (see Chapter 13). Thus, if the mutation rate were 100 times higher, 10^{-8} per cell division, we would immediately go extinct.

Keeping a mutation rate that low must be a formidable task: try to type 10 000 000 000 letters (3 000 000 pages) and make only one typo. Cells perform a number of complex,

Crumbling Genome: The Impact of Deleterious Mutations on Humans, First Edition. Alexey S. Kondrashov.

coordinated jobs which together ensure an amazing fidelity of handling of their DNA. Many proteins involved in these jobs are similar throughout all life, indicating their very early origin and extreme importance. Human genome encodes at least 16 different DNA polymerases, enzymes which synthesize new DNA strands (see Chapter 1), at least 31 proteins that detach the two DNA strands from each other, called helicases, and many other DNA-handling proteins which work in the course of DNA replication, repair, and recombination. Mechanisms of fidelity of these processes are still not fully understood, but even what is already known can fill volumes. Hopefully, even a short summary will be sufficient to appreciate how much a cell cherishes its genotype and at what lengths it goes in order to preserve its integrity and to avoid mutation.

Let us first consider fidelity of DNA replication, a process which must take place before every cell division (Fig. 1.13). Replication is performed by the replisome, a molecular machine composed of a topoisomerase (see below), helicase, DNA polymerases, and many other proteins. A replisome moves along an old, template DNA molecule, leaving behind two new DNAs, each consisting of an old and a new strand, and thus produces a structure known as a replication fork (Figure 5.1).

The time interval during which DNA replication occurs in the eukaryotic cell is called the S phase of its cycle. In mammals, replication starts from many initiation sites, and proceeds in both directions from each one, resulting in two "sister" forks. Because each genotype segment must be replicated exactly once, replication involves precise coordination of multiple replisomes. In particular, complex molecular mechanisms are employed to handle collisions of replisomes which move toward each other from adjacent initiating sites.

Let us now take a look inside a replisome. One might think that it has an easy job: two complementary strands of a DNA molecule make it perfectly suitable for being replicated. Use a helicase to detach the strands from each other and a DNA polymerase to make a new, complementary strand for each old one, and the two new double-stranded DNA molecules, with nucleotide sequences identical to that of the original one, are ready (Figure 1.3).

In fact, because DNA is a double helix, there is one unavoidable physical problem with its replication, not evident from the simplified Figure 1.3: the two DNA strands must not only be detached from each other, but also unwound. There is one turn of a double helix for each 10.5 nucleotides, and a eukaryotic replisome replicates ~100 nucleotides per second. It would be impossible for the two strands of a long DNA molecule to rotate around each other ~10 times per second. The replisome solves this problem by employing proteins, called topoisomerases, which allow DNA strands to pass through each other and, thus, to unwind without rotating (Figure 5.2).

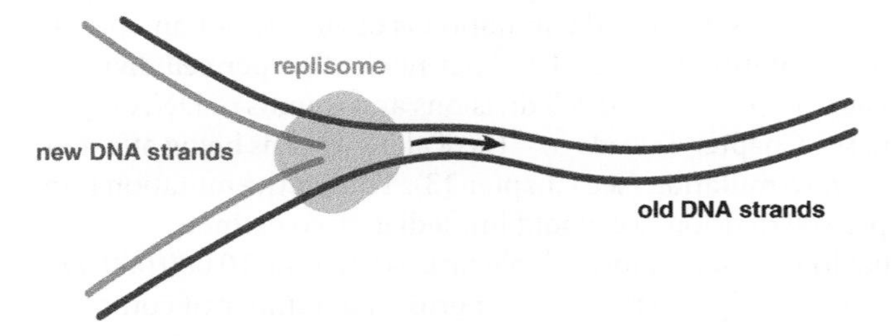

Figure 5.1 A replisome, moving rightward, at the replication fork.

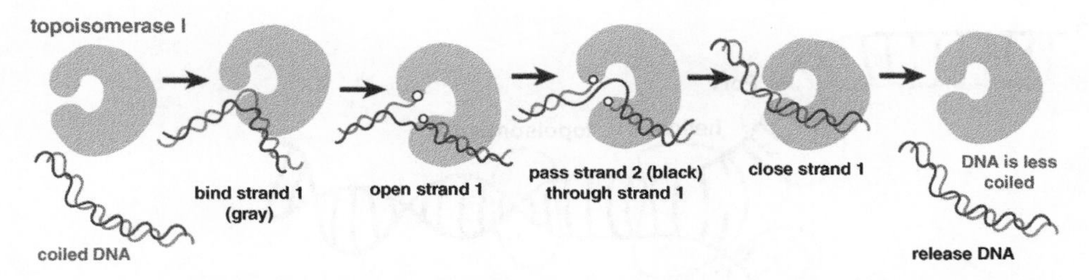

Figure 5.2 A DNA topoisomerase type I allows the two DNA strands to pass through each other, by temporarily cutting one strand, moving the other strand through the cut, and sealing the cut. A DNA topoisomerase type II (not shown) cuts both strands.

On top of this one, a replisome also has to deal with two "self-inflicted" problems. First, DNA polymerases that work within the replisome cannot start their job from scratch. Most likely, this limitation of these enzymes evolved together with their proofreading activity (see below). Instead, a replicative DNA polymerase needs a short RNA "primer", synthesized by a specialized RNA polymerase, called a primase. This primase, which can start working from scratch, forms a complex with DNA polymerase alpha, and together they synthesize composite RNA-DNA primers consisting of ~20–30 nucleotides. These primers are later removed, and replaced by "permanent" DNA.

Secondly, all DNA polymerases can synthesize a new DNA strand only in one direction, 5' > 3' (thus reading the template strand in the 3' > 5' direction). Because the two DNA strands are antiparallel, this allows for straightforward replication of only one, leading, strand, which faces the advancing replisome by its 3'-end. By contrast, replicating the other, lagging, DNA strand, which faces the replisome by its 5'-end, is complicated. The nascent DNA strand, complementary to the lagging strand, emerges as many short pieces, called Okazaki fragments, each initiated with a primer and growing in the direction opposite to that of advancement of the replisome. These short pieces, of the average length of 165 nucleotides in eukaryotes, are assembled into one continuous strand by an enzyme called DNA ligase. In eukaryotes, leading and lagging strands are replicated by different DNA polymerases (Figure 5.3).

One can naively think that, provided that all the problems just described are somehow solved, high fidelity of DNA replication is all but guaranteed. However, this is not so. Instead, it depends on at least four processes, which may be viewed as mechanisms of replication fidelity.

First, a high quality of raw materials for DNA replication must be ensured. DNA polymerases utilize deoxyribonucleotides (A, T, G, and C) with two extra phosphates – their removal provides energy for attaching the nucleotide to the nascent strand. These small molecules can undergo a variety of chemical modifications, and several enzymes work incessantly to sanitize their pool within the nucleus, by degrading modified molecules. Fidelity of replication of those segments of the genome that are replicated late in the S phase of the cell cycle is reduced, apparently because by this time the pool of raw materials becomes depleted.

Secondly, fidelity of attachment of nucleotides is directly controlled by the DNA polymerases. Despite its beauty, Watson–Crick complementarity is absolutely insufficient to ensure an acceptable fidelity of replication, even with perfect raw materials. Although complementary nucleotide pairs A:T and G:C fit nicely into the DNA double

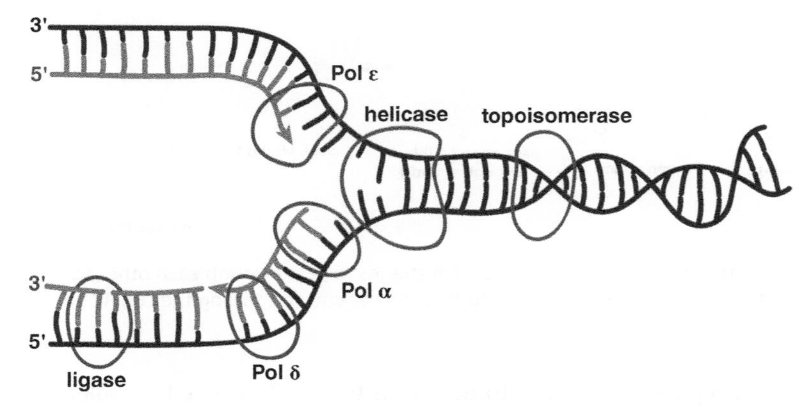

Figure 5.3 A crude scheme of processes that take place in the course of DNA replication. Two strands of the original DNA are unwound by topoisomerase and detached from each other by helicase, composite RNA-DNA primers are synthesized by primase which forms complex with DNA polymerase alpha, only to be soon removed and replaced with "permanent" DNA, leading strand is replicated continuously by DNA polymerase epsilon, and lagging strand is replicated discontinuously by DNA polymerase delta, with the resulting Okazaki fragments being joined by ligase (Lujan *et al.*, *Trends in Cell Biology*, vol. 26, p. 640, 2016).

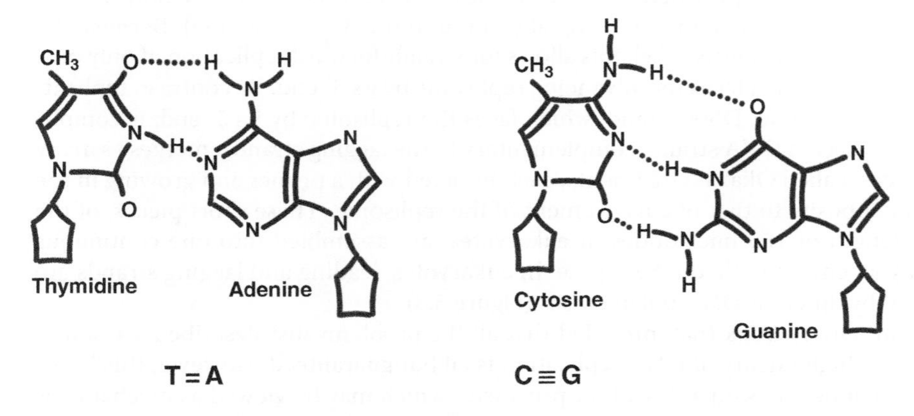

Figure 5.4 Watson–Crick A:T and G:C nucleotide pairs have very similar overall shapes and, thus, fit well into the double-stranded DNA molecule, regardless of its sequence.

helix (Figure 5.4), and all other, noncomplementary, pairs do not, the difference in the amount of energy released in the course of formation of complementary versus non-complementary pairs is not large. It may be calculated that if DNA polymerases relied on complementarity alone, the probability of attaching a wrong nucleotide to the nascent DNA strand would be ~0.01. With such a fidelity, 100 rounds of DNA replications in the course of a human generation would turn the genotype of an offspring into garbage, essentially uninformed by genotypes of the parents.

Fortunately, fidelity of nucleotide attachment depends not on complementarity, but on active involvement of DNA polymerases. When a polymerase senses, say, an A in the template strand, it tries hard to make sure that T – and not any other nucleotide – is attached to the nascent strand. DNA polymerases are complex proteins, which can do the job of discriminating between right and wrong nucleotides rather efficiently. As a result, the actual error rate in the course of nucleotide attachment is only $\sim 10^{-5}$, or 1000 times lower than what chemical properties of nucleotide pairs alone could provide. Still, even this error rate is way too high for humans, as it would lead to the per nucleotide per generation mutation rate of 10^{-3}, and to millions of *de novo* mutations in the genotype of an offspring, which is clearly inconsistent with life.

How could the replisome reduce its error rate even further? This can be done by employing the third mechanism of fidelity, which is called proofreading, although this name could be misleading. Indeed, an error within a text written in a human language can usually be recognized by a spellchecker, because only some short sequences of letters – words – are allowed. Surely, a spellchecker would not prevent you from writing, in a job application, "I have an extensive experience in ruining complex operations". But the situation is much worse for DNA texts, where there are no spelling rules and, thus, no sequence of nucleotides can be rejected *a priori*.

Thus, DNA proofreading has to work differently. A DNA polymerase immediately probes every attached nucleotide, while the template strand is still available, and tries to detach it, using an enzyme called 3'-to-5' exonuclease (Figure 5.5). In eukaryotes, this enzyme is incorporated into replicative DNA polymerases epsilon and delta, forming a separate domain within each of them. In rare instances when the nucleotide has been attached erroneously, and thus is not complementary to the template strand, 3'-to-5' exonuclease removes it with a probability of ~99%. Thus, the replisome makes an error only if a nucleotide was attached incorrectly and then escaped proofreading, and the probability of this is $10^{-5} \times 10^{-2} = 10^{-7}$. In humans, this would result in the per nucleotide per generation mutation rate of $10^{-7} \times 100 = 10^{-5}$, or tens of thousands of *de novo* mutations in a genotype. This is definitely an improvement, although still way too high to produce a viable offspring.

DNA proofreading comes at a cost: it removes not only almost every nucleotide that was attached incorrectly, but also a substantial proportion of those nucleotides that

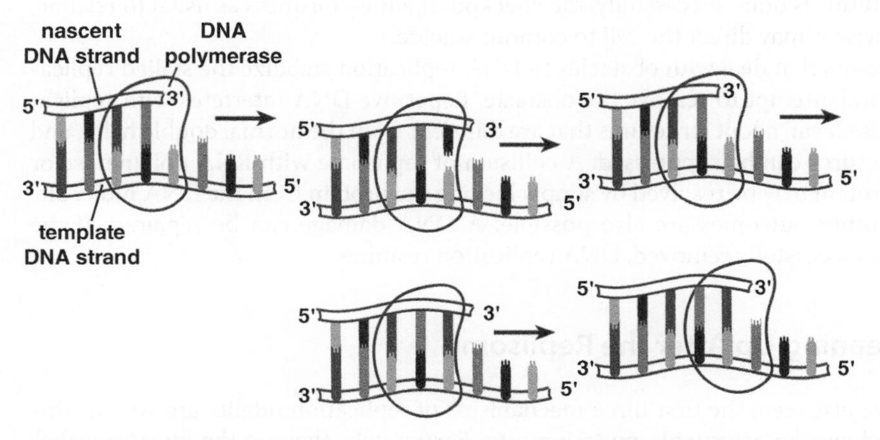

Figure 5.5 Proofreading 3'-to-5' exonuclease activity of a DNA polymerase. A nucleotide which was incorrectly attached to the growing 3'-end of the nascent DNA strand is usually removed, and another nucleotide is attached.

were attached correctly. This is unavoidable, because no process can achieve a perfect selectivity. Thus, any attempt to improve fidelity of DNA replication much further by increasing proofreading activity would disable the replisome, which obviously cannot remove more nucleotides than it attaches.

Also, there is a hole in the third mechanism of DNA replication fidelity. While polymerases epsilon and delta, responsible for the bulk of DNA replication, perform proofreading, DNA polymerase alpha, as well as RNA-synthesizing primase, does not. Unless a damage-inducing error occurs, RNA portions of RNA-DNA composite primers are completely replaced by DNA synthesized by high-fidelity polymerases. In contrast, some fragments of DNA portions of these primers, synthesized by polymerase alpha, are routinely retained. As a result, ~3% of the length of lagging DNA strand is replicated without proofreading. This fundamental imperfection (see Chapter 10) of DNA replication, which likely results from constraints imposed by the overall organization of the replisome, is responsible for a substantial proportion of mutations.

So far we discussed fidelity of DNA replication in the course of a normal, uneventful progression of a replisome, when errors could mostly lead to minor mutations. However, a replisome can also be stalled after encountering an obstacle. There are least four kinds of them. First, some DNA sequences, in particular those that contain short repeats, are unusually difficult to replicate. Secondly, a replisome may collide with DNA-dependent RNA polymerase while it transcribes DNA (see Chapter 1). Thirdly, a replisome may bump into a particularly stable DNA-protein complex. Finally, a replisome may encounter a DNA damage (see Chapter 4) of an "intolerable" kind (see below). If not handled properly, an obstacle may lead to collapse of the replication fork and to a major mutation, including deletion of a whole chromosome.

Dealing with a stalled replisome follows a pattern that appears in many cellular processes, when something goes so wrong that attempts to continue business as usual would only do harm. Mechanisms that deal with such situations are called checkpoints. A checkpoint consists of a sensor, which detects that something went wrong, and a signal which halts business as usual and activates a process that attempts to rectify the problem. If this is done successfully, the checkpoint allows business as usual to resume, but otherwise it may direct the cell to commit suicide.

Checkpoints that deal with obstacles to DNA replication stabilize the stalled replication fork and attempt to remove the obstacle. Repetitive DNA interferes with replication because it can adopt structures that are different from the normal double helix, and these structures can be suppressed. A collision of replisome with RNA polymerase or another protein may be resolved by simply kicking the protein from the DNA molecule, although other outcomes are also possible. A DNA damage can be repaired. If the obstacle is successfully removed, DNA replication resumes.

5.2 Cleaning Up After the Replisome

As we have just seen, the first three mechanisms of replication fidelity are still insufficient to achieve an acceptable mutation rate. Fortunately, there is the fourth mechanism, which works right after the replisome completed its work. Indeed, an error in the course of semiconservative DNA replication does not immediately lead to a mutation.

Instead, it only produces a mismatch between the two strands of a novel DNA molecule. Thus, it is not too late to prevent a mutation even after the passage of a replisome.

One can say that a mismatch is a soft damage to the DNA: although complementarity of its two strands is violated, each strand is perfectly normal individually. Such damages are dealt with by mismatch repair (MMR) machinery, which consists of a number of interacting DNA-handling proteins, some of which are present in all life. First, a mismatch is recognized, after which a segment of one of the strands which overlaps the mismatch is removed and replaced with the sequence that is complementary to the other strand. Thus, MMR involves synthesis of only one DNA strand, by the same DNA polymerase delta that replicates the lagging strand as a part of the replisome (Figure 5.6).

A priori, MMR can do two opposite things to a mismatch – reverse it, thus undoing the error made in the course of replication, or convert it into a mutation. Which outcome takes place depends, of course, on which strand is used as the template to correct the other strand and thus to get rid of a mismatch. If the old, template strand is used, and the new strand is corrected, the error is undone, but otherwise a mutation appears (Figure 5.7).

By preferentially choosing the old DNA strand as a template for repairing mismatches, MMR serves as the fourth mechanism of DNA replication fidelity. Indeed, MMR machinery may be physically coupled to the replisome, working right behind it. Prokaryotes recognize the old DNA strand by its methylation, attachment of $-CH_3$ groups to some nucleotides. Eukaryotes apparently use a different and still poorly characterized mechanism(s), which seems to rely on the presence of nicks in the nascent DNA strands. In line with this hypothesis, MMR acts on the discontinuously synthetized lagging strand ~3 times more efficiently than on the leading strand. In any case, in all life MMR prevents mutations much more often than creates them.

MMR fails to correctly undo a mismatch produced by an error made by the replisome with probability $\sim 10^{-3}$. This results in $10^{-7} \times 10^{-3} = \sim 10^{-10}$ per nucleotide per replication mutation rate (Figure 5.8), or in $\sim 10^{-8}$ per nucleotide per generation mutation rate, or

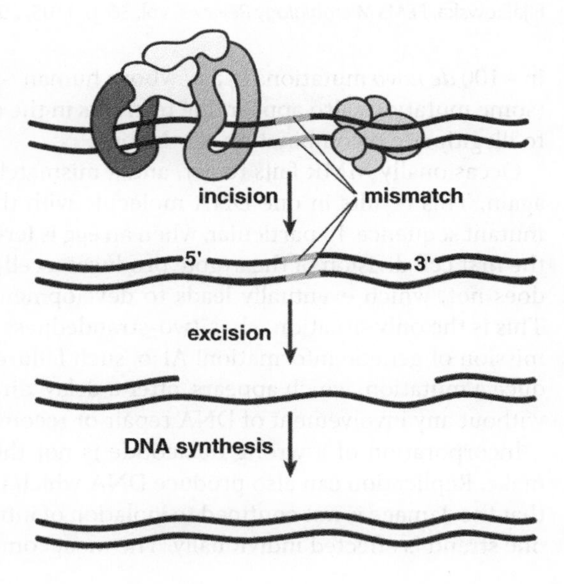

Figure 5.6 Scheme of mismatch repair. Proteins that perform MMR bind the DNA molecule around a mismatch and replace a segment of one of its two strands.

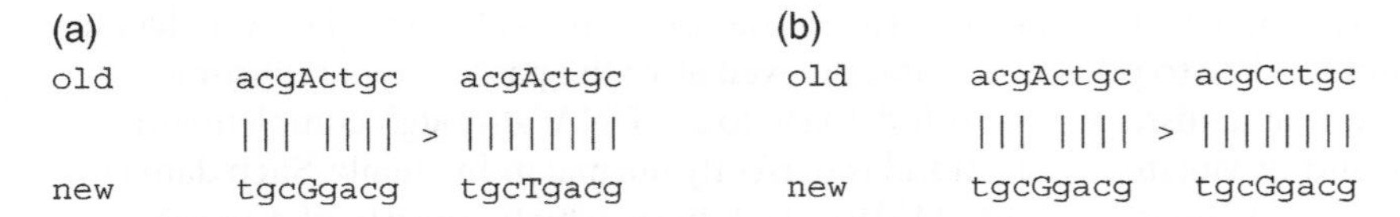

Figure 5.7 Correct (a) and mutagenic (b) repair of a mismatch (site in upper case) between DNA strands.

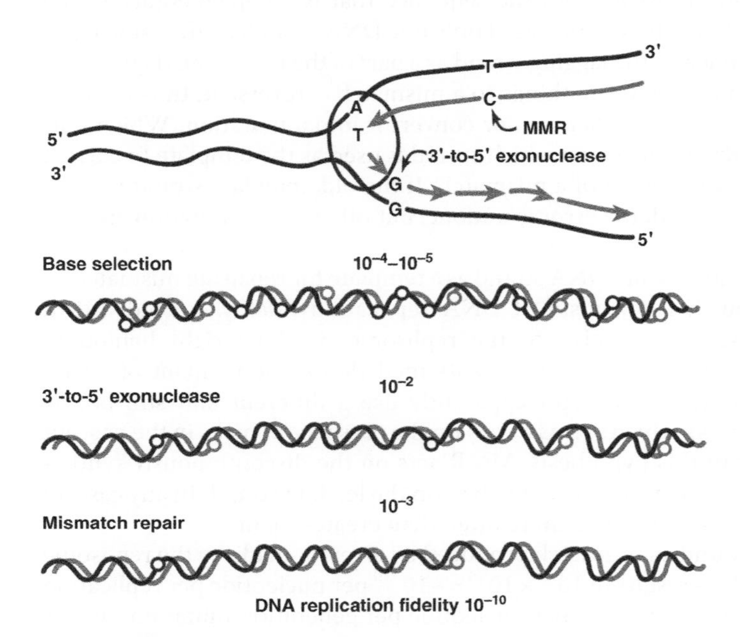

Figure 5.8 Contributions of steps 2, 3, and 4 in the overall fidelity of DNA replication (after Fijalkowska, *FEMS Microbiology Reviews*, vol. 36, p. 1105, 2012).

in ~100 *de novo* mutations in a newborn human – something with which we have to live (some mutations also appear due to errors in the course of repair of DNA damages and to illegitimate recombination, see below).

Occasionally MMR fails to act, and a mismatch persists until the DNA is replicated again. This results in one DNA molecule with the old sequence and the other with a mutant sequence. In particular, when an egg is fertilized by a mismatch-carrying sperm, the first cell division of the zygote produces a cell that carries a mutation and a cell that does not, which eventually leads to development of a half-body mosaic (Figure 5.9). This is the only situation when two-strandedness of DNA makes a difference for transmission of genetic information! Also, such failure of MMR is the simplest way to produce a mutation, which appears, after a delay, directly as a result of a replication error, without any involvement of DNA repair or recombination.

Incorporation of a wrong nucleotide is not the only kind of error a replisome can make. Replication can also produce DNA which is damaged in a hard way, in the sense that the damage is not confined to violation of interstrand complementarity, but at least one strand is affected individually. The most common kind of hard damages left by the

Figure 5.9 After replication of a DNA molecule which contains a mismatch (site in upper case) between an old nucleotide a and a new nucleotide g, two DNA molecules emerge, an unchanged and a mutant. Thus, fertilization of an egg by a mismatch-carrying sperm can produce a half-body mosaic (see cover).

```
                                  5'acgtAagtc3'
                                   |||||||||
                                  3'tgcaTtcag5'
old   5'acgtAagtc3'
        |||| ||||         >
new   3'tgcaGtcag5'
                                  5'acgtCagtc3'
                                   |||||||||
                                  3'tgcaGtcag5'
```

replisome are ribonucleotides, which belong to RNA only, wrongly incorporated into DNA strands instead of deoxyribonucleotides. They can be the remnants of RNA portions of primers that were not properly removed. Alternatively, DNA polymerases (most commonly polymerase epsilon) can erroneously incorporate ribonucleotides, which is not surprising given that their concentration within the nucleus is much higher than that of deoxyribonucleotides, As a result, ~1000 ribonucleotides become misincorporated into a human genotype after each round of replication. They do not distort genetic information, but make DNA, if it still can be called so, hard to use. Thus, there is a special mechanism for detecting misincorporated ribonucleotides and replacing them with deoxyribonucleotides.

5.3 Dealing with DNA Damages

On top of soft mismatches and misincorporated ribonucleotides, DNA can be damaged in many other ways. There are three broad classes of hard DNA damages: single-strand damages (SSDs) that are confined to one strand, leaving the other strand intact; interstrand crosslinks (ICLs) that involve a strong, covalent chemical bond between the two strands (normally connected only by weak hydrogen bonds between paired nucleotides, see Chapter 1); and double-strand breaks (DSBs), such that both strands are broken close to each other (Figure 5.10). There is a lot of chemical heterogeneity within each of these three classes. For example, a SSD can consist of a misincorporated ribonucleotide, a base chemically modified (in over 100 ways), a base detached from the strand, two adjacent bases being conjoined, or a strand being broken.

Why does DNA become damaged? Buddha's wisdom provides the general answer, but we can be more specific. A DNA damage can be spontaneous, induced, or deliberate.

Spontaneous DNA damages are due to processes that unavoidably occur within every living cell. Errors made in the course of DNA replication, considered above, is one prominent cause. Also, topoisomerases, which unwind the DNA helix in the course of its replication and transcription, occasionally fail to reseal the cuts they make (Figure 5.2), thus producing strand breaks. Equally important, DNA becomes damaged, regardless of its replication or transcription, as a result of chemical modification of its bases, which is often facilitated by reacting with a variety of aggressive chemical compounds (in particular, with the so-called reactive oxygen species) that are produced in the course of normal cell functioning.

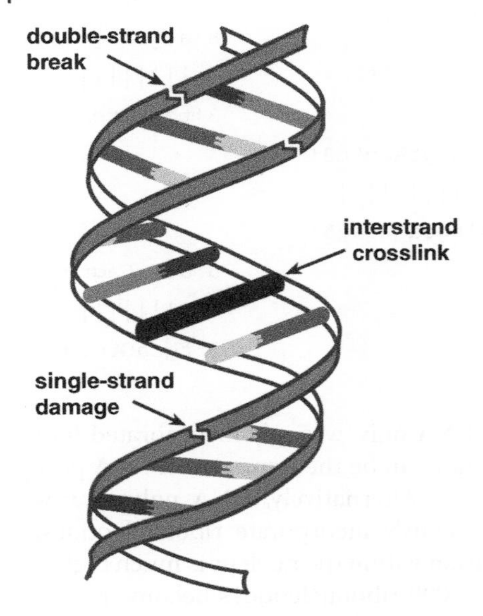

Figure 5.10 Three general kinds of hard DNA damages.

Induced DNA damages are caused by UV light, high-energy ionizing radiation, and a variety of chemical compounds that are normally absent within cells. However, such compounds are present in many natural environments, being produced by bacteria, fungi, and plants, and can penetrate cellular membranes. Some of the most potent DNA-damaging compounds are synthesized artificially.

Deliberate DNA damages are introduced by cells as a necessary step in their normal functioning. Several functions cannot be performed without a cell damaging its own genotype. In particular, meiotic recombination (crossing-over) between maternal and paternal chromosomes, which occurs in almost all eukaryotes, starts from the cell deliberately making a DSB and ends with its making four single-strand breaks (SSBs; Figure 5.11). Meiotic recombination likely evolved from a mechanism of repair of spontaneous DSBs which involves transfer of information from an intact double-stranded DNA and is present in all life (see below).

Normally, when the cell is not subject to any damage-inducing treatments, the vast majority of damages to its DNA is spontaneous. The genotype of every cell sustains a large number of such damages every day (Table 5.1). This should not be surprising, because the total length of DNA molecules which constitute the human genome is ~1 m, and such enormous molecules are bound to be extremely fragile.

Any DNA damage is a very serious event. Still, some damages can be tolerated for a while, in the sense that a cell can continue using damaged DNA. In particular, DNA that is damaged in some ways can, nevertheless, be replicated. A regular replisome can work through a DNA segment where in one strand two adjacent pyrimidine nucleotides are conjoined. When encountering a more serious, but still tolerable, SSD a replisome may recruit specific translesion DNA polymerases, which, however, work with reduced fidelity. Still, some kinds of SSDs, as well as all ICLs and DSBs, are absolute obstacles to DNA replication and transcription. In other words, they cannot be tolerated and, instead, must be immediately repaired.

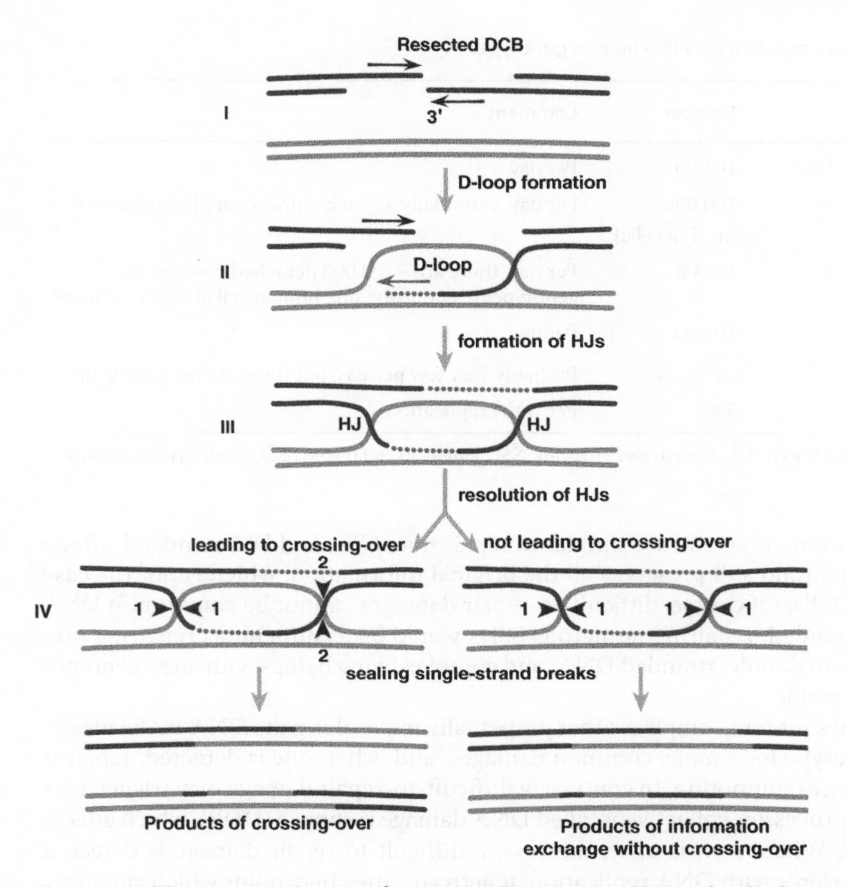

Figure 5.11 The major mechanism of interaction between two DNA molecules (maternal and paternal, shown in black and gray) which results in their recombination. (i) A DSB is introduced into one molecule, after which its strands that face this DSB by their 5'-ends are shortened, producing 3′ single-stranded ends. (ii) One single-stranded DNA end invades into the other, intact DNA molecule, by pairing with its (approximately) complementary strand. Double-stranded DNA that consists of two strands of different origin is called a heteroduplex, and the resulting structure is called a D-loop. (iii) Gaps and strand discontinuities are repaired by DNA synthesis and ligation. As a result, the two original DNA molecules become connected by two structures, called Holliday junctions (HJs). (iv) Each HJ can be resolved by strand cleavage and ligation in orientation 1 or 2. When the two junctions are resolved in opposite orientations (e.g., left HJ in orientation 1 and right HJ in orientation 2, as shown), crossing-over occurs. In contrast, when both junctions are resolved in the same orientation (e.g., in orientation 1, as shown), crossing-over does not occur (after Cromie *et al.*, *Cell*, vol. 127, p. 1167, 2006).

Even when possible, damage tolerance is only a temporary solution. Damages of any kind cannot accumulate indefinitely, and, therefore, every damage must be eventually repaired, if a cell is to remain viable. Because the ability of damaged DNA to function, if any, is limited, it is better to repair a damage early (Figure 5.9). Indeed, although damages appear incessantly (Table 5.1), at any particular moment they are rare within a normal genotype, indicating their prompt repair. Clearly, DNA repair must be one of the most amazing adaptations (see Chapter 7) of living beings.

Table 5.1 Spontaneous damages to a human genotype.

Kind of damage		Number	Comment
SSD	Modified bases	10 000	Per day
	Conjoined bases	100 000 (in skin cells)	Per day; essentially all such damages are light-induced
	Detached bases	10 000	Per day; there are ~100 000 detached bases in the genotype of a non-dividing human cell at every moment
	SSB	10 000	Per day
ICL		~1 (guess)	Probably very few per day, but there are no reliable data
DSB		50	Per DNA replication

DSB, double-strand break; ICL, interstrand crosslink; SSB, single-strand break; SSD, single-strand damage.

SSDs occur more often and are simpler to repair than ICLs and DSBs. Indeed, after a SSD the intact strand still preserves all the original information, which is not the case for ICLs and DSBs. Of course, difficult-to-repair damages cannot be common. If DNA were single-stranded, repairing numerous SSDs would be as difficult as repairing rare ILCs and DSBs in double-stranded DNA, and complex living beings with large genomes would be impossible.

A cell employs protein complexes that perpetually move along the DNA molecules to survey its genotype for simple, common damages, and, when one is detected, repair it without causing a commotion. In contrast, a difficult-to-repair damage may trigger a set of interacting processes, collectively called DNA damage response (DDR), which affects the whole cell. When, as it is often the case, a difficult-to-repair damage is detected because it interferes with DNA replication, it activates the checkpoint which stabilizes the stalled replisome, temporarily arrests preparation of the cell for division, and launches the necessary mechanism of repair.

DNA repair consists of a number of different mechanisms, dealing with damages of different kinds. Still, there are overlaps, and a damage can often be repaired by more than one mechanism. Also, the same protein can participate in multiple repair mechanisms, sometimes dealing with very different kinds of DNA damages. Many mechanisms of DNA repair are present in all life, but some others are confined to particular groups of organisms.

Although mechanisms of DNA repair involve very diverse proteins that catalyze a wide variety of chemical reactions, they possess some broad similarities. First of all, a damage must be detected. This involves binding of specific proteins to the damaged DNA. After this, components of the machinery which is responsible for dealing with damages of the detected kind are assembled around these proteins. Then, this machinery does its job, which usually involves a number of steps. Finally, if the damage interfered with replication or transcription, the stalled process resumes.

Repair of soft damages, or mismatches, has been described above. A few kinds of SSDs can be reverted directly, in a simple single-step reaction. Such simple repair mechanisms very often work perfectly. In particular, conjoined bases can be separated by a protein complex which, like the process of photosynthesis in plants, utilizes the energy of light. It is unfortunate that this neat mechanism has been lost in the common ancestor of placental mammals.

Direct reversal is the only mode of SSD repair which does not utilize information from the intact strand. All its other mechanisms use this stand as a template, which requires excision of a segment of the affected stand. This process is generally similar to what happens in the course of MMR, but here the excised segment carries a hard damage. Excision repair of SSDs leads to restoration of the original DNA as long as the information stored on the undamaged strand is transferred to the repaired strand faithfully. Two modes of excision repair, base excision repair (BER) and nucleotide excision repair (NER), are of particular importance.

BER acts when the damage does not disturb the overall spatial conformation of the DNA. It consists of the following steps: (1) detachment of the damaged base from the strand backbone (performed by enzymes called glycosylases; different glycosylases detach bases damaged in different ways, although their targets overlap); (2) cutting the damaged DNA strand near the resulting abasic site; (3) removal of deoxyribose at the abasic sites (sometimes several adjacent nucleotides can also be removed); (4) filling the gap by a DNA polymerase which uses the intact strand as a template; and (5) sealing the strand by DNA ligase (Figure 5.12). Quite often, the primary spontaneous damage consists of base detachment or of a SSB, in which cases this pipeline starts working from steps 2 or 3, respectively.

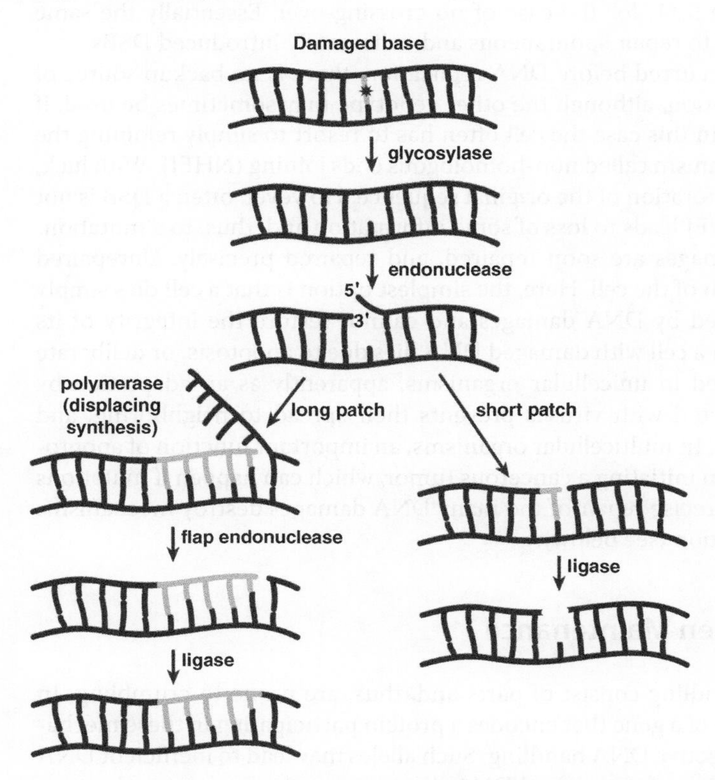

Figure 5.12 Pipeline of the base excision repair which can involve synthesis of a long (left) and a short (right) patch. Newly synthesized DNA is shown in gray.

NER acts if a SSD disturbs conformation of the DNA, for example, when a large molecule became attached to it or a damage is not confined to one nucleotide. In begins from cutting the damaged DNA strand at both sides of the damage, followed by removal of the damaged segment, usually ~30 nucleotides long. After this, the resulting gap is filled and sealed, as it is the case with BER. Damages that need to be repaired by NER may be recognized by proteins that survey the genome for drastic alterations of the overall DNA conformation; however, their recognition in the course of transcription also plays a major role. Many kinds of DNA damages can be repaired by both BER and NER and, whichever arrives first, does the job.

An easy way to repair an ICL would be to directly reverse it, by dissolving the abnormal interstrand bound. However, as far as we know, evolution failed to produce a mechanism for this. Instead, ICLs are repaired by a variety of complex mechanisms, which involve obtaining information either from a damaged strand, which is liable to mutations, or from another double-stranded DNA molecule, through recombination. An ICL detected in the course of DNA replication may be particularly hard to repair, as it leads to formation of a DSB.

Repairing a DSB cannot be easy, because such a break completely destroys the integrity of a DNA molecule and leaves neither strand intact. If a DSB occurred after DNA replication but before cell division, the recently produced copy of a broken molecule can serve as backup, from which missing information can be transferred. Often, this is done as shown in Figure 5.11, for the case of no crossing-over. Essentially the same mechanisms can be used to repair spontaneous and deliberately introduced DSBs.

In contrast, if a DSB occurred before DNA replication, there is no backup source of exactly the same information, although the other genotype may sometimes be used, if the cell is diploid. Thus, in this case the cell often has to resort to simply rejoining the ends of a DSB, by a mechanism called non-homologous ends joining (NHEJ). With luck, this may lead to exact restoration of the original sequence. However, often a DSB is not chemically clean, and NHEJ leads to loss of some information and, thus, to a mutation.

A vast majority of damages are soon repaired, and repaired precisely. Unrepaired damages can lead to death of the cell. Here, the simplest option is that a cell dies simply because it is overwhelmed by DNA damages and cannot restore the integrity of its genotype. However, often a cell with damaged DNA dies due to apoptosis, or deliberate suicide. Apoptosis evolved in unicellular organisms, apparently as an adaptation by which an organism infected with viruses prevents their spread to neighboring, and tightly related, organisms. In multicellular organisms, an important function of apoptosis is to prevent a cell from initiating a cancerous tumor, which can happen if mutations that may result from imprecise repair of too many DNA damages destroy mechanisms that control cell proliferation (see below).

5.4 Harms of Broken Maintenance

Mechanisms of DNA handling consist of parts and, thus, are prone to crumbling. In particular, a mutant allele of a gene that encodes a protein participating in these mechanisms often results in defective DNA handling. Such alleles may lead to inefficient DNA replication, to accumulation of unrepaired DNA damages, and/or to increased rate of spontaneous mutation (mutator alleles, see Chapter 4), and are a powerful tool for studying all these processes. Indeed, a natural way to find out what something is doing is to see what happens when this something is broken.

Before we consider impaired handling of DNA by cells, it makes sense to say a few words about RNA viruses, which lack some of the key mechanisms of genetic fidelity. In contrast to cells, which rarely replicate any RNA molecules, RNA viruses use RNA as a carrier of their hereditary information and, thus, must replicate it. However, with the exception of coronaviruses, replication of viral RNA does not involve any proofreading. Also, viruses do not perform any repair of mismatches or any other damages in their RNA. As a result, mutation rates in most RNA viruses are very high, $\sim 10^{-4}$–10^{-5} per nucleotide per replication. Because genomes of most RNA viruses are short (of the order of 10^4 nucleotides), such mutability does not lead to their extinction. Instead, it increases the rate of evolution, which is often good for the virus, which must evolve to survive. Rapid evolution is the necessary condition for persistence of the influenza virus within the human population, because a case of influenza leads to lifelong immunity against the virus of the genotype which caused it. Rapid evolution of HIV within an individual appears to be a major factor which allows the virus to evade the immune system and eventually leads, in the absence of treatment, to progression of the infection to AIDS. Beneficial mutations, in this case for a virus, do happen, and natural selection may favor high mutation rates when the genome is small (see Chapter 4).

In contrast, every cell performs DNA proofreading and repair, and their defects, as well as those of DNA replication, can lead to severe problems. In humans, which have been studied in this respect better than any other species, inherited defects of DNA handling lead to a number of Mendelian diseases. Most of these diseases are individually very rare, because defective alleles which cause them are strongly deleterious, and negative natural selection (see Chapter 7) prevents them from reaching high frequencies. Some of the diseases of DNA handling are dominant, manifested in heterozygotes, in which case homozygotes likely die well before birth (Figure 2.14). Others are recessive, with overt symptoms appearing only in homozygotes.

Still, impairments of many components of the DNA handling machinery do not lead to any known diseases. Some of them must still be awaiting discovery. However, impairments of other components may never lead to disease, due to two opposite reasons. On the one hand, redundancy makes loss of some functions clinically mute. Of course, any function is important, and loss of (almost) any function reduces fitness under a tough, natural environment (see Chapter 7). Nevertheless, a human, especially one living in a modern, benign environment (see Chapter 10), may not need medical attention if a non-critical, duplicated function is broken. On the other hand, a loss-of-function allele of a non-redundant gene involved in DNA handling may cause early death of homozygous embryos, but no overt symptoms in heterozygous individuals. In such cases, a disease is sometimes seen in homozygotes for alleles that cause only partial loss of function.

Despite their diversity, diseases of DNA handling present three pervasive themes. First, they often primarily affect the nervous system. Neurons are particularly vulnerable to many dysfunctions of DNA repair because they do not divide, and their DNA does not replicate, thus depriving them of a number of means of DNA repair that are associated with its replication. Also, neurons are very metabolically active, which leads to high intracellular concentrations of aggressive DNA-damaging molecules.

Both development and maintenance of the nervous system can be affected. Impaired development of the nervous system may lead to microcephaly (growing brain pushes cranial bones), ataxias (poor coordination of the muscles), apraxias (difficulties in performing purposeful movements), and/or intellectual disability, evident from early age. Impaired maintenance leads to neurodegeneration, which may also start rather early. Of course, other tissues can also be impaired by defective DNA handling, which, in particular, may cause cardiovascular diseases, with DNA damages accumulating in atherosclerotic plaques.

Secondly, defective DNA handling often leads to progeria, premature general aging. This should not be surprising, as accumulation of unrepaired DNA damages must impair or kill cells. Neurodegeneration is just one of the facets of progeria, which may also involve the cardiovascular system, connective tissues, and so on.

Thirdly, many, but not all, diseases of DNA handling are involved with increased incidence of cancer, in particular, at early age, when cancer in healthy people is very rare. Such diseases are known as cancer predisposition syndromes. Apparently, cancer predisposition appears when a defect in DNA handling leads not only to unrepaired DNA damages, but also to the increased rate of somatic mutation.

Because the relationship between genotypes and phenotypes is still understood very poorly (see Chapter 1), it is currently impossible to predict what pathology will be caused by a particular defect of DNA handling, and even the known connections are often hard to explain. For example, many such defects affect the cerebellum, part of the brain responsible for coordinating movements, more than its other parts. Thus, a person with severe ataxia may have normal intelligence. We do not know why it is so.

Let us consider several examples. Humans, heterozygous for alleles of DNA polymerase delta which are incapable of performing its primary function of replicating the lagging DNA strand, have a syndrome that includes a host of impairments: hardened skin, dilated small blood vessels, ligament contractures, reduced mass of limb muscles, underdeveloped jaws, undescended testes in males, and sensorineural deafness. Of course, a zygote carrying two copies of such an allele would be unable to develop. Apparently, these alleles do not increase the mutation rates and the incidence of cancer. In contrast, heterozygous alleles that abolish only the proofreading activity of replicative DNA polymerases are mutators and lead to multiple adenomas, which often progress to cancer.

Homozygous incomplete loss-of-function alleles of several genes that encode proteins responsible for removing misincorporated ribonucleotides from the DNA cause a disease known as Aicardi–Goutières syndrome, which involves inflammation of the central nervous system, skin, and small blood vessels, and intellectual disability. Complete loss of function of these proteins likely leads to death well before birth.

Heterozygous loss-of-function alleles of several genes that encode proteins participating in MMR drastically increase incidence of several cancers, in particular, of non-polyposis colorectal cancer. Affected people have an 80% life-time risk of this cancer, which is rare in the general population. Very rare homozygotes die from multiple childhood cancers.

Complete loss of function of one of the glycosylases that detach damaged bases in the course of BER leads to polyposis and increased risk of colorectal cancer. In contrast, loss of function of enzymes which participate in processing of ends of SSBs is known to cause three neurological diseases, none of which is involved with increased risk of cancer. One of them, ataxia oculomotor apraxia 1, leads to difficulties with walking and the inability to control eye movements.

Xeroderma pigmentosum is a collective name for seven diseases caused by loss of function of proteins that participate in NER. It is characterized by severe hypersensensitivity of the skin to sunlight, and by a 2000-fold increase of the incidence of skin cancers. Some forms of xeroderma pigmentosum, as well as two other diseases caused by defective NER, Cockayne syndrome and trichothiodystrophy, also involve neurological symptoms.

Fanconi anemia is a collective name of 16 diseases caused by loss of function of proteins that participate in ICL repair. Affected people suffer from progressive bone marrow failure, developmental disabilities, and have a very high incidence of various cancers, including leukemias. The average life span of patients is ~30 years.

Impairments of DSB repair can cause a wide variety of diseases. Dysfunctions of NHEJ usually lead to immunodeficits, because this process plays the key role in the generation of the diversity of antibodies (see below). Some defects in recombination-dependent DSB repair drastically increase the risk of breast and ovarian cancer.

Diseases caused by defects in DDR are also rather diverse. Complete loss of function of a protein called ATM, which initiates the cell's response to DNA breaks, leads to a diseases called ataxia telangiectasia. Among its symptoms are poor coordination of muscles due to impairment of the cerebellum, dilated small blood vessels, weak immunity, and increased risk of lymphomas and leukemias. People heterozygous for loss-of-function alleles of a gene that encodes a protein known as p53 have Li–Fraumeni syndrome, which drastically increases the risk of many cancers, in particular, of sarcomas, at early age. This protein plays a central role in regulating many activities of the cell, including two crucial components of DDR – arrest of the cell cycle after a DNA damage is detected, and initiation of apoptosis if DNA repair fails. Some other defects of DDR cause several forms of microcephalic dwarfism.

All these diseases result from defects in DNA handling that an individual inherits from their parents. However, in a multicellular organism, a mutant allele can be not only inherited (zygotic), and, therefore, present in all cells, but also somatic (postzygotic), present only in those cells that are descendants of the somatic cell in which the mutation occurred (see Chapter 4). A postzygotic defect of DNA handling that only impairs cell functioning will likely cause death of the mutant cell or of its descendants, without a noticeable impact on the organism as a whole. By contrast, postzygotic mutator alleles play a major role in cancer.

Cancer is a diverse collection of diseases which begin from the origin, within the organism, of a cell with a pathological genotype that produces a "cancerous" phenotype. The key features of this phenotype are uncontrollable division, immortality, abolition of suicide mechanisms, and the ability to disperse within the organism which leads to metastases. A cancerous phenotype would be lethal in a zygote. Thus, it can appear only as a result of somatic mutation. It takes several mutations to convert a normal genotype into one that makes the cell cancerous. That is why inherited mutator alleles strongly increase the risk of cancer.

Still, a majority of cancers appear in people who do not suffer from any inherited cancer predisposition syndromes. However, even in such cases the mutation rate in tumor cells is elevated in a majority of human cancers, due to acquisition of somatic mutator alleles. A notable exception is "liquid tumors", that is leukemias and lymphomas, which usually have normal rates of somatic mutation.

A mutator allele may appear either in a healthy cell and facilitate its malignization or in an already-cancerous cell and eventually take over the whole tumor. Indeed, a cancerous cell that acquired a mutator allele may gain an advantage over its peers in the course of pathological intercellular selection (see Chapters 6 and 7) which favors more aggressive cells. Some tumor cells accumulate thousands and even millions of somatic mutations, and the nature of these mutations constitutes a "signature", which often makes it possible to determine what defect of the DNA handling led to the increased mutation rate. For example, transversions are disproportionally common in genotypes of cells in which a particular component of BER is broken, and short insertions and deletions within DNA segments with repetitive sequences (microsatellites, see Chapter 6) is a hallmark of MMR defects. Elevated mutation rates allow cancerous cells to rapidly evolve resistance to chemotherapies and represent a major obstacle to successful treatments.

5.5 Mechanisms of Mutation

Even when all the exquisite mechanisms of DNA handling are in good order, they fail occasionally, because nothing can work without a glitch. These glitches can lead to DNA damages, followed by cell death or by mutations. Unless the cell dies, any DNA damage is eventually repaired, precisely or not. Thus, mutations are the only possible long-lasting consequence of DNA mishandling. A wide variety of molecular events that can lead to a mutation can be subdivided into five broad classes (Table 5.2).

Small-scale errors in handling of the existing DNA. Such errors are not informed by any other DNA sequence, and, thus, produce truly novel mutations. There are at least three ways in which a novel small-scale mutation can appear.

The first is malfunctioning of the mechanisms of replication fidelity, in the course of routine replication of undamaged DNA. A mutation occurs if all these mechanisms fail to prevent it. A majority of small-scale spontaneous mutations appear in this way.

Table 5.2 Mechanisms of spontaneous mutation.

Small-scale errors in handling of the existing DNA	Errors in DNA replication and repair is by far the most common mechanism of mutation, mostly producing small-scale mutations
Non-template DNA synthesis	Apparently, this mechanism does not play a major role in germline mutation and only produces short novel sequences
Gene conversion	Gene conversion leads to replacement of a genotype segment with another one, which can be the other allele at the same locus in a diploid cell or a different segment of the genotype, usually with a similar sequence
Reassortment of DNA segments	This involves errors in recombination, repair, and replication, and is often facilitated by repetitive segments of the genotypes. The key mechanism producing large-scale mutations
Gain or loss of complete chromosomes	This occurs mostly due to malfunctioning of meiosis

The second is replication of damaged DNA, which allows the cell to tolerate the damage for some time. Translesion polymerases, which do not have a luxury of relying on a sound template, are much less accurate than polymerases which replicate undamaged DNA. Thus, prompt repair of DNA damages is likely to be one of the mutation-preventing adaptations, although repairing them too hastily could also be mutagenic.

The third is imprecise repair of DNA damages, which restores chemical soundness, but not the original sequence, of the DNA. Imprecise repair of a SSD should initially lead only to a mismatch, because the damage is confined to one strand, and a mutation eventually occurs only if MMR fails to revert this mismatch. In contrast, repair of a ICL or a DSB readily leads to mutation. In particular, NHEJ appears to be a recipe for mutation, although, surprisingly, even this mechanism of repair often restores the original DNA sequence. We still do not know what causes rare very complex small-scale mutations (Figure 4.5).

Non-template DNA synthesis. Although DNA replication, which uses a pre-existing molecule as a template (see Chapter 1), is by far the most common mode of DNA synthesis, some DNA polymerases can produce more or less random DNA molecules without any template. Mutant sequences resulting from non-template DNA synthesis are truly novel. Still, it appears that this mechanism of mutagenesis is of a rather limited importance; in particular, DNA segments produced without a template are always quite short, because non-template DNA polymerases are prone to abandoning their tasks.

Gene conversion. The nucleotide sequence of a DNA segment can be replaced with some other sequence, "pasted" instead of the original one. This mechanism of mutation is called gene conversion. In a diploid cell, sources of the pasted sequence can be of four kinds (Figure 5.13).

Gene conversion occurs as a result of physical interaction between the "donor" and "recipient" DNA segments. Such a contact always involves the temporary formation of a heteroduplex – a segment of double-stranded DNA in which two strands are of different origin (Figure 5.11). When these strands came from segments with non-identical sequences, there will be mismatches in the heteroduplex, and MMR often repairs them, leading to gene conversion. Also, gene conversion may result from the synthesis of a segment of one chromosome that uses the other chromosome as a template (Figure 5.11).

Regular, meiotic recombination involves physical interaction between the corresponding segments of maternal and paternal genotypes and often leads to gene conversion between them. Such within-locus gene conversion usually results only in minor changes of the converted sequence, because the differences between maternal and paternal alleles are small (see Chapter 2). The other three kinds of gene conversion, shown in Figure 5.13, occur as a result of "ectopic", unintended interactions between different loci. Still, sequences of such loci usually are over 90% identical (over the length of at least ~100 nucleotides), because otherwise a stable heteroduplex is not formed and interaction does not occur. The probability of an ectopic interaction is higher if the two DNA segments are located close to each other on the same chromosome. When a genome contains many similar segments (repetitive elements) gene conversion between them can be rather common.

Typically, a gene conversion event involves a sequence segment (conversion track), of 100–1000 nucleotides. However, as long as the two interacting segments have rather similar sequences, such an event leads to only one or a small number of small-scale changes.

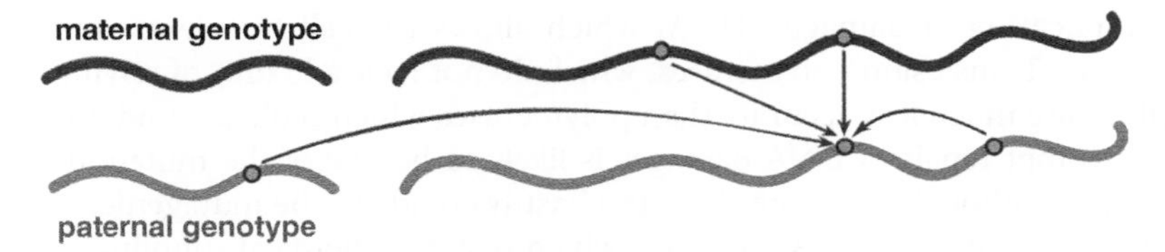

Figure 5.13 Possible sources of information for gene conversion in a diploid cell: another allele of the same locus, another locus of the same genotype, situated either on the same or on a different chromosome, and another locus of the other genotype.

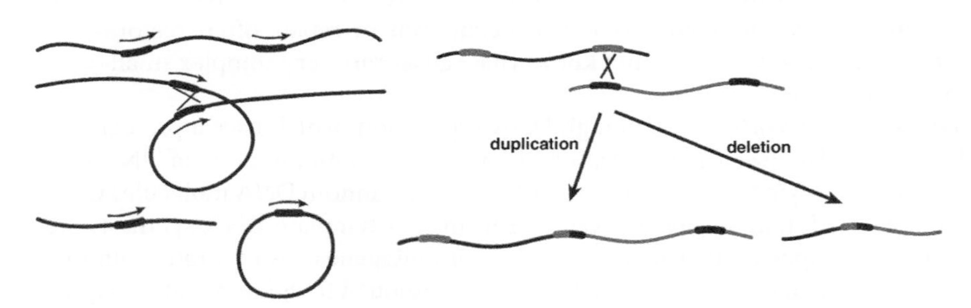

Figure 5.14 Origin of a long deletion (and a circular piece of DNA, to be lost in the course of the next cell division) and of a long deletion and a long insertion due to ectopic recombination between similar sequence segments within one genotype (left) and in two haploid genotypes of a diploid cell (right). Similar segments are shown by thick line segments.

A mutant sequence that appears through conversion has already being tested by natural selection and, therefore, may be not as disruptive to function as a truly novel one.

Reassortment of DNA segments. Reciprocal reassortment of the corresponding segments of the maternal and paternal genotype is the essence of meiotic crossing-over. However, it is convenient to consider crossing-over, which only recombines alleles at pre-existing polymorphic loci, as a separate phenomenon, distinct from mutation (Figure 4.3). In contrast, outcomes of all other modes of segment reassortment are considered mutations. Reassortment of DNA segments is the only mechanism that can generate large-scale mutations. Such mutations are relatively rare (see Chapter 6) but disproportionally important, due to their large individual deleterious effects (see Chapter 13). Large-scale mutations are sometimes called CNVs (copy number variations) or SVs (structural variations), although neither of these terms is logical. A wide variety of molecular processes can lead to a reassortment of DNA segments.

In the simplest case, segment reassortment involves only one breakpoint, so that a new, mutant DNA sequence consists of only two segments of the original sequences. In particular, ectopic recombination between non-allelic sequences easily produces duplications and deletions (Figure 5.14). As it was the case with gene conversion, recombination requires physical contact between DNA molecules, mediated by formation of a heteroduplex (Figure 5.11) and mostly occurs between very similar DNA sequences. In fact, mutations of both these kinds may be produced by the same interaction. Ectopic recombination, as well as gene conversion, between similar DNA segments that are located close to each other occurs more often (see Chapter 6). Also, a breakpoint and resulting segment reassortment can appear as the result of an error in the course of repair of a DSB by NHEJ. A large-scale mutation often involves small-scale sequence changes at breakpoints, which may provide clues about the mechanism of its origin.

Inversions and insertions (see Chapter 4) are large-scale mutations that involve two breakpoints. An inserted DNA sequence can be of foreign origin or can originate somewhere else in the same genome. Most genomes contain sequences, called transposable elements (TEs), that are prone to get inserted into new places and, as a result, are present in many similar copies. There are many families of similar TEs in the human genome, which together comprise at least 50% of it; however, most of them are no longer capable to insert themselves. The only exception is a family of TEs called L1, which also facilitate insertions of short (~300 nucleotides) sequences known as Alu elements.

More complex segment reassortments, which involve three or more breakpoints (Figure 4.6) also happen occasionally. Usually, malfunctioning of DNA replication plays an important role in their origin. In particular, when a stalled replication fork resumes working from the different site of the same template or from another template, this results in a nascent DNA which consists of reassorted segments.

Gain or loss of complete chromosomes. Malfunctioning of meiosis can lead to loss or gain of complete chromosomes (aneuploidy), or even to a whole-genome duplication (WGD). Not surprisingly, such mutations are most deleterious. In humans, loss of any autosome is lethal. Gains of autosomes, which result in three, instead of two, copies of an autosome, is usually consistent with life only in the case of chromosomes 21 (Down syndrome), 18 (Edwards syndrome), and 13 (Patau syndrome). Humans are more tolerant to gains and losses of X and Y chromosomes (see Chapter 13). WGD is lethal in humans and almost all mammals, with the only known exception being the red viscacha rat from South America, whose genome appears to be the product of a WGD.

All the paths to mutation considered above involve failure of some mechanism that ensures fidelity of DNA handling and, therefore, fit into the general paradigm that mutations are undesirable but unavoidable accidents. Indeed, this is likely to be generally true at least for multicellular eukaryotes (see Chapter 4). However, there are exceptions, when cells introduce mutations deliberately, in order to perform some biological function – very much like DNA damages are introduced in order to perform crossing-over.

The most striking among such exceptions is a set of genetic processes that produces mature immunoglobulin-encoding genes in jawed vertebrates, including, of course, ourselves. Immunoglobulins (antibodies) are proteins whose function is to recognize and bind foreign molecules (antigens). Antibodies are produced by specialized cells called lymphocytes, and are the key component of the system of acquired immunity. Each lymphocyte possesses a unique immunoglobulin-encoding gene and, thus, produces a unique antibody. Totally, our adaptive immune system includes a repertoire of ~10 billion different antibodies, which obviously cannot be directly encoded by the human genome. Two mechanisms are responsible for each lymphocyte acquiring a unique antibody-encoding gene. First, a mature gene is assembled randomly from many segments of the genome, which involves deliberate introduction of DSBs and their repair by NHEJ. Secondly, in the course of this assembly, a template-independent DNA polymerase introduces short pieces of random DNA between these segments.

The diversity of antibodies ensures that, when an antigen enters human organism, there will be antibodies that can bind it, perhaps imperfectly. Then, lymphocytes that bind the antigen start to proliferate, and those that bind better proliferate faster. In the course of the resulting cell divisions, the region of the immunoglobulin-encoding gene of length ~1000 nucleotides which encodes the parts of an antibody that directly interact with the antigen undergoes somatic hypermutation, being replicated by DNA polymerases with low fidelity. These polymerases make errors up to 10^6 times more often than high-fidelity replicative polymerases epsilon and delta. This leads to efficient natural selection (see Chapter 7) of lymphocytes that produce antibodies which bind the antigen stronger.

Apparently, in multicellular organisms deliberately introduced mutations never occur in the germline and, thus, are never transmitted to the offspring. From this perspective, mutation is radically different from meiotic recombination, whose only function is to alter genotypes of the offspring, although it is still not clear why natural selection favored this process. However, deliberate introduction of mutations into small segments of the genome is also known for some unicellular organisms that infect long-lived hosts, and in this case mutations are transmitted to the offspring. For example, a unicellular eukaryote that causes sleeping sickness, *Trypanosoma brucei,* often alters the gene that encodes its surface protein, which allows it to evade the immune response of the host. Gene conversion which involves copying information from other segments of the genome plays the key role in this case of deliberate mutagenesis. Clearly, if doing so increases fitness (see Chapter 7), cells can fiddle with their DNA and modify their genotypes very efficiently.

Further Reading

Caldecott, KW 2014, Ribose – an internal threat to DNA, *Science*, vol. 343, pp. 260–261.
Contamination of DNA by ribose in the course of DNA replication.

Carvalho, CMB & Lupski, JR 2016, Mechanisms underlying structural variant formation in genomic disorders, *Nature Reviews Genetics*, vol. 17, pp. 224–238.
A review of mechanisms of large-scale mutations.

Chen, J-M, Cooper, DN, & Chuzhanova, N 2007, Gene conversion: mechanisms, evolution and human disease, *Nature Reviews Genetics*, vol. 8, pp. 762–775.
A review of mechanisms of gene conversion.

Friedberg, EC 2003, DNA damage and repair, *Nature*, vol. 421, pp. 436–440.
A review of DNA repair.

Hogg, M, Osterman, P, & Bylund, GO 2014, Structural basis for processive DNA synthesis by yeast DNA polymerase ε, *Nature Structural and Molecular Biology*, vol. 21, pp. 49–56.
An incredibly complex spatial structure of DNA polymerase epsilon.

Hombauer, H, Srivatsan, A, & Putnam, CD 2014, Mismatch repair, but not heteroduplex rejection, is temporally coupled to DNA replication, *Science*, vol. 334, pp. 1713–1716.
Mismatch repair that corrects errors of the replisome occurs immediately after DNA replication.

Li, Z, Woo, CJ, & Iglesias-Ussel, MD 2004, The generation of antibody diversity through somatic hypermutation and class switch recombination, *Genes and Development*, vol. 18, pp. 1–11.

Deliberate production of somatic mutations in the course of development of the immune system.

O'Driscoll, M 2012, Diseases associated with defective responses to DNA damage, *Cold Spring Harbor Perspectives in Biology*, vol. 4: a012773.
A review of human diseases caused by broken maintenance of DNA.

Wang, W 2007, Emergence of a DNA-damage response network consisting of Fanconi anaemia and BRCA proteins, *Nature Reviews Genetics*, vol. 8, pp. 735–748.
DNA-damage response network in human cells.

Zakharyevich, K, Tang, S, & Ma, Y 2012, Delineation of joint molecule resolution pathways in meiosis identifies a crossover-specific resolvase, *Cell*, vol. 149, pp. 334–347.
Deliberate production of DNA damages in the course of meiotic crossing-over.

6

Mutation Rates

I was born when my father was seventy-one years old and was hence a rather sickly child.

Yamamoto Tsunetomo, *Hagakure, 1716.*

Mutation rates can be measured directly by sequencing mother–father–child trios and counting alleles in the genotype of a child that could not be inherited from their parents. In a variety of multicellular eukaryotes, the rate of mutations of all kinds is $\sim 10^{-8}$ per nucleotide per generation. Thus, each newborn human carries about 100 *de novo* mutations. Most mutations are small-scale, with single-nucleotide substitutions being the most common of all. Mutation rate varies between nucleotide sites, due to both known and unknown reasons. Large-scale mutations are promoted by repeats in the genotype sequence. In mammals, the mutation rate increases with paternal age, primarily because of ongoing cell division in the male germline. In some cases, drastic mutations can be detected by studying phenotypes of offspring. In particular, the per genome rate of dominant drastic mutations in humans is ~0.004, and the per genome rate of recessive lethals in fruit flies is ~0.02. Phenotypic effects of mild mutations, which cannot be detected individually, can be described by mutational pressure, the steady-state rate of change of some characteristic of the population due to unopposed accumulation of mutations. Mutation always pushes fitness-related traits in the direction which reduces fitness.

6.1 Measuring Mutation Rates

Direct detection of germline mutations is based on comparison of genotypes of parents and offspring. Because per nucleotide mutation rates (see Chapter 4) are usually very low, $\sim 10^{-8}$ per generation in multicellular organisms, measuring them directly requires a lot of data and, thus, depends on sequencing technologies that are cheap and precise. Fortunately, several years ago next-generation sequencing (NGS) became good enough for the job. This brought our knowledge of spontaneous mutation to a new level. As it happened many times in the history of biology, dramatic progress resulted not from clever thinking but from applying the brute force of a new technology. Of course, inventing this technology required a lot of clever thinking.

In the simplest case of asexual haploids, an offspring has only one parent, each organism carries only one haploid genotype, and without *de novo* mutations genotypes of parent

Crumbling Genome: The Impact of Deleterious Mutations on Humans, First Edition. Alexey S. Kondrashov.

P cacatcgactgatacgatcgatcagtac

O cacatcgactgatGcgatcgatcagtac

M cacat*c*gactgatacgatcgatc*a*gtac

cacat*g*gactgatacgatcgatc*a*gtac

F cacat*c*gactgatacgatcgatc-gtac

cacat*c*gactgatacgatcgatc*a*gtac

O cacat*g*gactgatacgatcgatc*a*gtac

cacat*c*gactgatGcgatcgatc-gtac

Figure 6.1 Detecting *de novo* mutations within a parent–offspring duo (asexual haploids) and a mother–father–offspring trio (sexual diploids). Sites that are polymorphic within parents are in italics, and mutations are in upper case.

and offspring would be identical. Then, each difference between them can be interpreted as a mutation (Figure 6.1). With sexual diploids, there are two parents and two genotypes in an organism, which makes it necessary to sequence mother–father–offspring trios, each comprising six haploid genotypes. Any allele that is present in the offspring but absent in both parents must be a new mutation (Figure 6.1). Of course, comparing DNA texts consisting of billions of nucleotides by eye would be impossible, but a laptop computer does it easily.

In fact, direct detection of mutations is still not free from technical problems. Some of them appear because precision of any sequencing technique cannot be absolute – occasionally, a sequencing error occurs, and a wrong nucleotide is called. Double-checking any suspected mutation effectively eliminates false positives, but some mutations may be missed due to such errors. Other problems appear because NGS works by producing huge numbers of relatively short reads (see Chapter 1). As a result, detecting small-scale mutations which affect DNA segments that are much shorter than the length of a read may be simpler than detecting large-scale mutations that do not fit within a read. Also, when mutations are studied in diploids, including humans, it may be difficult to determine in which parental genotype a particular mutation occurred. Still, even today these problems can be addressed, with some care. In several years, further reduction of the frequency of sequencing errors and increased length of reads will likely make them obsolete.

Data produced by NGS did not come as a revelation: by the end of the 20th century, old, indirect methods already produced good estimates of mutation rates. One of these methods relies on detecting mutations through their effects on phenotypes. For example, if *de novo* mutations of a particular gene that cause dominant Mendelian disease occur at a rate of 10^{-5}, 10% of these mutations are nonsense substitutions (see Chapter 2), and a particular nucleotide substitution can produce a stop codon at 300 nucleotide sites within the gene, we can conclude that the per nucleotide rate of all substitutions is $10^{-5} \times 0.1/(300/3) = 10^{-8}$ (a factor of 3 appears because each nucleotide can be substituted in three ways).

Another indirect method infers mutation rates from data on differences between genomes of different species. For example, if the last common ancestor of humans and chimpanzees lived 400 000 generations ago (assuming that the length of a generation was the same in the ancestral lineage of both species), and the alignments of selectively neutral (see Chapter 8) segments of their genomes contain 1.3% of isolated mismatches, we can conclude that the per nucleotide rate of all substitutions in the course of their divergence was 0.013/800 000 = 1.6×10^{-8} (in fact, this figure is a little bit too high, because variation within the common ancestral population can also contribute to differences between modern species). Here, selective neutrality is essential, because only in this case mutation rate and evolution rate are equal (see Chapter 8).

Nevertheless, direct data are very important, because they are free from uncertainties inherent to all indirect estimates. Per gene rates of *de novo* mutations causing Mendelian diseases are usually known only rather approximately. The number of generations on the evolutionary path that connected two modern species also cannot be know precisely, and ruling out natural selection (see Chapter 7) within a genome segment is difficult. In contrast, one cannot argue with a DNA sequence, properly determined.

6.2 Data on Mutation Rates

Mutations are changes in DNA sequences. Thus, they must be, first of all, studied and characterized at the level of genotypes. Table 6.1 presents data on the rates of small-scale germline mutations (affecting DNA segments shorter than 50 nucleotides) in humans and four species of so-called model organisms, which play particularly prominent roles in research. Over 99% of all mutations belong to this category. Of course, the cut-off at 50 is as arbitrary as any other. The future will bring some refinements; however, it seems likely that true mutation rates do not differ from the figures presented in the table by more than a factor of two or three. Because the Main Concern is about humans, I consider only multicellular eukaryotes and ignore prokaryotes and unicellular eukaryotes.

Table 6.1 shows that the per nucleotide per generations rates of small-scale mutations are higher in species with larger numbers of cell divisions from zygote to zygote.

Table 6.1 Data on rates of small-scale mutations in some multicellular eukaryotes.

Species	Per nucleotide per generation rate of single-nucleotide substitutions	Per nucleotide per generation rate of other small-scale mutations	Per diploid genome per generation rate of all small-scale mutations	Number of cell divisions per generation	Per nucleotide per cell division rate of all small-scale mutations
Homo sapiens	1.3×10^{-8}	0.15×10^{-8}	95	220	0.6×10^{-10}
Mus musculus	0.5×10^{-8}	0.03×10^{-8}	30	45	1.2×10^{-10}
Drosophila melanogaster	0.5×10^{-8}	0.05×10^{-8}	1.5	35	1.6×10^{-10}
Caenorhabditis elegans	0.3×10^{-8}	0.1×10^{-8}	0.8	9	4.4×10^{-10}
Arabidopsis thaliana	0.7×10^{-8}	0.1×10^{-8}	2	30	2.7×10^{-10}

However, this correlation is not very strong, because the per cell division mutation rate is higher in species with smaller numbers of cell divisions. These patterns can be explained if selection always favors reduction of the per generation mutation rate, but very low per cell division mutation rates are involved with high costs of fidelity (see Chapter 4). Under this scenario, when, for some reason, the number of cell divisions per generation increases, selection for lower mutation rate gets stronger, but leads to only a modest reduction of the per cell division mutation rate, insufficient to prevent some increase of the per generation mutation rate. Wide variation of the per genome per generation mutation rates apparently suggest that this rate is not a direct target of natural selection. This is not surprising, because larger genomes carry higher proportions of junk DNA (see Chapter 8).

Relative rates of mutations of different kinds are also mostly uniform. In all species, ~90% of mutations are single-nucleotide substitutions, and deletions are two to three times more common than insertions. Among all single-nucleotide substitutions, substitutions of a purine with another purine (A < > G), or of a pyrimidine with the another pyrimidine (T < > C), collectively called transitions, are two to three times more common that transversions, substitutions of a purine with a pyrimidine or vice versa (A < > T, A < > C, G < > T, or G < >C). Note, that if all the 12 possible substitutions occurred with the same probability, transversions would be two times more common than transitions. Frequencies of insertions and deletions rapidly decline with their length. In fact, ~50% of all insertions and deletions affect just one nucleotide. An insertion is often a duplication of a short sequence flanking the site of insertion. Complex mutations (see Chapter 4) constitute ~2% of all small-scale mutations.

Mutation rates presented in Table 6.1 are only averages, because rates of mutation of a particular kind are not uniform. At some sites, called mutational hotspots, mutation rates can be 10 or even 100 times above the average. Mutagenic small-scale properties of DNA sequences can be of two kinds. First, there are mutagenic contexts, specific short sequences where mutation rates are elevated. In humans and other mammals, the most important mutagenic context is a sequence of just two nucleotides, CG. The reason for this is that within this context, nucleotide C is often methylated ($-CH_3$ group attached to it), which makes it more difficult for the cell to handle it correctly. On average, rates of transitions and transversions at nucleotide sites that belong to CG contexts are increased by factors of 15 and 3, respectively.

Secondly, repetitive DNA sequences, such as AAAAAAAAAA, AGAGAGAGAGAG or ACTACTACTACTACTACT, called microsatellites, are mutagenic, because the replisome often errs replicating them. Most of the resulting mutations are insertions or deletions of a repetitive unit (A, AG, and ACT, in the above examples), and the total mutation rate per microsatellite can be as high as 10^{-4} or even 10^{-3}. As a result, microsatellites rapidly appear and disappear in evolution. The number of microsatellites in the human genome is rather high, so that the per genome rate of mutations in them, mostly not reflected in Table 6.1, may be as high as 100 or even more.

Indirect evidence suggests that, on top of heterogeneity of the mutation rates caused by obvious small-scale features of DNA sequences, there is also a substantial cryptic heterogeneity which cannot currently be explained. If, at a particular site, we find a derived allele within the population of one species, the chances that the same allele will be found in the population of another, related, species are much higher than at other similar sites, even when the possibility that these alleles were inherited from the common ancestor of the two species can be ruled out. Thus, mutations at some sites appear more often than at others, although we do not know why.

To directly measure mutation rates at individual nucleotide sites, billions of mother–father–offspring trios need to be analyzed, which is not going to happen soon. In contrast, patterns of within-population variation will likely produce some indirect estimates in the near future, when the numbers of studied genotypes will become 10–100 times higher than today (see Chapter 11). This will hopefully shed more light on the intriguing cryptic small-scale heterogeneity of mutation rates.

There is also a large-scale heterogeneity of rates of small-scale mutation. Mutation rates in genome segments that are replicated late are higher than in segments that are replicated early (see Chapter 5). There are also substantial differences between per nucleotide mutation rates in different chromosomes.

An extreme example of large-scale heterogeneity of mutation rates is provided by the circular chromosomes of length of only 16 569 nucleotides (in humans) that reside within mitochondria (see Chapter 1). Mitochondria evolved from bacteria, which ~2 billion years ago became endosymbionts of organisms from which all modern eukaryotes eventually evolved. Modern mitochondrial chromosomes are remnants of the genome of these bacteria, which, however, lost almost all of their genes, as they migrated over time into the major, nuclear portion of eukaryotic genomes. Replication of mitochondrial DNA is performed by a DNA polymerase which is not part of the nuclear replisome, so that mitochondrial mutation rate could evolve independently of the mutation rate in the rest of the genome. In humans, the per nucleotide per generation mutation rate in mitochondrial chromosome is $\sim 2 \times 10^{-5}$, which is ~1000 times higher than in nuclear chromosomes. Even with this rate, there is only a ~30% chance that a mutation will appear in the mitochondrial chromosome in the course of a generation. Thus, selection for further reduction of the mitochondrial mutation rate is probably weak.

Most small-scale mutations appear one-by-one, independently of other mutations. Still, a non-trivial proportion of them are clustered, in two rather different ways. First, ~2% of mutations arrive as parts of groups of mutations that occur, likely as parts of a single event, close to each other within the genotype. Most such groups consist of two or more single-nucleotide substitutions, located within ~1000 nucleotides of each other ("mutational showers"), but some are spaced more tightly and together constitute complex mutations (Figure 4.5). It seems that mutational showers appear due to specific mechanism(s), different from that responsible for isolated mutations. Secondly, instead of being singletons, many small-scale mutations are parts of clusters and, thus, may be present in multiple siblings (Figure 4.7). In humans, if an individual carries a *de novo* mutation, their sibling carries the same mutation with probability ~2% (see Chapter 14).

Let us now consider large-scale mutations. They are relatively rare: only ~0.4 mutations that affect >50 nucleotides, versus ~100 of small-scale mutations, occur per human genome per generation. As a continuation of the trend that is present in small-scale mutations, the longer the DNA segment affected by large-scale mutations, the rarer such mutations are. As a rough approximation, the mutation rate declines with the length of the affected DNA segment x as $\sim 1/x^2$. Nevertheless, the average total number of nucleotide sites in a human genome that are affected by large-scale mutations in the course of a generation is 8000, which is almost 100 times more than the number of sites affected by small-scale mutations (or the number of such mutations, which is only slightly less). Of course, this contrast appears because some large-scale mutations are truly large.

A substantial proportion of large-scale mutations is caused by ectopic recombination between similar non-allelic sequences (Figure 5.14). Movements of transposable elements are responsible for ~0.1 large-scale mutation per human genome per generation.

All quantitative traits are genetically variable (see Chapter 3), and mutation rates are no exception. So far, there are no data on Mendelian variation of the rates of germline mutation in humans due to mutator alleles with large effects (see Chapter 13). Still, substantial variation in rates of germline mutation was reported between human families, which is likely due to alleles with small or moderate individual effects.

Under usual circumstances, when no mutagens (see Chapter 4) are used, germline mutation rates in multicellular organisms do not depend strongly on the environment. Still, in 1928 Hermann J. Muller demonstrated that when *D. melanogaster* is maintained at a temperature close to the maximum that they can tolerate, their per generation mutation rate increases by a factor of ~2. We will encounter Muller, one of the founders of genetics and a namesake of the Haldane–Muller principle (see Chapter 9), Muller's ratchet, and Dobzhansky–Muller incompatibilities (see Chapter 15), many times in the pages to follow. Later, similar results were obtained for *C. elegans*. However, in humans and other mammals no features of their ordinary environments are known to have any mutagenic – or antimutagenic – effects.

6.3 Guilty Older Men

In contrast, human germline mutation rate strongly depends on the sex – men transmit to children many more *de novo* mutations than women. Moreover, the number of *de novo* mutations transmitted to an offspring rapidly increases with the paternal age. Thus, Table 6.1 in fact presents average mutation rates, observed in children whose fathers are ~30 years old. The effects of both sex and the paternal age are important for the Main Concern, and I want to briefly review the history of their discovery.

The first step was made very soon after the phenomenon of mutation was described by de Vries (see Chapter 4). In 1912, the German obstetrician Wilhelm Weinberg, a codiscoverer of the Hardy–Weinberg law (see Chapter 2), noticed that in large families the last-born children suffer from genetic diseases more often than first-born children. Weinberg presciently noticed that this pattern may imply that such diseases are caused by mutations.

The second step was made by one of the founders of modern evolutionary biology, John B. S. Haldane, who proposed, *inter alia*, Haldane's dilemma, Haldane's rule, Haldane's sieve, and the Haldane–Muller principle (see Chapter 9). In 1947, he studied pedigrees of boys with hemophilia. Two forms of hemophilia, A and B, are caused by recessives alleles of different X chromosome genes. Because, until recently, severe hemophilia was fatal, disease-causing alleles are rare. Thus, an affected individual usually carries an allele which appeared not long ago in his ancestry. A girl carries two X chromosomes (that is why she is a girl), and the chances that both her alleles are disease-causing are very low. In contrast, a boy carries only one X chromosome and one Y chromosome (that is why he is a boy), and there are no hemophilia genes on the Y chromosome. Thus, it takes only one disease-causing allele for a boy to have hemophilia, and he obtains this allele from his mother, because his father gives him the Y chromosome (otherwise, he would be a girl). Thus, hemophilias, as well as many other X-linked recessive conditions (see Chapter 12), are almost exclusively diseases of males.

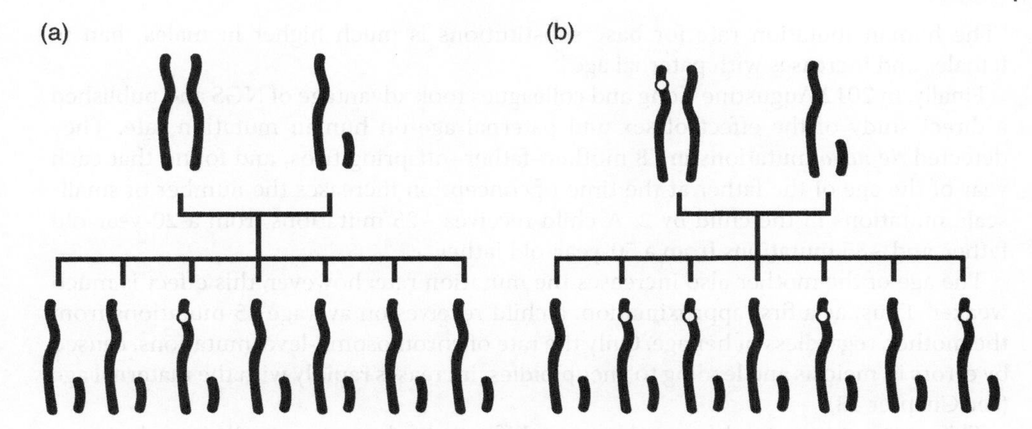

Figure 6.2 If hemophilia in an affected boy is due to a *de novo* mutation, his brothers would be mostly unaffected (a). In contrast, if hemophilia is inherited from the mother, who is, therefore, a heterozygous carrier, a brother of an affected boy has 50% chance to also have the disease (b). X and Y chromosomes are shown by long and short rods, and disease-causing alleles are shown as open circles.

All these facts had already been well-established by 1947. Haldane, however, went a step further, by asking "What kind of a disease-causing allele does a hemophilic boy receive from his mother – a *de novo* mutation or an older allele which the mother received from her parents?". In the first case, the mother would have both her alleles normal and, because most *de novo* mutations in humans appear as singletons, brothers of an affected boy would have only a low risk of having hemophilia. In contrast, in the second case the mother would be a heterozygous carrier of hemophilia, and a brother of an affected boy would have 50% chance to also have hemophilia (Figure 6.2). Haldane determined that the risk of hemophilia in brothers of affected boys is indeed close to 50%. Thus, in a vast majority of cases hemophilia is inherited from a heterozygous mother. From this fact, Haldane concluded that mutations occur more often in a male than in a female germline: a mutation that causes hemophilia usually appears in the maternal grandfather of the affected boy, and not in his mother.

In 1987 Takashi Miyata and colleagues studied the difference between mutation rates in female versus male germline from a rather different perspective. They compared rates of evolution of junk DNA (see Chapter 8) in mammalian autosomes and sex chromosomes and observed that X chromosome junk evolves slower, and Y chromosome junk evolves faster, than autosome junk. An autosome spends equal times in the germline of females and males, an X chromosome spends two-thirds of the time in females, and a Y chromosome is present only in males. Because the rate of evolution of junk DNA is equal to the mutation rate (see Chapter 8), Miyata and colleagues concluded that mutations occur more often in the male germline.

Efficient methods of DNA sequencing, developed in 1977 (see Chapter 1), made it possible to determine the origin of *de novo* mutations that cause autosomal dominant Mendelian diseases (see Chapter 2). Data on a number of such diseases confirmed Haldane's discovery: in most cases, pathogenic mutations are of paternal origin. These data also demonstrated that pathogenic mutations appear more often in the offspring of older fathers. Thus, in 1997 James F. Crow confidently stated that

"The human mutation rate for base substitutions is much higher in males than in females and increases with paternal age".

Finally, in 2012 Augustine Kong and colleagues took advantage of NGS and published a direct study of the effect of sex and paternal age on human mutation rate. They detected *de novo* mutations in 78 mother–father–offspring trios, and found that each year of the age of the father at the time of conception increases the number of small-scale mutations in the child by 2. A child receives ~25 mutations from a 20-year-old father, and ~85 mutations from a 50-year-old father.

The age of the mother also increases the mutation rate; however, this effect is much weaker. Thus, as a first approximation, a child receives on average 15 mutations from the mother, regardless of her age. Only the rate of chromosome-level mutations, caused by errors in meiosis and leading to aneuploidies, increases rapidly with the maternal age (see Chapter 13).

This contrast is certainly caused by very different life histories of cells in male versus female germline. In the germline of a woman, mitoses cease before she is born, and only one last, meiotic cell division occurs right before ovulation. Thus, regardless of the woman's age, there are ~35 cell divisions between the zygote from which she developed and an egg which she produced. Male germline goes through approximately the same number of cell divisions before the onset of puberty, after which cells start to divide more often. As a result, a sperm produced by a 20-year-old man goes through 150 cell divisions, and a sperm produced by a 55-year-old man goes through 800 cell divisions, each of which is preceded by error-prone DNA replication (see Chapter 5).

The paternal age effect is not unique to humans. Apparently, it is even more pronounced in chimpanzees. In this species, testes are ~3 times larger than in humans, presumably because an ovulating chimpanzee female usually mates with multiple males, which leads to sperm competition. Thus, it is likely that in chimpanzee male germline cells divide even faster than in human.

On top of accumulation of replication-produced mutations, paternal age can occasionally increase the mutation rate due to a very different mechanism, natural selection between germline cells. Such selection increases the proportion of germline mutations that facilitate propagation of cells which carry them. When transmitted to an offspring, some of these mutations cause Mendelian diseases, including Apert syndrome and Multiple endocrine neoplasia type 2B. Thus, results of pathological selection between germline cells are very different from that of pathological selection between somatic cells, which plays a major role in cancer (see Chapter 5).

6.4 Rates of Phenotypically Drastic Mutations

Although every mutation changes the genotype, phenotypically mute mutations are irrelevant to the Main Concern. Due to our poor knowledge of the relationship between genotypes and phenotypes (Figure 1.21), phenotypic effects of germline mutations cannot be currently deduced from any genotype-level data. Thus, these effects need to be studied independently, which requires accomplishing two feats. First, phenotypic effects of *de novo* mutations must be observed. Secondly, these effects must be distinguished from possible effects of pre-existing mutant alleles or of non-genetic factors.

Let us first consider phenotypically drastic mutations, visible or lethal (see Chapter 2), which can individually cause easily identifiable phenotypes. Even detecting such mutations may be involved with problems. Only in the simplest case of visible mutations that are dominant and fully penetrant (see Chapter 3), each *de novo* mutation can be properly recorded: every time when an offspring of two wild-type parents has a mutant phenotype, this is due to a *de novo* mutation, and every *de novo* mutation is manifested in this way. By screening offspring of wild-type parents for mutant phenotypes, one measures the rate of such mutations in the whole genome.

In *D. melanogaster*, the per genome rate of dominant visible mutations is ~0.002. In humans, the per genome rate of mutations that lead to dominant Mendelian diseases is ~0.004. These figures are subject to some uncertainty, because there is no definite boundary between Mendelian and complex phenotypes (see Chapter 3). Still, one could expect the genomic rate of visible mutations to be higher in humans that in fruit flies: there are more protein-coding genes in the human genome, and the human per nucleotide mutation rate is higher (Table 6.1).

Unfortunately, such a simple screening cannot be used to study recessive visible mutations, even if they always produce a detectable phenotype when homozygous. First, a *de novo* mutation will usually appear as a heterozygote and, thus, will not be immediately observable. Secondly, a phenotypically detectable homozygote can appear in the offspring of two wild-type individuals due to not a *de novo* mutation, but to Mendelian segregation, if both parents carry pre-existing heterozygous mutant alleles (see Chapter 2). Thus, different approaches are needed in this case.

One of them is to cross wild-type individuals with individuals homozygous for recessive alleles of some genes (analyzing cross, in Mendel's terminology, see Chapter 2). Visible *de novo* mutations that occurred in wild-type individuals will be manifested in the offspring from this cross. Unfortunately, this method, known as the specific-locus (gene) test, can detect mutations in only a small number of genes, because an individual cannot be made homozygous for too many visible alleles (14 is a record, for *D. melanogaster*). Also, it is important to distinguish clusters of *de novo* mutations (see Chapter 4) from pre-existing visible alleles, which can be present in wild-type individuals. This can be achieved by screening offspring from many wild-type siblings (Figure 6.3).

The specific-locus test was employed by William L. Russell and Liane B. Russell to measure the rate of recessive visible mutations of seven genes that affect coat color or ear morphology in mice. These studies produced the spontaneous mutation rate per gene of 5×10^{-5}. This rate is likely to be above the average mutation rate of a murine gene, because the seven genes used in the test appear to be exceptionally mutable. In *D. melanogaster*, several studies, each based on a small number of loci, produced average per gene mutation rates between 5×10^{-6} and 1×10^{-5}. If the combined length of protein-coding exons of an average gene is 1500 nucleotides, and 100% of indels and, say, 20% of nucleotide substitutions within these exons affect the function of the protein drastically, the per nucleotide mutation rates presented in Table 6.1 imply that an average per gene rate of visible mutations in *D. melanogaster* is $1500 \times (0.05 \times 10^{-8}) + 1500 \times (0.5 \times 10^{-8}) \times 0.2 = 2.3 \times 10^{-6}$.

As an alternative to the specific-locus test, one can perform a mutation-accumulation experiment and allow *de novo* recessive visible mutations to accumulate in the heterozygous state for some number of generations, after which they are made homozygous and,

Figure 6.3 Detecting recessive visible *de novo* mutations. A number of wild-type siblings, produced by wild-type parents (white) are crossed with individuals homozygous for recessive alleles of one or, better, several genes (black). A singleton mutation that occurred in one of these siblings is manifested as one mutant offspring (gray) from one cross (left). A cluster of mutations is manifested as several mutant offspring from one cross (center). If no mutations occurred in the genes under study, which is by far the most common case, all the offspring in the cross are wild-type (right). A pre-existing mutant allele (inherited by some wild-type siblings from one of their parents) would lead to ~50% mutant offspring from ~50% of crosses (not shown).

thus, observable. Homozygotes can be produced through inbreeding or, in the case of *D. melanogaster* only, by using balancer chromosomes (see below). This way, mutations at a large set of loci can be scored. Lev Y. Yampolsky and colleagues scored recessive mutations that produce eye phenotypes in ~200 genes of *D. melanogaster* after they accumulated for 10 generations, and estimated the per genome rate of recessive visible mutations at ~0.001. Unfortunately, there are no data on the rate of mutations that lead to human recessive Mendelian diseases. If, however, we assume that such diseases can be caused by mutations of 2000 genes, and that the average per gene mutation rate is 10^{-5}, we get an estimate of 0.002 – 2 times less than what has been directly measured for dominant diseases.

Obviously, the rate of lethal, as opposed to visible, drastic mutations cannot be studied by a simple phenotypic screening. Death is not a specific phenotype and one cannot perform genetic analysis of the dead. We still know essentially nothing about rates of both dominant and recessive lethals in humans and other mammals, because studying them is complicated by a very high prenatal mortality (see Chapter 11). Obviously, mortality of offspring from the moment of conception provides an upper bound for the rate of dominant lethals. However, a developing organism can also die due to many other reasons, so that this upper bound is useful only if low. This is, indeed, the case for some organisms with external development. In fishes, under favorable conditions, egg mortality can be as low as 1%, implying that the rate of early-acting dominant lethals is even lower. However, in *D. melanogaser* mortality of eggs is ~10%, and it does not seem likely that dominant lethals are the main cause of it. The same must be true for prenatal death of ~70% of human zygotes (see Chapter 11).

Studying recessive lethals is also difficult. The specific-locus test obviously cannot be used in this case. Thus, mutation-accumulation experiments remain the only generally available option. Surprisingly, such experiments have never been performed, although scoring accumulated lethals in, say, fish (but not in mammals) would be straightforward.

Fortunately, for *D. melanogaster* a yet another method for studying recessive lethals is available – balancer chromosomes, discovered by Alfred H. Sturtevant and developed into a powerful tool by Muller. A balancer chromosome has three properties: (i) it carries one or several large inversions which prevent it from recombining with its ordinary counterpart; (ii) it carries a dominant visible allele, making its presence in a fly easily observable; and (iii) it carries a recessive lethal allele and, thus, is lethal when homozygous (autosome) or hemizygous (X chromosome).

Let us see how balancers can be used to study recessive lethals on the X chromosome (Figure 6.4). First, a male is crossed with a female which has one balanced and one normal chromosome. Secondly, daughters from this cross that carry a maternal balancer X chromosome, revealed by the dominant allele on it, are collected. Of course, each of these daughters also carries a paternal X chromosome. Finally, the offspring from each such daughter is examined. Here, it does not matter with whom she mated. Sons that inherited the balancer X chromosome from their mother do not survive. Thus, as long as the other, paternal, X chromosome confers normal viability, there will be one son for two daughters among these offspring. In contrast, if the paternal X chromosome carries a recessive lethal, there will be no sons at all.

Muller used this method to demonstrate that X-rays are mutagenic. However, that study also involved measuring, as a control, the natural rate of appearance of recessive lethals on the X chromosome. Muller found that on average two F_1 daughters out of 1000 produce no sons. In other words, an X chromosome which must be lethal-free, being the only X chromosome in a male, acquires a recessive lethal, in the course of one generation, with probability 0.002. Because one X chromosome constitutes ~10% of the diploid genome of *D. melanogaster*, this result implies that the per genome rate of

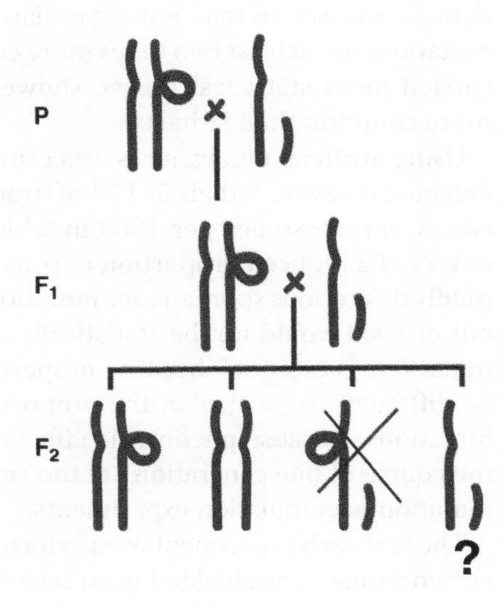

Figure 6.4 Detecting recessive lethals on the X chromosome (see text). Balancer chromosome is shown by a loop.

recessive lethals is ~0.02. This estimate was confirmed by studying recessive lethals on autosomes. Thus, at least in *D. melanogaster*, recessive lethal mutations appear with a rate much higher than that of dominant and recessive visible mutations.

6.5 Mild Mutations and Mutational Pressures

Drastic derived alleles, individually detectable through their effects on phenotypes, constitute only a small proportion of genetic variation (see Chapter 2), because polymorphic loci which harbor mild alleles that affect only complex traits are much more common (see Chapter 3). The same is true for *de novo* mutations, although the proportion of drastic *de novo* mutations is higher than that of drastic polymorphisms, because negative selection eliminates drastic alleles from populations more rapidly than mild alleles (see Chapter 7).

Still, drastic mutations were discovered first, and for some time the very existence of mild mutations was not obvious. This is not surprising, because mild mutations, which cannot be observed individually at the level of phenotypes, even when homozygous, are much more difficult to study than drastic mutations. Mild mutations were discovered in the 1930s by Nikolay V. Timofeev-Ressovsky and Yulij Ya. Kerkis.

Timofeev-Ressovsky performed the same crosses as shown in Figure 6.4, but introduced two modifications. First, before crossing males from the parental generation he irradiated them by X-rays, which substantially increased the rate of *de novo* mutations. Secondly, he scored the F_2 generation quantitatively, going beyond simply recording whether sons are present or absent in a particular sibship. Timofeev-Ressovsky found that in the offspring of many F_1 daughters, sons were present, but their numbers were well below 50% of the number of daughters. He correctly concluded that irradiated X chromosomes that are present in genotypes of these daughters carry one or several mutations which cause death only in some proportion of hemizygous males. One can view such mutations as incompletely penetrant lethals, or as mild mutations affecting the probability of survival. Statistical analysis of his data demonstrated that, among radiation-induced mutations, such mildly deleterious mutations are at least two times more common than lethals. Kerkis, whose experiments carried more statistical power, showed that mild mutations are at least four times more common than lethals.

Using artificial mutagenesis was critical for these experiments. Timofeev-Ressovsky detected recessive lethals in 12% of irradiated X chromosomes, which is 60 times above two X chromosomes per 1000 in which recessive lethals appear spontaneously, and observed a reduced proportion of sons in 25% of sibships. Clearly, he could not detect mildly deleterious spontaneous mutations: a reduced proportion of sons in four sibships out of 1000 would not be statistically significant. And yet studying mild spontaneous mutations is essential, because properties of spontaneous and induced mutations can be different. In particular, the proportion of drastic alleles is higher among induced mutations. Because phenotypic effects of mild spontaneous mutations that appear in the course of one generation are too small, they need to be studied in multigeneration mutation-accumulation experiments.

The first such experiment was performed by Terumi Mukai in 1964. By using a balancer chromosome 2, he shielded from selection (see Chapter 7) all mutations that appeared on all copies of one ancestral chromosome 2 in the course of many generations. Then, Mukai recorded the rate of decline in the mean of viability and of increase of its variance, which resulted from free accumulation of mutations. If we take into account that one chromosome 2 constitutes ~20% of the diploid genome of *D. melanogaster*, Mukai's data imply that free accumulation of mutations in the whole genome reduced mean viability by ~2% per generation. On top of this, Mukai used an elaborate statistical method which takes into account changes of both mean and variance of viability to estimate that this decline was due to ~1 viability-reducing mild mutation appearing on average in the genome every generation. If so, the rate of mildly deleterious mutations is ~50 times higher than that of recessive lethals, and the average effect of these mutations on viability is about −2%.

The estimate of the actual per genome per generation number of *de novo* mild mutations depends on a number of assumptions, some of which are hard to test, and should thus be treated with caution. In contrast, the rates of decline of the mean viability, as well as of increase of its variance, have been observed directly. Later, my laboratory at Cornell University confirmed this part of Mukai's results using a rather different method of shutting down selection (see Chapter 9). Indeed, simply knowing the collective effect of *de novo* mutations at the level of phenotypes can already be very interesting, even if no attempt is made to infer their numbers and individual properties.

This purely phenotypic approach to mutation rates is based on the concept of mutational pressure, which is the phenotype-level analog of the genotype-level concept of mutation rate. The mutational pressure P on some characteristic of the population is the steady-state relative rate of change of this characteristic due to unopposed accumulation of *de novo* mutations. P may depend on the environment, being higher under harsh conditions, under which the effects of mutations are often stronger (see Chapter 10). One can say that Mukai inferred the per genome rate of deleterious mutations on the basis of his measurements of the mutational pressures on the mean and the variance of viability.

In the case of a two-state trait, Mendelian or complex, the mutational pressure on the frequency of its rare, disease state is simply the per generation rate of increase of the frequency of this state, assuming that mutations are not eliminated. In this case, $P = \mu b$, where μ is the total rate of mutations producing alleles that may cause the disease state and b is the probability that a mutant allele actually causes this state. Here, b can be broadly understood as penetrance of mutant alleles, although one possible reason why a mutant allele may not immediately lead to the disease is its recessivity. Clearly, the mutational pressure is the highest when a mutant allele always causes the disease and it cannot exceed the total mutation rate at the target of the trait. In all cases, the mutational pressure is equal to the frequency of those cases of the disease which would not occur without *de novo* mutations (see Chapter 13).

In the case of a quantitative trait, the mutational pressures on both its mean and variance are important. Mutation increases mean values of some quantitative traits (e.g., of development time), but decreases mean values of others (e.g., of fertility). In contrast, the mutational pressure on the variance of a quantitative trait is always positive. To make a mutational pressure on a quantity dimensionless (see Chapter 3), it is often convenient to normalize it over the current value of the quantity. The variance-normalized mutational pressure on variance is sometimes called mutational heritability.

However, mutational pressure on variance can be also normalized over the squared mean, in which case it can be called mutational evolvability (see Chapter 3).

Some modes of incomplete penetrance of disease-causing *de novo* mutations could lead to systematic short delays in their manifestation. For example, in the case of an X-linked recessive disease a paternal mutation will not cause a disease in the next generation, when it resides in a female genotype. Similarly, in the case of an autosomal recessive disease in a population where moderate inbreeding (such as first cousin marriages) is common but incest is avoided, a *de novo* mutation is likely to cause the disease several generations, but not immediately, after its appearance. In contrast, incomplete penetrance of *de novo* mutations that affect quantitative traits is unlikely to lead to such delays. Anyway, as long as delays are short, unopposed accumulation of mutations should soon lead to constant-rate changes of the population.

Since the pioneering work by Mukai, a lot of data on mutational pressures have been obtained. So far, mutational pressures on two-state traits are available almost exclusively for humans. These data will be considered in Chapter 13. In contrast, all the available direct data on the mutational pressures on quantitative traits were obtained on non-human species. Indeed, neither artificial mutagenesis nor mutation-accumulation experiments, the two most straightforward methods of measuring such mutational pressures, can be performed on humans.

Artificial mutagenesis was used in my laboratory at Cornell University to study mutational pressures on nine quantitative traits in *D. melanogaster*. We studied flies whose grandfathers were treated with ethyl methanesulfonate (EMS; see Chapter 4). Mutation exerted substantial pressures on the mean values of four traits that were tightly related to fitness: it increased developmental time, and reduced viability, fertility, and longevity. In contrast, we did not detect any mutational pressure on the mean values of traits that were not tightly related to fitness – metabolic rate, motility, body weight, and numbers of bristles on two segments of adult flies. One treatment with EMS produces as many recessive lethals as ~100 generations of spontaneous mutations. Assuming that a fly whose both grandfathers were treated with EMS carries as many *de novo* mutations as produced spontaneously in the course of 50 generations, our data imply that spontaneous per generation mutational pressures (normalized by the mean values of the corresponding traits) on developmental time, viability, fertility, and longevity were 0.0004, −0.014, −0.002, and −0.0008, respectively. Of course, *de novo* mutations behind all these pressures, except the one on viability, must be mild.

Still, estimating mutational pressures by means of artificial mutagenesis is not a perfect substitute for measurements of effects of spontaneous mutation. A number of multigeneration mutation-accumulation experiments has been performed on species with short generation times. As it was the case for artificial mutagenesis, mutational pressures on mean values of quantitative traits that are directly related to fitness are negative. In contrast, mutational pressures on variance are always positive (Table 6.2).

These patterns make sense. Most non-neutral mutations are deleterious and, thus, should shift a fitness-related trait in the undesired direction, although there is no reason why mutations should disproportionately increase or decrease the value of a trait that is not tightly related to fitness. Still, mutations somehow affect all traits and, thus, mutation always increases their variances.

Table 6.2 Mutational pressures on some quantitative traits.

Species	Trait	Mutational pressure on the mean	Mutational evolvability	Comment
Drosophila melanogaster	Viability of larvae	−0.01	0.0004	Estimates obtained in different experiments vary widely
Caenorhabditis elegans	Fitness	−0.003	0.001	This species mostly reproduces by self-fertilization, and mutational pressures in related outcrossing species are several times higher
Amsinckia gloriosa (a flowering plant)	Number of flowers	−0.005	0.0001	This species mostly reproduces by self-pollination; however, mutational pressures in a related outcrossing species are similar

Further Reading

Conrad, DF, Keebler, JE, & DePristo, MA, 2011, Variation in genome-wide mutation rates within and between human families, *Nature Genetics*, vol. 43, pp. 712–714.
Variation of the human germline mutation rate.

Haldane, JBS 1947, The mutation rate of the gene for haemophilia, and its segregation ratios in males and females, *Annals of Eugenics*, vol. 13, pp. 262–271.
Indirect, pedigree-based evidence for paternal bias in human mutation.

Halligan, DL & Keightley, PD 2009, Spontaneous mutation accumulation studies in evolutionary genetics, *Annual Reviews of Ecology and Systematics*, vol. 40, pp. 151–172.
A review of spontaneous mutational pressures.

Miyata, T, Hayashida, H, & Kuma, K 1987, Male-driven molecular evolution: a model and nucleotide sequence analysis, *Cold Spring Harbor Symposia on Quantitative Biology*, vol. 52, pp. 863–867.
Genome comparison-based evidence for paternal bias in mammalian mutation.

Mukai, T, Chigusa, SI, & Mettler, LE 1972, Mutation rate and dominance of genes affecting viability in *Drosophila melanogaster*, *Genetics*, vol. 72, pp. 335–355.
A classical mutation-accumulation experiment.

Muller, HJ 1928, The measurement of gene mutation rate in *Drosophila*, its high variability, and its dependence upon temperature, *Genetics*, vol. 13, pp. 279–357.
A classical study of mutation in *Drosophila*.

Rahbari, R, Wuster, A, & Lindsay, SJ 2016, Timing, rates and spectra of human germline mutation, *Nature Genetics*, vol. 48, pp. 126–133.
A large-scale direct measurement of human mutation rate.

Willems, T, Gymrek, M, & Poznik, GD 2016, Chromosome-wide characterization of Y-STR mutation rates, Retrieved January 9, 2017, *bioRxiv*, http://dx.doi.org/10.1101/036590.
A high total rate of mutations in short repetitive segments of the human genome.

Wong, WSW, Solomon, BD, & Bodian, DL 2016, New observations on maternal age effect on germline *de novo* mutations, *Nature Communications*, vol. 7: 10486.
A weak maternal age effect in human mutation.
Yang, HP, Tanikawa, AY, & Van Voorhies, WA 2001, Whole-genome effects of ethyl methanesulfonate-induced mutation on nine quantitative traits in outbred *Drosophila melanogaster*, *Genetics*, vol. 157, pp. 1257–1265.
Measurements of artificial mutational pressures on several quantitative traits.

7

Natural Selection

Nothing in biology makes sense, except in the light of evolution.
Theodosius G. Dobzhansky, 1973.

Adaptations, parts of phenotypes that perform various functions contributing to survival and reproduction of organisms, are products of long evolution, which proceeds through natural selection of spontaneous mutations. Selection occurs when the ability of an individual to survive and reproduce depends on its genotype or, in other words, when different genotypes confer different fitnesses. It is productive to think of selection in terms of a fitness landscape, the graph of function that attributes a fitness to each genotype. Distribution of fitnesses of genotypes can be characterized by imperfection and evolvability of fitness. Measuring selection is difficult, due to a variety of reasons. Evolvability of actual, instead of inherited, fitnesses of individuals provides an upper limit for the strength of selection. Positive selection favors initially rare alleles and promotes changes, and negative selection favors common alleles and maintains status quo. The strength of selection at a locus can be described by a selection coefficient. Selection acting on a quantitative trait can be directional or stabilizing, and can be described by the selection differential, the difference between mean values of the trait after and before selection.

7.1 Vulnerable Adaptations and Their Evolutionary Origin

We learned a lot about mutation, but why should we care? We should care because a mutation of the genotype of an organism may affect its phenotype. Phenotypic traits that are important for the organism are called adaptations. In order to survive and reproduce, an organism must do a lot of things – or, in scientific parlance, perform a lot of functions – and a part of its phenotype which performs a particular function is an adaptation. Examples of adaptations and functions are countless: a DNA-binding protein makes sure that the gene it regulates works only when needed, factor IX of blood coagulation takes part in activating factor X, an artery directs the flow of blood, eyes see, and so on.

Partitioning all activities of an organism into separate functions, and of the whole phenotype into traits and adaptations is not without problems. Heart pumps blood, and also emits quiet sounds, which a physician, armed with a stethoscope, may find very useful – so, does it make sense to say that pumping blood is a function but emitting

Crumbling Genome: The Impact of Deleterious Mutations on Humans, First Edition. Alexey S. Kondrashov.

sounds is not, despite these sounds being unavoidably produced by blood turbulence? Spatial arrangement of atoms in the active center of an enzyme, being crucial to its catalytic function, is definitely an adaptation, but what about other parts of the molecule, which play mostly supportive roles and whose structures are not prescribed by the function? These are important issues; however, simplistic, naive understanding of functions and adaptations will be sufficient for our purposes.

Improbability, complexity, and vulnerability are hallmarks of adaptations. The human brain (Figure 7.1) possesses all these qualities to the extreme degree. Most human genes are expressed in the brain. In particular, thousands of different proteins work in synapses, structures responsible for transmission of signals between neurons. There are ~100 billion neurons and about the same number of glial cells in the brain, and a neuron can form up to 10 000 synaptic contacts with other cells. An entity that consists of so many interacting parts must be prone to crumbling. Not surprisingly, the genomic targets (see Chapter 3) of neurodevelopmental diseases are particularly large, and both pre-existing (see Chapter 12) and *de novo* (see Chapter 13) mutations disproportionally impair the brain.

Deleterious mutations, the essence of our Main Concern, are those mutations that impair adaptations and, as a result, reduce fitness, the ability of an organism to reproduce. Such mutations also reduce wellness of modern humans (see Chapter 12). How did adaptations of modern organisms come into being? Biology is still very far from a comprehensive understanding of the origin of any complex adaptation. Nevertheless, the general mechanism of adaptive evolution is already well-understood – it occurs through natural selection of spontaneous mutations.

In a better world, where Buddha's words would not apply, mutation alone would be sufficient for organisms to evolve in the direction which improves their adaptations and, thus, their ability to thrive under the current environment. This would be the case if organisms could change genotypes of their offspring into the right direction. Suppose you are a giraffe, and have trouble reaching to leaves high above the ground. Of course, you wish your offspring to live more comfortably – and, thus, modify your genotype, before transmitting it, in such a way that it will cause the development of longer necks.

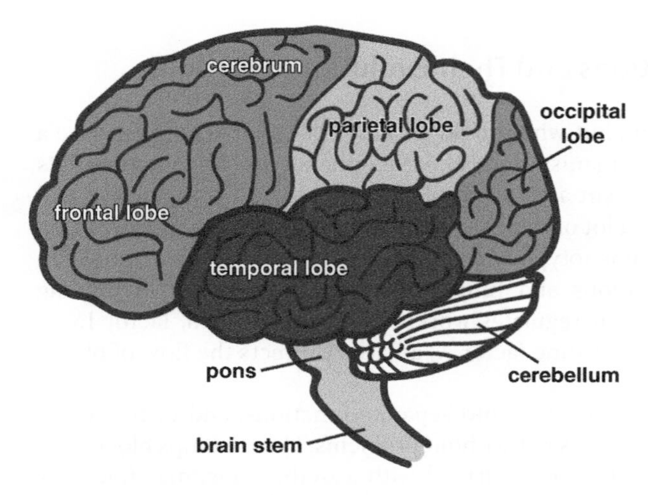

Figure 7.1 Lateral view of the human brain.

The trouble is, this never – or almost never – happens. There is simply no biological mechanism to alter genotypes sensibly, because mutation is not informed by which changes would be good for the organism. Thus, another force is needed to push evolution in the direction that creates and improves adaptation. This force is called natural selection, or simply selection.

The key role of selection in evolution was discovered by Charles Robert Darwin and Alfred Russel Wallace, who worked independently. On reading Darwin's *On the Origin of Species by Means of Natural Selection*, first published in 1859, it is hard to avoid a feeling that he had direct access to some of the deepest mysteries of nature. Indeed, Darwin, who knew nothing not only about the molecules of heredity, but even about its discrete, Mendelian nature, nevertheless proposed a framework for understanding evolution which has not changed much since his time and is not going to change much in the future. Contrary to a common misconception, Darwin and Wallace were not the first to claim that modern life is a product of long evolution in the past, although only Darwin proved this beyond reasonable doubt. In contrast, their discovery of selection had almost no forerunners, except for a Scottish landowner and farmer, Patrick Matthew, who in 1831 outlined this concept, of all places, in an Appendix to his book titled *On Naval Timber and Arboriculture*.

In the narrow, genetic sense, selection is differential reproduction of genotypes. Usually, it is enough to characterize reproductive success of an individual by just one number, which takes into account its viability, ability to attract mates, fertility, and longevity. Below, the term fitness will refer to this number. Often, the fitness of an individual may be simplistically understood as the number of its offspring. From the point of view of a geneticist, fitness is just another quantitative phenotypic trait – although of a unique complexity and importance. Fitness of a genotype is the inherited value (see Chapter 3) of fitness of an individual of this genotype. Inherited values, and not transmittable values, are relevant here because selection occurs within a generation.

By definition, selection occurs as long as different genotypes confer different fitnesses. If no other force is interfering, selection leads to increased frequency, and eventual fixation, of the genotype which confers the highest fitnesses, and to eventual elimination of all other genotypes. Offspring of those who reproduce faster eventually take over the population and offspring of those who reproduce slower eventually disappear; this banal truth is behind adaptive evolution by natural selection.

A productive way of thinking about selection is to perceive fitness as a surface flying over the space of all possible genotypes (Figure 7.2), called the fitness, or adaptive, landscape. The altitude of this surface over a genotype coincides with the fitness of this genotype, so that the fitness landscape is the graph of the function (in the mathematical sense) that relates fitness to genotypes. In a genetically variable population, selection is absent only when the fitness landscape is flat over the region of the space of genotypes which harbors all the present genotypes, which means that the available genetic variation is mute, fitness-wise. The notion of fitness landscape was introduced, in a slightly different form, in 1931 by Sewall Wright, one of the founders of modern evolutionary biology. In fact, it is possible – and often useful – to consider landscapes of all complex traits, and not only of fitness. Figure 3.13 presents such trait landscapes, which could be fitness landscapes, as long as the "trait" shown there is fitness.

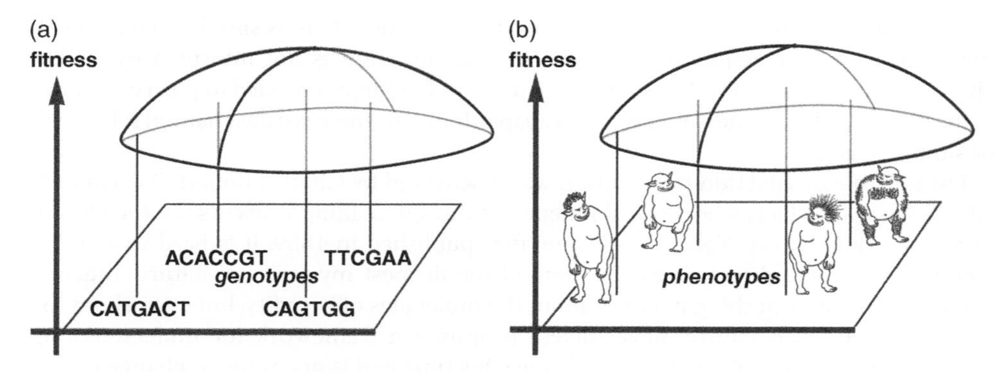

Figure 7.2 Fitness landscapes over the spaces of (a) genotypes or (b) phenotypes.

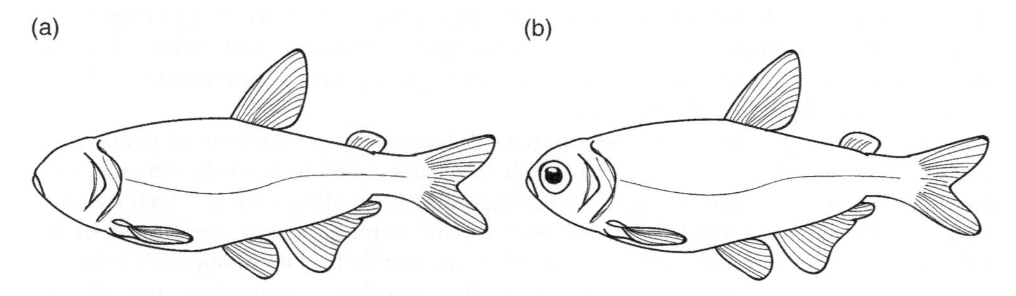

Figure 7.3 Individuals from cave populations of fish *Astyanax mexicanus* possess only vestigial, functionless eyes or are totally eyeless (a), in contrast to individuals from surface populations (b).

Often, it is convenient to expand the notion of selection on phenotypes and to consider fitness landscapes flying over the space of phenotypes. In this case, the fitness of a phenotype is the average fitness of individuals with the phenotype. Because genes, and not phenotypic traits, are transmitted from parents to offspring, differential reproduction of phenotypes affects the next generation and, thus, could lead to evolution, only if it causes differential reproduction of genotypes. This is often the case, because phenotypic variation is usually affected by genotypic variation (see Chapter 3).

Natural selection is responsible not only for the origin of adaptations, but also for their preservation. Sometimes a change in the environment renders useless a part of the phenotype that used to perform an important function. Such used-to-be adaptations, no longer protected by selection, disappear soon, at the evolutionary time scale. This process can be seen both at the level of phenotypes and genotypes. Cave animals lose their eyes (Figure 7.3). The genome of *Mycobacterium leprae*, which causes leprosy, possesses about as many broken, functionless remnants of genes, called pseudogenes (see Chapter 8), as functional genes. Most of these pseudogenes correspond to very similar but functional genes in the genome of a closely related organism, *M. tuberculosis*. A recent loss of functions of many previously crucial genes in the ancestry of *M. leprae* likely occurred due to deepening of its adaptation to parasitism. As a result, *M. leprae* became an obligate intracellular parasite. In contrast, *M. tuberculosis* can grow on artificial media.

Natural selection is an incredibly cruel force, and it gives one chills to think that this force was responsible for the origin of modern species, including ours. Think, for example, about selection which drove the evolution of a wider birth canal, necessary to deliver big-brained babies of modern humans – it certainly led to an enormous amount of suffering and death of *Homo erectus* women. Similarly, selection that protected our adaptations against the never-ending onslaught by deleterious mutations led to suffering and death of countless mutant individuals who failed to become our ancestors in the course of millions of generations. The brutality of natural selection may be one of the reasons for the persistent popularity of creationism, despite it becoming intellectually untenable over 150 years ago. Unfortunately, nature could not care less whether we like the ways it works.

7.2 Two Basic Characteristics of Selection

Selection is a population-level phenomenon. Ignoring, for a while, the issue of the exact nature of genetic variation which leads to variation of fitness, we can describe selection in the population by distribution $q(w)$, where w is the fitness of a genotype. Selection is absent only if all the available genotypes possess the same fitness, in which case $q(w)$ is concentrated in one point.

Being a function, $q(w)$ is a rather complex entity. Thus, it is desirable to describe selection by something simpler, best of all, by a number. One may think that the mean value of $q(w)$, called the mean population fitness W, is the most important property of selection. However, this is not the case, because it does not really matter whether the fitness of a genotype is absolutely high or low. Instead, everything depends on how a genotype fares in comparison with other genotypes.

Imagine two populations, each consisting of, say, 1000 individuals, 80% of which are of genotype A and 20% are of genotype a. In one population, individuals of these genotypes produce 5 and 3 offspring, respectively, and in the other 50 and 30 offspring, respectively. Then, in the first population there will be $800 \times 5 = 4000$ offspring of genotype A out of $(800 \times 5) + (200 \times 3) = 4600$ offspring of both genotypes. Analogously, there will be 40 000 offspring of genotype A out of 46 000 offspring totally in the second population. We can see that, although the mean fitness is 10 times higher in the second population than in the first one, selection leads to exactly the same change in the genetic compositions of both populations: the frequency of genotype A among the offspring is $4000/4600 = 40\,000/46\,000 \sim 87\%$. Thus, W *per se* is not an informative characteristic of selection.

Still, there are useful numerical characteristics of selection. Two of them, evolvability of fitness E and its imperfection I, which describe selection from rather different perspectives, are of particular importance (Figure 7.4). Formally, E is the variance of $q(w)$ normalized by its squared mean (see Chapter 3). If, for example, the variance of fitness within the population is 0.25 (so that its standard deviation is 0.5), and its mean is 2, $E = 0.25/4 = 1/16$. We can say that E characterizes the mean-normalized width of scattering of fitness.

In contrast, I characterizes the reduction of fitness relative to its highest possible value. Formally, $I = (w_{max} - W)/W = w_{max}/W - 1$, where w_{max} is the maximal fitness, possessed by the best possible ("perfect", see Chapter 10) genotype. Thus, if the maximal

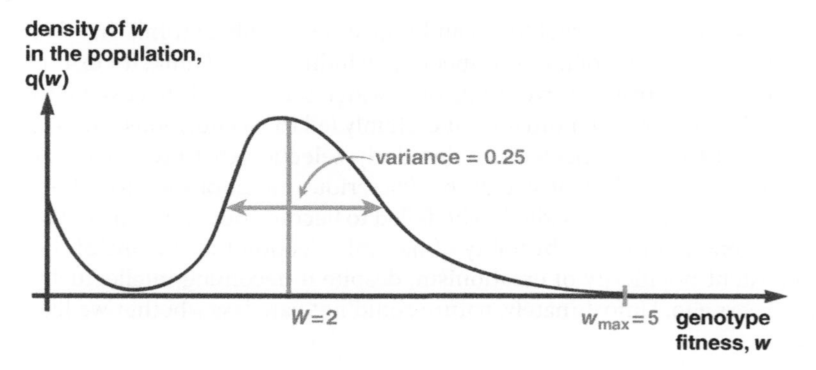

Figure 7.4 Distribution of fitness q(*w*) can be characterized by evolvability of relative fitness $E = 0.25/4 = 1/16$ and by imperfection $I = (5-2)/2 = 1.5$.

fitness is 5 but the mean population fitness is only 2, the imperfection is $(5-2)/2 = 1.5$, because the excess of the maximal possible fitness over the mean population fitness constitutes 150% of this mean fitness. Traditionally, instead of imperfection *I*, load $L = (w_{max} - W)/w_{max}$ is used to characterize deviation to the mean population fitness from the maximal one. These two quantities are mutually interchangeable; however, to normalize this deviation by *W* is more convenient for us.

Both *E* and *I* are dimensionless, so that the values of each of them are the same in the two populations considered above. If, and only if, selection is absent, both *E* and *I* are equal to zero. Indeed, when all genotypes have the same fitness, this fitness is both the mean one, implying that $E = 0$, and the maximal one, implying that $I = 0$.

When different genotypes possess different fitnesses, so that selection is operating, knowing *E* is not enough to determine *I* (and vice versa), as long as we do not know q(*w*). Still, *E* and *I* are not completely independent of each other. It can be shown that *I* is always equal to or larger than *E*, and that $I = E$ only when selection works in the simplest way possible: some genotypes are lethal, and all others possess the same non-zero fitness (Figure 7.5).

The importance of *E* is obvious: selection is synonymous to scattering, around the mean population fitness, of fitnesses of genotypes that are actually present in the population. Thus, $E > 0$ means that selection is at work. In contrast, the notion of *I* may appear to be vague, because it compares the mean fitness of actually present genotypes to the fitness of some obscure perfect genotype. It is even possible that, despite $I > 0$, selection does not operate, if all the available genotypes have the same fitness which is below that of the perfect genotype which is effectively absent from the population. Still, because the Main Concern appears due to the differences between properties of actual genotypes and of the "weakly perfect" (see Chapter 10) deleterious mutations-free genotype, we will use the concept of imperfection extensively.

Another reason why *I* is important is that it imposes the lower limit on the maximal fecundity of individuals. If we consider females only, the long-term (geometric) mean fitness of a population must be exactly 1: if each female fails to produce, on average, one successful daughter, the population will go extinct, and if she produces more than one, the population will expand indefinitely, which is impossible. Therefore, the mean fecundity of females of the perfect genotype must be at least $I + 1$. For example, if $I = 4$,

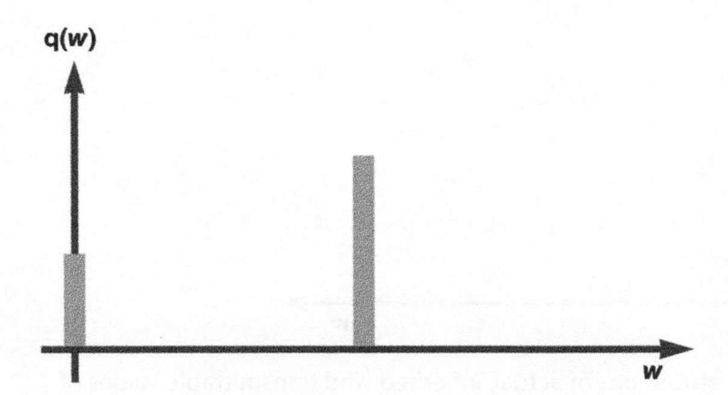

Figure 7.5 Selection which leads to the minimal I, under a particular E. If, as it is shown in the figure, 40% of genotypes do not reproduce at all and 60% have the same, non-zero fitness, $I=E=2/3$. Under any other mode of selection that produces the same value of E, I is higher.

a female of the perfect genotype must produce, on average, exactly five successful daughters, or five times more than an average female in the population. Of course, the actual average fecundity of females of the perfect genotype must be even higher, as long as there is any mortality of their daughters. Thus, any analysis which leads to a conclusion that $I=999$ in the human population must be wrong, unless we are willing to believe that a deleterious mutations-free woman could have 1000 daughters (see Chapters 11 and 15).

7.3 Measuring Natural Selection

Quantitative studies of selection are very difficult, due to several reasons. First, even the actual fitness of an individual is much harder to measure than almost any other quantitative trait. It is easy to weight a frog, but determining the number of offspring it produced during the lifetime requires years of careful observation.

Secondly, there is no substitute for studying selection in natural populations living under their native environments. Indeed, in contrast to mutation rates, which usually do not depend much on the environment and, thus, can be measured in the laboratory (see Chapter 6), strength of selection can be deeply reduced under a benign laboratory environment (see Chapters 9 and 10). Of course, counting offspring of an individual is especially difficult in the wild.

Thirdly, selection is variation of inherited, and not of actual, values of fitness of individuals (see Chapter 3). Thus, to measure selection directly, we need to determine, for each individual in our sample, the expected fitness of an individual with its genotype. This is usually unrealistic, unless the species can be propagated clonally (see Chapter 3). Balancer chromosomes (see Chapter 6) make it possible to extract and "clone" intact chromosomes of *Drosophila melanogaster*, but this technique is not available to any other species, and, again, cannot be applied in the wild.

Fortunately, even data on actual, instead of inherited, values of fitnesses of individuals shed some light on selection, by providing an upper limit to its strength. Indeed, there could be no selection in a hypothetical population where every individual produces the same number of offspring. Of course, in every real population, including that consisting

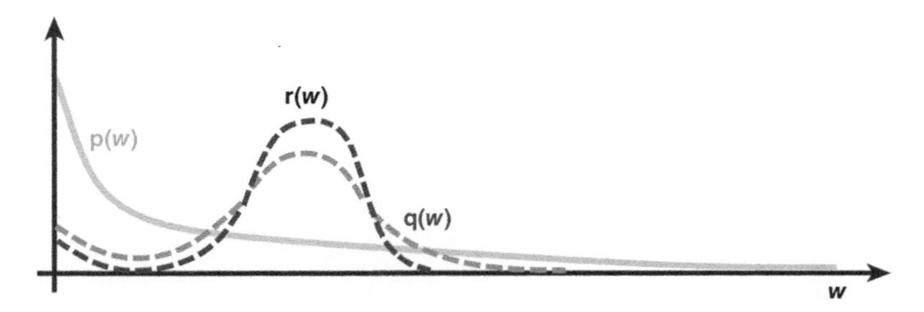

Figure 7.6 A likely relationship between distributions of actual, inherited, and transmittable values of fitness of individuals, p(*w*), q(*w*), and r(*w*). Densities of inherited and transmittable values of fitness, which cannot be observed directly, are shown by broken lines.

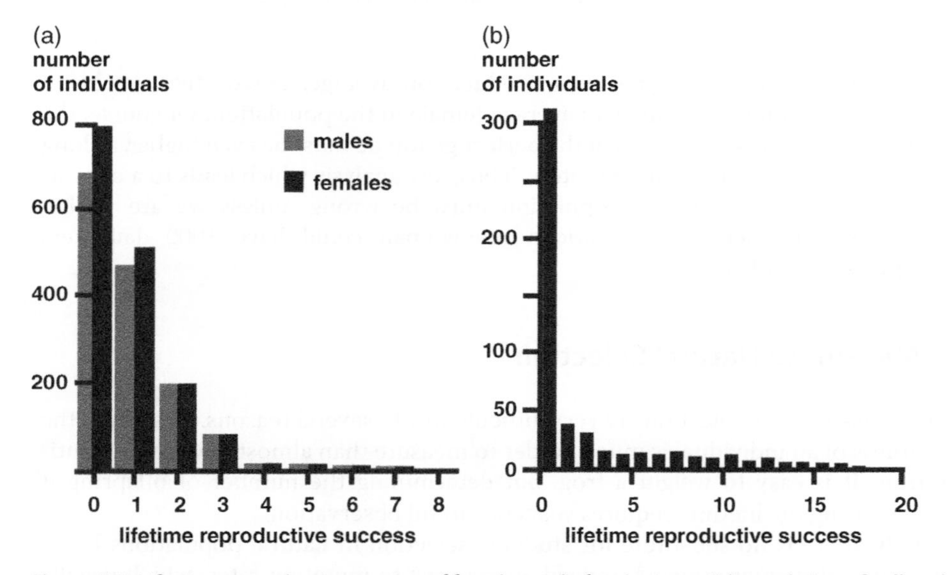

Figure 7.7 Life-time reproductive success of females and of males in a wild population of collared flycatcher (a) and of females in a semi-wild population of rhesus macaques (b) (after Merilä and Sheldon, *American Naturalist*, vol. 155, p. 301, 2000 and Blomquist, *Evolutionary Ecology*, vol. 24, p. 657, 2010).

of genetically identical individuals, where selection is impossible, there is still a substantial variation in actual fitnesses of individuals, due to chance (see Chapter 3). As a result, distribution of actual fitness p(w) always has a wider scattering than that of inherited fitness q(w) (Figure 7.6), and the difference depends on the magnitude of the chance scattering of actual fitnesses of individuals.

Therefore, evolvability of p(w) is called opportunity for selection (or Crow's index). Good data on densities of actual values of fitness are available for a limited number of wild or semi-wild populations. In all the studied populations, there was a substantial opportunity for selection, in particular, because a sizeable proportion of individuals do not reproduce at all. Figure 7.7 shows data on life-time reproductive success in two populations. A wild population of collared flycatcher, *Ficedula albicollis*, that inhabits the island of Gotland (Sweden) was studied over the course of 15 years. A semi-wild population of rhesus macaque, *Macaca mulatta,* that inhabits the island of Cayo Santiago (Puerto Rico) was studied for over 30 years. In the first case, the number of offspring which were entering adulthood (recruits) was recorded, and in the second case all live births were counted. The first measure is probably closer to fitness, because it includes reproduction and survival over the whole life cycle.

Unfortunately, it is impossible to say *a priori* by how much the real selection, described by q(w) (Figure 7.6), is weaker than what the opportunity for selection suggests. Directly ascertaining q(w) is also essentially impossible in a sexual population. In contrast, the variance of the distribution of transmittable values of fitness r(w) can be estimated (see Chapter 3). In the study of collared flycatcher parent–offspring regression was used for this purpose, and the study of rhesus macaques used a more complex technique which takes complete pedigrees into account. For females of collared flycatcher, evolvability and heritability of transmittable values of their lifetime reproductive success were 0.08 and 0.2, respectively. For females of rhesus macaque, they were 0.8 and 0.4, respectively.

There are at least two problems with such estimates. On the one hand, in the presence of epistasis data on transmittable values of fitness are bound to underestimate selection, which occurs due to scattering of inherited values of fitness. On the other hand, similarity between relatives, especially between parents and offspring, can be inflated by non-genetic factors, which can lead to overestimation of selection (see Chapter 3). Still, it is likely that real evolvability and heritability of inherited values of fitness are not too different from estimates obtained for its transmittable values.

In contrast to high opportunity for selection, which may simply reflect vulnerability of individuals to random insults, high evolvability of fitness of genotypes, and, thus, strong selection within a natural population needs an explanation(s). Indeed, simple theory predicts that, as a result of selection, eventually only one, the fittest, genotype will remain in the population, after which there will be no more selection. Several factors may prevent this from happening. One is continuous changes of the environment and, thus, of relative fitnesses of genotypes, which can prevent fixation of any genotype, because no genotype retains the status of the most fit one for long. However, there are serious doubts on whether rapid enough changes of the environment are common. In particular, a number of lineages, such as ginkgo and horseshoe crab, changed very little, at least morphologically, in the course of tens and even hundreds of millions of years. It is hard to imagine that environments of such living fossils kept changing rapidly during all these times, without eliciting any noticeable morphological response. Another factor that can be responsible for persisting selection is deleterious mutation (see Chapter 9).

Still, it should not be surprising that some studies of natural populations do not detect strong selection. On top of differences in methodology and experimental errors, natural populations may be genuinely different from each other in this respect. We will consider data on selection in humans in Chapter 11.

Whatever direct data on fitnesses of genotypes could be collected, they are likely to shed light only on those aspects of selection that can be characterized by E. In contrast, determining I directly is next-to-impossible, because it depends on fitness of the perfect genotype, which is very hard to estimate. On the one hand, this fitness can be lower that the maximal actual fitness of a lucky individual. On the other hand, it can be higher than the fitness of the best available genotype, if the perfect genotype is effectively absent from the population (see Chapter 10). We will consider this issue in Chapter 15.

7.4 Selection at a Polymorphic Locus

Description of selection by distribution of fitness of genotypes in the population $q(w)$ is mute on how selection acts at individual polymorphic loci. However, this action is of key importance, because it determines what kind of heritable changes are produced by selection in the population. In the simplest case of a diallelic locus with alleles A and a, selection acts at this locus when average fitnesses of diploid individuals of genotypes AA, Aa, and aa at this locus, denoted by w_{AA}, w_{Aa}, and w_{aa}, are not the same. The two most important modes of selection at a locus are positive and negative selection.

Positive selection favors an allele which is currently rare and, thus, is likely to be derived (see Chapter 2). If we assume that allele a is rare (and, thus, allele A is common), selection is positive if $w_{AA} < w_{Aa} < w_{aa}$. Positive selection can lead to an allele replacement, the increase of the frequency of an initially rare beneficial mutation, until it eventually reaches fixation in the population, ousting the less-fit ancestral allele (Figure 7.8).

An A > a mutation which will be eventually fixed by positive selection disturbs genetic variation at polymorphic loci that are close to it in the genome. The frequency of a "passenger" allele at such a locus that was lucky enough to be present in the genotype in which this "driver" mutation occurred will increase, until crossing-over separates them. This phenomenon is called hitch-hiking (Figure 7.9). A passenger allele can be selectively neutral or even mildly deleterious; however, in cannot be severely deleterious, because in this case the tightly linked beneficial mutation will not even start spreading.

Positive selection-driven allele replacement is the key population mechanism of adaptive evolution. Functional parts of the genome of any contemporary species can be thought of as a record of such allele replacements in the past. However, quantitatively our knowledge of the genetic basis of adaptive evolution is still rudimentary. For example, we cannot list all allele replacements which together led, in the course of the last ~8 Ma, to the evolution of modern humans from the last common ancestor of humans and chimpanzees. We even have no idea how many allele replacements would

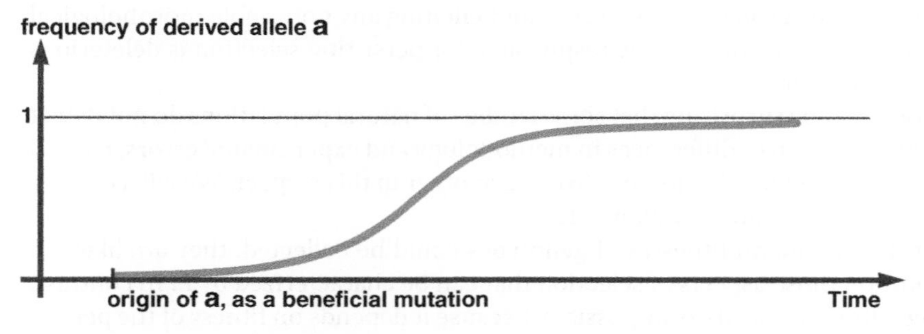

Figure 7.8 The frequency of an initially rare beneficial allele a, produced by a mutation, increases and eventually reaches 100% in the course of a positive selection-driven allele replacement.

● **strongly beneficial allele**
• **neutral or mildly deleterious allele**

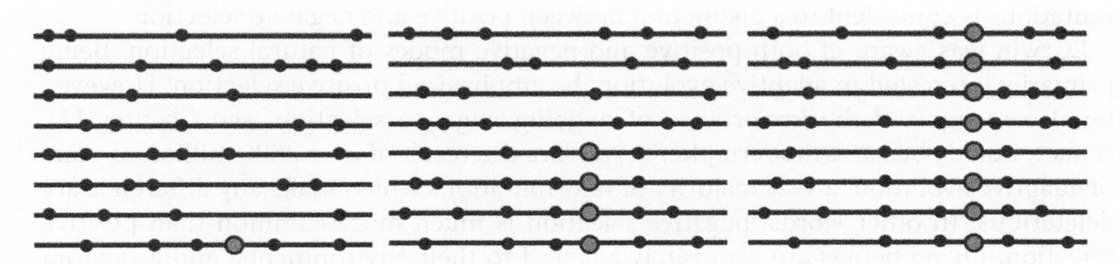

Figure 7.9 Phenomenon of hitch-hiking. Neutral or mildly deleterious alleles located close to the site of the *de novo* mutation that produced a strongly beneficial allele will attain higher frequencies or, in the case of very tight linkage, even fixation. While the beneficial allele goes toward fixation, recombination chips away the edges of the hitch-hiked genotype segment. From left to right: genetic compositions of the population at the moment when the beneficial allele appeared, half-way through the process of positive selection-driven allele replacement, and soon after fixation of the beneficial allele.

it take to convert the genotype of a chimpanzee into that of a passable human. Any number between, say, 100 and 10^5 cannot be ruled out.

Still, many individual cases of positive selection are well-understood, at the levels of both genotypes and phenotypes. For example, in mammals synthesis of an enzyme lactase, which metabolizes milk sugar lactose, normally ceases after the offspring stops consuming the mother's milk. However, in humans two alleles which lead to continuation of lactase synthesis through the whole lifespan appeared independently, in Africa and in Europe, after domestication of the cow ~10 000 years ago. Each of these alleles is a single-nucleotide substitution in a genome segment responsible for regulation of transcription of the lactase-encoding gene. These alleles reach high frequencies locally, but are not fixed within any human population, perhaps due to interbreeding of milk-consumers with people from adjacent populations who do not have dairy cows. Rapid local increases of frequencies of these two alleles have definitely been driven by positive selection.

Positive selection increases the rate of evolution but reduces the level of genetic variation. Methods of detection of positive selection are mostly based on these two facts. However, we are not going to consider these methods, because positive selection is not directly relevant to the Main Concern.

Indeed, negative, and not positive, selection acts against deleterious alleles produced by mutation. Selection is negative if $w_{AA} > w_{Aa} > w_{aa}$, so that the currently common, and likely ancestral, allele A is favored. In other words, instead of promoting evolution, negative selection maintains the status quo. Formally speaking, after the frequency of a derived, beneficial allele exceeds 50%, in the course of an allele replacement (Figure 7.6), selection that favors this allele stops being positive and becomes negative. Later, this negative selection protects the fixed beneficial allele against new deleterious mutations.

We have already implicitly introduced negative selection in Chapters 2 and 4, by stating that rare alleles and *de novo* mutations are more often deleterious than beneficial. By definition, a beneficial (deleterious) allele or mutation increases (decreases) fitness of its carriers. Thus, a distinction between beneficial and deleterious mutations is equivalent to a distinction between positive and negative selection.

Darwin was aware of both positive and negative modes of natural selection. Being primarily interested in adaptive evolution, he emphasized positive selection. However, he also recognized the importance of ongoing negative selection (see Chapter 11). In fact, exactly because modern phenotypes are the result of over 3500 million of years of adaptive evolution, a vast majority of new mutations which make any difference are deleterious. In other words, negative selection is much more common than positive selection. Living beings are exquisitely adapted to their environments, and a random change of a genotype damages adaptation much more often than improves it. One may wonder how a small proportion of mutations turns out to be beneficial, increasing the ability of their carriers to function and reproduce.

Preponderance of negative selection leads to a radical difference between properties of within-population genetic variation and of interspecific genetic differences. Within-population variation that affects adaptations and fitness is mostly due to deleterious alleles which have no chance to become fixed and to contribute to between-species differences, because negative selection keeps them rare. In contrast, almost all important interspecific differences emerge due to positive selection. Thus, most functionally important genetic variation which is present within a population at any given moment of time has nothing to do with its long-term evolution.

Although estimates vary widely, it may be that in a typical population positive selection causes only one beneficial allele replacement every ~100 generations. In contrast, negative selection simultaneously acts against millions of deleterious alleles, most of which having low frequencies (see Chapters 2 and 11). Negative selection reduces both the rate of evolution and the level of within-population variation, and these facts can be used to study it quantitatively. The recent deluge of data on genomes and genotypes has shed some light on the parameters of negative selection. We will consider these parameters in the following chapters.

The distinction between positive selection which favors a rare allele versus negative selection which favors a common allele is a qualitative one. However, we also need to characterize the action of selection at a polymorphic locus quantitatively, because selection can be anything from very strong to very weak. Let us do this for a diallelic locus under negative selection against a rare allele a. Ignoring very rare homozygotes aa, the strength of selection can be described by one number, the coefficient of selection against a, $s = 1 - w_{Aa}/w_{AA}$. If the fitness conferred by the heterozygous allele a comprises 0.99999, 0.99, or 0.9 of the fitness conferred by the homozygous common allele A, the selection coefficient is 0.00001, 0.01, or 0.1, respectively. If a is a dominant lethal so that $w_{Aa} = 0$, $s = 1$. In the other extreme case of $w_{Aa} = w_{AA}$, selection is absent and $s = 0$.

A careful reader will notice that this definition of selection coefficient coincides with the definition of a relative effect of the derived allele at a complex trait locus (CTL) on the quantitative trait, introduced in Chapter 3. Indeed, loci that harbor alleles that affect fitness and, thus, are subject to selection, are fitness CTLs. For a variety of phenotypic traits, mild mutations and alleles are much more common than drastic ones (see Chapters 2, 3, and 6), and fitness is no exception. We will consider data on strengths of selection against individual deleterious alleles in Chapter 11.

7.5 Selection on a Quantitative Trait

Genetic composition of any natural population is extremely complex (see Chapter 11). Thus, selection acting on a natural population must also be a very complex phenomenon, and simplifications are needed to comprehend and study it. So far, we have simplified the description of selection in two opposite ways. One was to ignore all relations between the available genotypes and, instead, to characterize them exclusively by their fitnesses and to consider the resulting distribution $q(w)$. The other was to ignore all properties of genotypes outside one polymorphic locus, and to consider fitnesses only of alleles at this locus. Both these extreme approaches are useful, but something in between is also needed.

In particular, it is productive to consider selection acting on one genotype-level quantitative trait (see Chapter 3). In the context of the theory of natural selection, this model was proposed in 1967 by a famous trio of papers published in the journal *Genetics*. In this way, we do not attempt to capture all complexity of the space of genotypes, but also do not restrict our attention to just one locus. Instead, every genotype is characterized by one number, which can be called the genotype score or fitness potential, after which we study how fitness depends on this number. The key assumption that a properly chosen genotype score alone determines fitness holds if epistasis that characterizes the fitness landscape, if any, is one-dimensional (see Chapter 3). Geometrically, we collapse the space of genotypes not on one locus, but on a one-dimensional space of fitness potential, and consider fitness landscape, which becomes the graph of a function of one variable, which flies over this space (Figure 7.10). Examples of scores that can characterize a genotype are the number of nucleotides A in it or, more sensibly, the (weighted) number of deleterious alleles (see Chapter 9). The first of these scores can hardly determine fitness, but the second one might, at least to a good approximation.

It can also make sense to consider selection acting on quantitative traits that describe not genotypes but phenotypes, such as height, blood pressure, or intelligence quotient. Because fitness is a phenotypic quantitative trait itself, in this case we are dealing with a relationship between two such traits. Regardless of the nature of the first, independent quantitative trait, the value of fitness ascribed to its particular

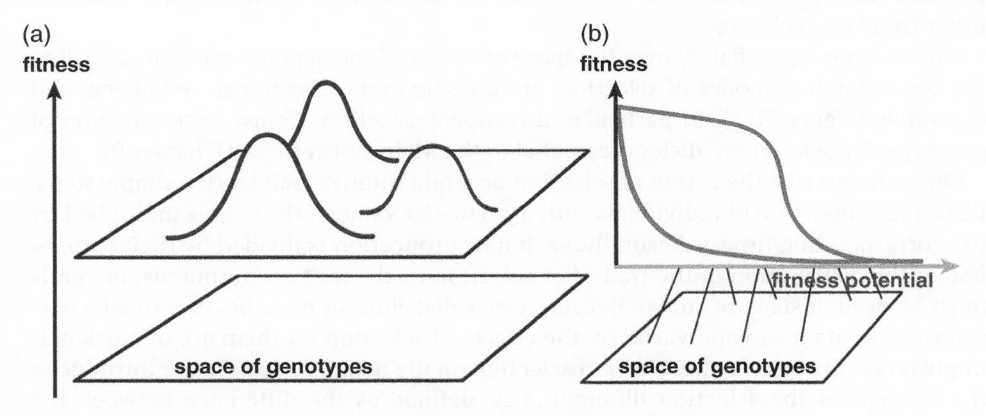

Figure 7.10 Generic (a) and fitness potential-mediated (b) fitness landscapes.

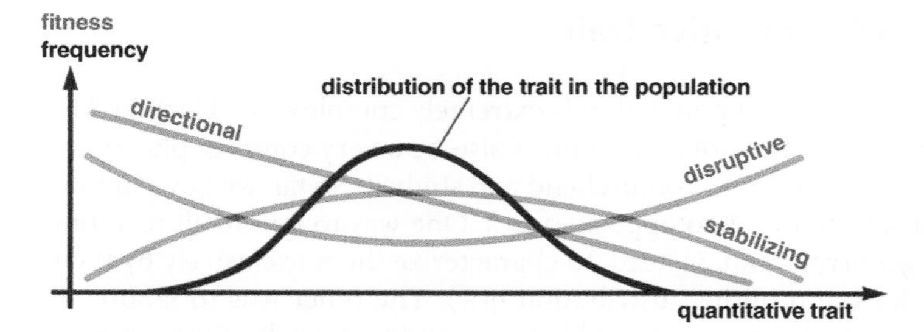

Figure 7.11 Directional, stabilizing, and disruptive fitness landscapes and modes of selection acting on a quantitative trait. A directional landscape is monotonic, a stabilizing landscape has a maximum close to the population mean, and a disruptive landscape has a minimum close to the population mean.

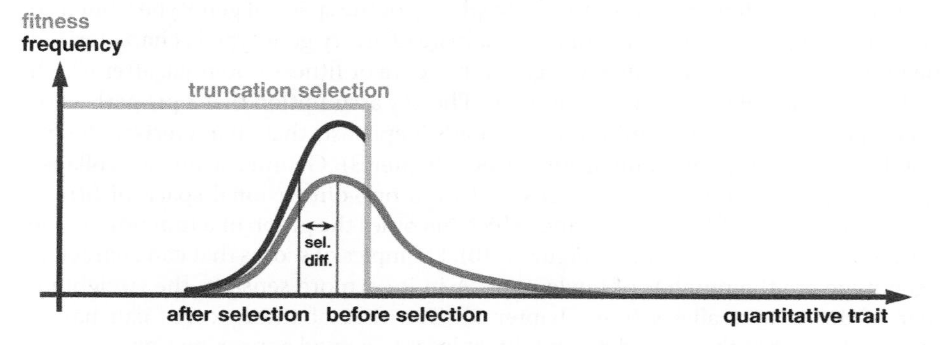

Figure 7.12 Action of truncation directional selection on a quantitative trait. The black and dark gray curves are densities of the quantitative trait after and before selection, respectively. The selection differential Δ is the difference between their mean values.

value is the expected fitness of individuals with this value of the trait. The word "expected" is essential here, because individuals of the same height may, nevertheless, produce different numbers of offspring or die at different ages, due differences in other traits or to chance.

Fitness landscapes that fly over the space of values of one quantitative trait, as well as the corresponding modes of selection, are classified into directional, stabilizing, and disruptive (Figure 7.11). In particular, directional selection against contamination of genotypes by deleterious alleles is essential to the Main Concern (see Chapter 9).

One can describe the action of selection on a quantitative trait by two simple steps. First, the proportion of individuals with a particular value of the trait is multiplied by the corresponding fitness. Secondly, each new proportion is divided by their sum, to obtain the distribution of the trait after selection (if the trait is continuous, integrals must be used, instead of sums). Because every distribution must be, first of all, characterized by its mean and variance, the effects of selection on them are of particular importance. To describe the effect of selection on the mean, Ronald Fisher introduced the concept of the selection differential Δ, defined as the difference between the mean values of the trait after and before selection (Figure 7.12). Of course, directional

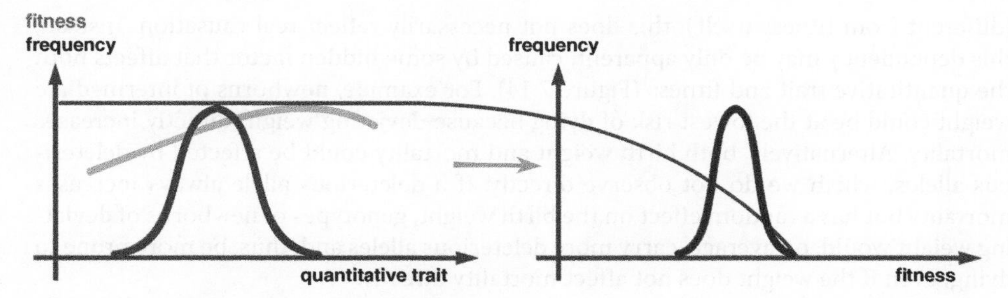

Figure 7.13 Action of selection (gray fitness landscape) on a quantitative trait (black distribution) produces a distribution of fitness.

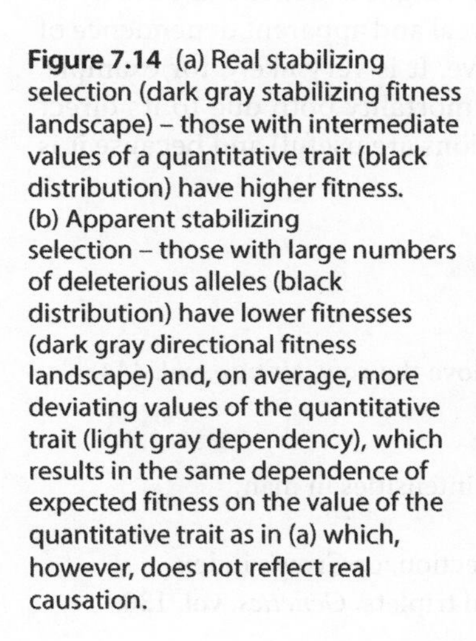
Figure 7.14 (a) Real stabilizing selection (dark gray stabilizing fitness landscape) – those with intermediate values of a quantitative trait (black distribution) have higher fitness. (b) Apparent stabilizing selection – those with large numbers of deleterious alleles (black distribution) have lower fitnesses (dark gray directional fitness landscape) and, on average, more deviating values of the quantitative trait (light gray dependency), which results in the same dependence of expected fitness on the value of the quantitative trait as in (a) which, however, does not reflect real causation.

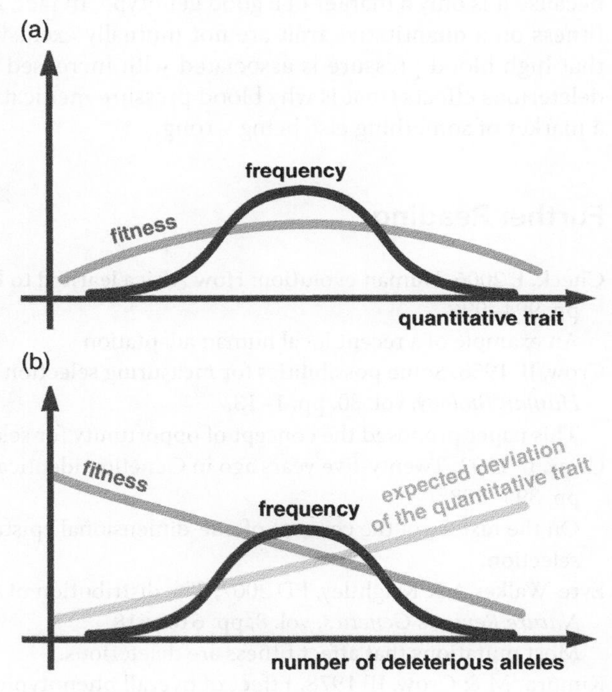

selection moves the mean value of the trait in the direction of increasing fitness. In Chapter 9, selection differential will be central for analysis of the mutation–selection equilibrium. The effect of selection on the variance of a quantitative trait will be briefly considered in Chapter 11.

When selection acts on a quantitative trait, this results, by definition, in different fitnesses of individuals that possess different values of the trait. Knowing the fitness landscape and the distribution of the trait, it is easy to infer the resulting distribution of fitness $q(w)$ (Figure 7.13). In Chapter 12, we will consider imperfection and evolvability of fitness that are induced in the population by selection acting on quantitative traits.

Obviously, one quantitative trait cannot provide a comprehensive description of a phenotype. Thus, even when we see that fitness depends on a quantitative trait (different from fitness itself), this does not necessarily reflect real causation. Instead, this dependency may be only apparent, caused by some hidden factor that affects both the quantitative trait and fitness (Figure 7.14). For example, newborns of intermediate weight could be at the lowest risk of dying because deviating weight directly increases mortality. Alternatively, both birth weight and mortality could be affected by deleterious alleles, which we do not observe directly. If a deleterious allele always increases mortality but has a random effect on the birth weight, genotypes of newborns of deviating weight would, on average, carry more deleterious alleles and, thus, be more prone to dying, even if the weight does not affect mortality directly.

The distinction between real versus apparent selection is also applicable to effectively directional fitness landscapes. People of above-average height (but not those exceptionally tall) may tend to have higher fitness because such height is beneficial directly or because it is only a marker of a good genotype. In fact, real and apparent dependence of fitness on a quantitative trait are not mutually exclusive. It is very likely, for example, that high blood pressure is associated with increased mortality both due to its direct deleterious effects (that is why blood pressure medications are useful) and because it is a marker of something else being wrong.

Further Reading

Check, E 2006, Human evolution: How Africa learned to love the cow, *Nature*, vol. 444, pp. 994–996.
An example of a recent local human adaptation.

Crow, JF 1958, Some possibilities for measuring selection intensities in man, *Human Biology*, vol. 30, pp. 1–13.
This paper proposed the concept of opportunity for selection, or Crow's index.

Crow, JF 1992, Twenty-five years ago in Genetics: identical triplets, *Genetics*, vol. 130, pp. 395–398.
On the history of the concept of one-dimensional epistasis in the studies of natural selection.

Eyre-Walker A & Keightley, PD 2007, The distribution of fitness effects of new mutations, *Nature Reviews Genetics*, vol. 8, pp. 610–618.
Most mutations that affect fitness are deleterious.

Kimura, M & Crow, JF 1978, Effect of overall phenotypic selection on genetic change at individual loci, *Proceedings of the National Academy of Sciences of the USA*, vol. 75, pp. 6168–6171.
On the connection between selection on a quantitative trait and at an individual locus.

Kondrashov, AS & Turelli, M 1992, Deleterious mutations, apparent stabilizing selection and the maintenance of quantitative variation, *Genetics*, vol. 132, pp. 603–618.
Discussion of the concept of apparent selection.

Reeve, HK & Sherman, PW 1993, Adaptation and the goals of evolutionary research, *Quarterly Review of Biology*, vol. 68, pp. 1–32.
An influential analysis of the concept of adaptation.

Sarkisyan, KS, Bolotin, DA, & Meer, MV 2016, Local fitness landscape of the green fluorescent protein, *Nature*, vol. 533, pp. 397–401.
An example of an application of the concept of fitness landscape.

Shnol, EE, Ermakova, EA, & Kondrashov, AS 2011, On the relationship between the load and the variance of relative fitness, *Biology Direct*, vol. 6: 20.
Connection between evolvability of fitness and its imperfection.
Singh, P & Cole, ST 2011, *Mycobacterium leprae*: genes, pseudogenes and genetic diversity, *Future Microbiology*, vol. 6, pp. 57–71.
A striking example of loss of obsolete adaptations.

8

Functioning DNA and Junk DNA

The amount of silent DNA is steadily shrinking. The question of how our species accommodates such [high deleterious] mutation rates is central to evolutionary thought.

James V. Neel, 1986.

Nucleotide sites and even long segments of genomes may be selectively neutral, if they do not participate in any molecular function, and, thus, are irrelevant to the phenotype and fitness of the organism. Dynamics of selectively neutral genetic variation are governed by mutation and random drift, chance fluctuations in the genetic composition of a population. Effective sizes of natural populations are usually much lower than their actual sizes, so that random drift proceeds much faster than one could expect. Evolving junk DNA accumulates changes of each particular kind at the rate that is equal to the rate of mutations of this kind. This simple property of selectively neutral evolution provides quantitative evidence for common ancestry of different species. Meaningful DNA segments can be identified through their low rates of evolution, and comprise ~10% of mammalian genomes. Thus, a newborn human carries about 10 *de novo* deleterious mutations.

8.1 Selective Neutrality and Random Drift

We have just spent a whole chapter considering natural selection – and for a good reason, because selection against deleterious mutations is crucial for the Main Concern. One might even think that selection – usually negative but occasionally positive – acts upon every mutation that enters a population. However, this is not so, because not every nucleotide site – and even not every long segment of a genome – is controlled by selection.

How could this be the case? Easily: if the same phenotype develops regardless of which nucleotide (A, T, G, or C) occupies a particular site, this site is phenotypically mute and selection cannot see it. Such a site is called selectively neutral, or just neutral, for brevity. Of course, a site can be neutral only if it does not affect any molecular function. If many adjacent sites are all functionally irrelevant, they form a selectively neutral segment of the genome. Such segments are called junk or silent DNA. Perhaps, exact neutrality is an abstraction – after all, different nucleotides are different molecules. Still, if one out of 3 billion sites of the human genome is not involved in any specific function,

Crumbling Genome: The Impact of Deleterious Mutations on Humans, First Edition. Alexey S. Kondrashov.

the choice of nucleotide at it must be irrelevant, fitness-wise, for all practical purposes. Of course, approximate neutrality, such that selection is very weak but not trivial, is also possible.

Even compact genomes of bacteria contain some approximately neutral sites. One reason behind this is degeneracy of the genetic code: all amino acids, except two, can be encoded by multiple codons. In particular, 8 out of 20 possible amino acids are encoded by 4-fold degenerate codon families, in which the third nucleotide does not matter at all (Table 1.2). In other words, the third site within such codons is fully synonymous. Selection usually does not care much which nucleotide occupies a fully synonymous site, although such sites are only approximately neutral: different synonymous codons work slightly differently as parts of mRNA during translation.

On top of synonymous protein-coding sites, bloated genomes of many multicellular eukaryotes, including mammals, possess long segments of non-coding junk DNA. Evolution can produce such segments in a number of ways. Insertions of transposable elements is one of them (see Chapter 6). Also, a protein-coding gene occasionally stops working, and begins to accumulate mutations which irreversibly destroy its ability to encode the protein, by introducing premature stop codons and frameshift deletions and insertions. This results in a pseudogene, a dysfunctional remnant of a gene. Some pseudogenes acquire new functions and, thus, become visible to selection again, but most of them are apparently functionless and neutral. With some exceptions (see Chapter 7) pseudogenes are rare in bacterial genomes. In contrast, genomes of mammals carry about as many pseudogenes as working genes. A junk pseudogene keeps accumulating differences from the original gene, and eventually becomes unrecognizable. It seems that long segments of junk DNA are often almost exactly neutral.

So, if selection, by definition, does not affect selectively neutral DNA sites and segments, what does? Of course, mutation, which proceeds regardless of whether it alters fitness. However, neutral DNA is also affected by another evolutionary force, known as random drift. Suppose that in a population (haploid, for simplicity) of fixed size of 100 at some site 60 individuals carry nucleotide A and the remaining 40 carry nucleotide T. Even if replacing A with T at the site is selectively neutral, because both A- and T-carriers produce, on average, the same number of offspring, the probability of seeing exactly 60 alleles A and 40 alleles T in the next generation is rather low. Much more likely, carriers of either allele will be luckier, and we will see 53 A's and 47 T's or, perhaps, 65 A's and 35 T's – just by chance. Of course, the expected number of A's in the next generation is still exactly 60: otherwise, A would be either advantageous or disadvantageous, which contradicts our assumption that selection does not act at the site. Such random, unbiased fluctuations of the genetic composition of the population are called random drift (Figure 8.1).

Random drift always operates in any population, even if selection is also present. Still, random drift plays the most prominent role when selection is absent or weak – otherwise, a systematic (dis)advantage of an allele trumps random fluctuations of its frequency, as long as the allele is present in a large enough number of individuals. Eventually, random drift acting alone would cause the population to lose all alleles but a lucky one. However, mutation either prevents this from happening or sooner or later reverses each fixation of a lucky allele, by creating a new, mutant allele.

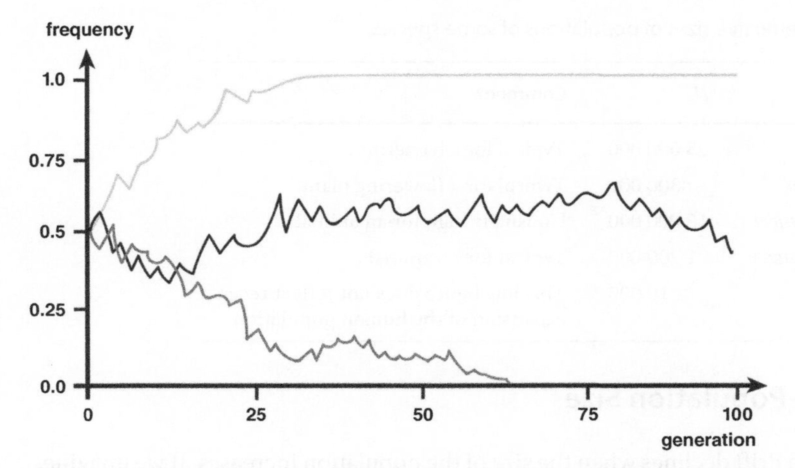

Figure 8.1 Three trajectories of allele A frequencies in a population of 100 individuals with no mutation or selection. The intial frequency of allele A is 0.5, and it changes due to random drift alone.

Darwin, with his usual insight, was the first to recognize and qualitatively describe the phenomenon of random drift:

> Variations neither useful nor injurious would not be affected by natural selection, and would be left either a fluctuating element, as perhaps we see in certain polymorphic species, or would ultimately become fixed…
>
> C. Darwin, *On the Origin of Species,* 1859.

In 1968, soon after it became obvious that natural populations harbor a lot of genetic variation (see Chapter 2), geneticist Motoo Kimura suggested that most of this variation is selectively neutral and, thus, is affected by random drift, but not by selection. Later, his student Tomoko Ohta argued that there also exists a large class of nearly neutral polymorphisms, which are under only a very weak selection, so that both selection and random drift affect them to about the same extent. We now know that both Kimura and Ohta were right. For example, software tools that take into account all kinds of evidence, such as PolyPhen, determined that among ~10 000 protein-altering derived alleles that are present in an average human genotype only ~1000 are subject to a substantial negative selection and most of the remaining ~9000 are nearly neutral or even completely (for all practical purposes) neutral. Polymorphic loci outside genes are affected by selection, on average, to an even lesser extent (see Chapter 11).

Random drift also plays a major role in accumulation of genetic differences in the course of independent evolution of ancestral lineages of modern species. Again, this role is particularly prominent in species with bloated genomes. Among ~ 30 million differences between genomes of humans and chimpanzees, accumulated since their lineages diverged ~8 Ma, the vast majority are phenotypically mute and selectively neutral or very nearly so.

Table 8.1 Estimated effective sizes of populations of some species.

Species	N_e	Comment
Escherichia coli	25 000 000	Typical for a bacterium
Arabidopsis thaliana	300 000	Typical for a flowering plant
Caenorhabditis brenneri	15 000 000	Unusually high for an animal
Drosophila melanogaster	1 200 000	Typical for an animal
Homo sapiens	10 000	This low figure does not reflect recent expansion of the human population

8.2 Effective Population Size

The rate of random drift declines when the size of the population increases. If we imagine, for a moment, a population of infinitely many individuals, allele frequencies in it would remain, in the absence of mutation and selection, strictly invariant, due to the law of large numbers. Thus, the rate of random drift in a real population can be characterized by its effective size N_e – the size of an idealized population with the equivalent rate of random drift. This idealized population is known as the Wright–Fisher population, to honor Ronald Fisher and Sewall Wright, who developed the concept of random drift in the 1920s and 1930s. For example, if in a population of 1 billion flies random drift proceeds as it would in a Wright–Fisher population of 100 000 individuals, we say that N_e of this fly population is 10^5. For those who are curious, the key property of a Wright–Fisher population is that the mean and variance of the number of copies of an allele transmitted to the next generation is 1, and, moreover, this number has Poisson distribution.

The larger the value of N_e of a population, the slower is random drift in it. Effective sizes of natural populations are much lower than their actual, census sizes, due to a wide random variation in reproductive successes of individuals. Thus, even very large natural populations are subject to a substantial random drift (Table 8.1).

There is a very useful simple rule, discovered by Wright in 1931, which makes it possible to determine which force, selection or random drift, dominates the dynamics of allele frequencies at a polymorphic locus. If selection coefficient s at a locus is such that $s > 1/N_e$, selection rules (and the polymorphism is non-neutral), if $s < 1/N_e$, random drift rules (and the polymorphism is effectively neutral), and if $s \sim 1/N_e$, both factors are about equally important (and the polymorphism is nearly neutral). Thus, in humans, polymorphisms with $s \sim 0.0001$ are nearly neutral and those with $s < 0.0001$ are effectively neutral.

8.3 Junk DNA Provides the Simplest Evidence for Evolution

Neutral evolution of junk DNA, affected by mutation and random drift but not by selection, proceeds at the rate which is equal to the mutation rate. For example, if $\mu = 10^{-8}$, there will be, on average, one allele replacement at a neutral site every 10^8 generations. This theorem, which makes it possible to estimate mutation rates through rates of neutral evolution (see Chapter 6) can be easily proven: without selection, a new mutation is no better and no worse than all other alleles. Among all alleles that are present in the population at a moment, eventually one will be fixed by random drift, and all others will be lost. In a population of census size N, $N\mu$ mutations appear at a site each generation, and each one of them will by fixed with probability $1/N$, thus leading to evolution with the rate $N\mu(1/N) = \mu$.

Such simplicity is unusual for biology and makes it possible to offer quantitative evidence for past evolution of species and their genomes. Indeed, the claim that humans, chimpanzees, flies, worms, flowers, and bacteria evolved from a common ancestor is so astonishing that it must not be accepted without strong evidence. Darwin was the first to systematically consider such evidence. However, until this day, most of evidence for past evolution of life is only qualitative, consisting, first of all, of data on similarities between species that cannot be explained rationally without assuming their common ancestry.

In this respect, biology is rather different from physics. In physics a hypothesis is accepted or rejected depending on whether its implications are in quantitative agreement with the data of observations and experiments. Famously, when in 1919 Arthur Eddington observed that the Sun's gravity bends light two times stronger that what Newtonian physics could explain but in agreement with the prediction of Einstein's general relativity, the latter was elevated to the status of an established theory.

Unfortunately, evolution of life, viewed as a hypothesis, is generally far too complex to generate such quantitative predictions. Still, neutral evolution provides a fortunate exception, if we consider mutations of at least two kinds. Indeed, if similar segments of junk DNA in two species originated from a common ancestor, the relative abundance of interspecific differences of kind 1 and kind 2 must be the same as the ratio of rates of *de novo* mutations of kind 1 and kind 2. This prediction is easy to test: current rates of mutations of all kinds can be measured directly (see Chapter 6), and interspecific differences can be counted after the corresponding genomes are aligned.

Results of such analyses are in perfect agreement with the hypothesis of common ancestry of different species. For example, data on human mother–father–offspring trios show that single-nucleotide substitutions occur ~10 times more often than short insertions and deletions. In alignments of human and chimpanzee introns, which are mostly selectively neutral, mismatches also occur ~10 times more often than short gaps. This result agrees with what a hypothesis of neutral evolution predicts and, thus, testifies to the evolution of these two species from a common ancestor. Of course, this pattern is observed only in junk DNA. If we compare protein-coding exons from humans and chimpanzee genomes, we would see practically no gaps of lengths 1 and 2, due to a strong negative selection against frameshift mutations (see Chapter 2).

8.4 Finding Functioning Genome Segments

Deleterious mutations and negative selection against them obviously cannot occur at neutral sites. If so, why do we care about junk DNA? The reason is because, in order to study deleterious mutations, we first need to recognize meaningful DNA segments and sites, which requires discriminating them from surrounding junk DNA. This is an important and challenging task, because most of the human genome consists of junk DNA. Still, there are ways to identify meaningful DNA.

A functional approach to this task takes advantage of the fragility of complex adaptations. In particular, a long protein-coding gene is likely to be functionally important, because without negative selection it would rapidly become a pseudogene due to accumulation of nonsense and frameshift mutations. In contrast, a short site of attachment of a DNA-binding protein, which consists of a particular sequence of, say, six nucleotides, may appear by chance and does not need to be functional. There is some selection against such spurious sites, but often the strength of this selection is too low to eliminate them all. In fact, even a long genomic region which is occasionally transcribed but does not encode any proteins may be devoid of function, because transcription can also be spurious.

Another, comparative approach, known as phylogenetic footprinting, consists of comparing genomes of different species and finding segments and sites that diverged less and, thus, evolved slower, than junk ones. This approach works because negative selection is much more common than positive selection, so that functioning DNA usually evolves slower than junk DNA. Very similar species pairs, such as humans and chimpanzees, are unsuitable for this analysis. Indeed, even junk segments of human and chimpanzee genomes are 98.7% identical, so that an extra similarity due to negative selection can be detected only in very long segments of functioning DNA. However, very distant species are also unsuitable, because the lack of similarity between their genomes could be due not only to benign neglect of selective neutrality, but also to positive selection-driven allele replacements. A functional DNA segment which contributes to us being humans and fruit flies being fruit flies must be different in the genomes of the two species.

Thus, moderately different species, such that junk segments of their genomes had time to lose (almost) all traces of similarity, but functioning segments still mostly do the same things, should be used for phylogenetic footprinting. One such species pair is human and mouse. A comparison of human and murine genomes, performed in 2001, revealed that ~10% of their length belong to segments where their similarity remains substantial (Figure 8.2). These segments include almost all exons of protein-coding genes; in fact, some algorithms of finding exons (genome annotation, see Chapter 1) use information on interspecific similarity.

Still, most segments whose evolution was constrained by negative selection are located outside protein-coding genes. Initially, this fact surprised some biologists, who naively believed that almost all non-coding DNA, which constitutes ~99% of mammalian genomes (see Chapter 1), is junk. However, regulation of gene expression in mammals is extremely complex and non-coding DNA plays a key role in it. Thus, it is only natural to

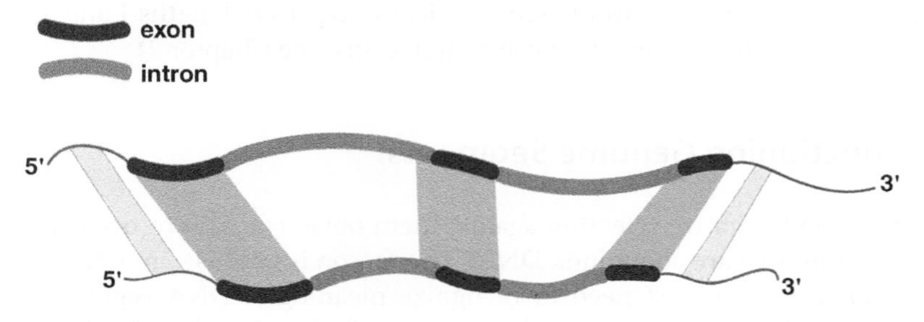

Figure 8.2 Alignment of sequences of the same protein-coding gene in human (top) and murine (bottom) genomes. Correspondence between highly similar genome segments is shown by dark gray shading, and between moderately similar segments by light gray shading.

find a lot of selective constraint outside genes, although negative selection at individual non-coding sites is usually weaker than at individual coding, non-synonymous sites (see Chapter 11).

Simultaneous comparisons of multiple genomes are even more sensitive, and make it possible to identify very short constrained DNA segments. Accumulating data on mammalian genomes keep refining estimates of their selective constraint. In contrast to large protein-coding genes, short functional units within non-coding DNA, such as sites of attachment of individual DNA-binding proteins, often disappear and reappear at a different place in the course of evolution. As a result, simple alignment of even not-too-distant genomes cannot reveal all sites under selective constraint. Still, the estimate of ~10% of meaningful sites (and, thus, ~90% of junk sites) has not changed much in the last 15 years and will likely stand. Of course, in species with lean genomes the proportion of functioning DNA is much higher than in mammals, and in species with superbloated genomes (Table 1.1) it is likely to be much lower.

One could think that, because deleterious mutations are much more common than beneficial mutations, functioning genotype segments should mutate slower than junk segments. However, the opposite is true, and short mutagenic contexts (see Chapter 6) are responsible for this evolutionary paradox. Mutagenic contexts are rare within junk segments, because mutations which destroy such contexts are much more common than mutations which create them. In contrast, when selection cares (in particular, within coding exons), it preserves sequences which perform important functions even when they contain mutagenic contexts, thus causing a higher mutation rate. In particular, a high proportion of cases of Mendelian diseases are due to mutations at highly mutagenic CG contexts within coding exons. This paradox provides another illustration of the fact that mutation is not informed by the effect of mutant genotypes on fitness.

8.5 The Genomic Rate of Deleterious Mutations

So, we have good reasons to think that in humans ~10% of small-scale mutations hit meaningful genome sites and segments. This figure must be even higher for large-scale mutations: a mutation which affects, say, 100 kb of human DNA is very likely to disturb some function, and for a deletion longer than 1 Mb this is almost certain, as continuous pieces of junk DNA longer than that must be very uncommon. However, large-scale mutations are relatively rare. Thus, we may conclude that among ~100 *de novo* mutations which appear in the genotype of a newborn (see Chapter 6) ~10 are substantially deleterious. In other words, per generation genomic rate of deleterious mutations, denoted U, is ~10 in our species. The actual figure may be a little lower or higher, but the order of magnitude appears to be certain. I will be using this figure (instead of, say, U ~7) to emphasize this point. Of course, this single value of U is, in fact, the average between a higher man and a lower woman rates, and depends on the ages at which individuals reproduce (see Chapter 6).

Unfortunately, it is not easy to find out what "substantially deleterious" means precisely. Theory only tells us that selection prevents fixations of deleterious mutations with $s > 1/N_e$. Thus, because our estimate of the proportion of meaningful DNA is based primarily on interspecific comparisons, we can only say, assuming that $N_e = 10^{-4}$, that 10% of human *de novo* mutations are subject to negative selection with $s > 0.0001$. Of course, there is a huge difference between deleterious mutations with $s = 0.001$ and

$s = 0.1$. In particular, even a free accumulation of mutations with $s = 0.001$ in the course of 10 generations should not be a major concern, but accumulation of mutations with $s = 0.1$ should (see Chapter 13). Methods which can determine the strength of negative selection more precisely will be considered in Chapter 11.

Further Reading

Charlesworth, B 2009, Effective population size and patterns of molecular evolution and variation, *Nature Reviews Genetics*, vol. 10, pp. 195–205.
A review of the concept of effective population size.

Kimura M & Ohta, T 1971, Polymorphism as a phase of molecular evolution, *Nature*, vol. 229, pp. 467–469.
A classical paper on evolution at selectively neutral loci.

Li, W-H, Gojobori, T, & Nei, M 1981, Pseudogenes as a paradigm of neutral evolution, *Nature*, vol. 292, pp. 237–239.
The first demonstration that selectively neutral genome segments evolve faster than functionally important segments.

Ohta, T 2002, Near-neutrality in evolution of genes and gene regulation, *Proceedings of the National Academy of Sciences of the USA*, vol. 99, pp. 16134–16137.
A review of evolution at loci under very weak selection.

Rands, CM, Meader, S, & Ponting, CP 2014, 8.2% of the human genome is constrained: variation in rates of turnover across functional element classes in the human lineage, *PLoS Genetics*, vol. 10: e1004525.
A sophisticated estimate of selective constraint in mammals.

Schmidt, S, Gerasimova, A, & Kondrashov, FA 2008, Hypermutable non-Synonymous sites are under stronger negative selection, *PLoS Genetics*, vol. 4: e1000281.
A positive correlation between the mutation rate and the strength of negative selection.

Shabalina, SA, Ogurtsov, AY, & Kondrashov, VA 2001, Selective constraint in intergenic regions of human and mouse genomes, *Trends in Genetics*, vol. 17, pp. 373–376.
A simple estimate of selective constraint in mammals.

Silva, JC, Shabalina, SA, & Harris, DG 2003, Conserved fragments of transposable elements in intergenic regions: evidence for widespread recruitment of MIR- and L2-derived sequences within the mouse and human genomes, *Genetical Research*, vol. 82, pp. 1–18.
Negative selection acts on some mammalian genome segments that originated from transposable elements.

Wasserman, WW & Sandelin, A 2004, Applied bioinformatics for the identification of regulatory elements, *Nature Reviews Genetics*, vol. 5, pp. 276–287.
Using phylogenetic footprinting for finding functional non-coding segments of genomes.

Wright, S 1931, Evolution in Mendelian populations, *Genetics*, vol. 16, pp. 97–159.
A classical paper which introduced, *inter alia*, the concept of random drift.

9

It Takes All the Running You Can Do

Does nature rank and truncate, or do something approximating this? It will have to be answered empirically.

James F. Crow and Motoo Kimura, 1979.

Unopposed accumulation of constantly appearing deleterious mutations leads to a rapid deterioration of the gene pool of the population and to the decline of fitness of individuals. However, negative selection against mutations operates in all natural populations, which thus exist in the state of the mutation–selection equilibrium. Pervasiveness of a deleterious mutation, the expected number of its copies produced before all of them are eliminated, is inversely proportional to the coefficient of selection against it. It is natural to assume that fitness of a genotype is determined by its contamination, the sum of coefficients of selection against all its deleterious alleles. As long as individual deleterious alleles remain rare, at the mutation–selection equilibrium the average contamination of a genotype is always equal to the genomic deleterious mutation rate. Deleterious alleles can act independently or can reinforce effects of each other. The importance of this second possibility, known as synergistic epistasis, remains obscure. Because with synergistic epistasis selection eliminates deleterious alleles in groups, mutation imperfection can be tolerable even when genotype contamination is very high. Fitness of inbred individuals is usually reduced, because many deleterious alleles are partially recessive. Selection is powerless to prevent fixations of slightly deleterious alleles, which can lead to profound imperfection.

9.1 Middle Class Neighborhood for *Drosophila*

The Red Queen told Alice in Wonderland: "Now, here, you see, it takes all the running you can do, to keep in the same place. If you want to get somewhere else, you must run at least twice as fast as that!". This is a fitting description of what happens when living beings face some relentless pressure.

Originally, the Red Queen metaphor entered biology in the context of coevolution of interacting populations. Often, such interactions are antagonistic – one's gain is somebody else's loss. Competitors, predators, and parasites keep evolving, becoming more and more efficient, and, as a result, the population also needs to permanently evolve by

Crumbling Genome: The Impact of Deleterious Mutations on Humans, First Edition. Alexey S. Kondrashov.

positive selection-driven allele replacements, just to stay alive. Such an ecological Red Queen can lead to endless arms races between coevolving populations.

However, the Red Queen metaphor is also applicable to a population pressed by an incessant influx of deleterious mutations. In this case, the only "running" the population can do in order to survive is to constantly remove these mutations, through the action of negative selection. If the rates of origination and of removal of mutations are equal, the mutation–selection equilibrium is reached and this running allows the population to keep in the same place. The population can get somewhere else, in the sense of reducing its burden of deleterious alleles, only if negative selection works faster than mutation.

A natural way to study how fast a population has to run is to see what happens if we tie its legs, by shutting down selection. Past evolution provides a lot of examples of sad fates of adaptations that are no longer needed (see Chapter 7). Still, such natural experiments do not tell us all we want to know about deleterious mutation and negative selection, for three reasons. First, they usually last for too long: we only observe a degraded adaptation very many generations after negative selection stopped protecting it and, thus, remain ignorant about the initial rate of degradation. Secondly, positive selection against costly former adaptations that are no longer used can also contribute to their demise. Thirdly, natural degradation of unneeded adaptations always involves only some parts of the phenotype, and, of course, never affects fitness. Comprehensive shutdown of selection is an absolutely unnatural thing, which can be attempted only in the laboratory.

Several methods can be employed for this purpose. In 1964 and 1972, Terumi Mukai used balancer chromosomes (see Chapter 6). However, this genetic trick is available only for *Drosophila melanogaster*. Also, even in this species only one chromosome, instead of the whole genome, can be shielded from selection by using a balancer chromosome.

Another method is to keep a population very small. Because when $s < 1/N_e$ (where s is the selection coefficient and N_e is the effective population size) random drift makes selection inefficient (see Chapter 8), only very deleterious mutations can be effectively eliminated from such a population. In hermaphrodites that are capable of self-fertilization, such as a worm *Caenorhabditis elegans* and many flowering plants, even a one-individual population can be maintained for many generations. In species with separate sexes, the minimal population size is two, and a female and a male that constitute such a population are bound to be (from generation 2) full siblings. Such very small populations, commonly referred to as mutation-accumulation (MA) lines, are widely used for studying spontaneous deleterious mutations (Figure 9.1). Indeed, all deleterious mutations with $s < 0.1$, that is, a vast majority of them, can become fixed within an MA line by random drift, with effectively no opposition from negative selection. The trouble is, after not too many generations, an MA line loses almost all genetic variation, again due to random drift. This results in emergence of homozygous genotypes, which are normal for selfers, but do not occur naturally in outcrossers where they suffer from inbreeding depression (see below).

In contrast, the third method of shutting down selection does not disturb the life of a population too much. If we can make sure that every individual contributes exactly two offspring to the next generation, opportunity for selection (see Chapter 7) disappears and, thus, selection, differential reproduction of genotypes, also disappears. This is what Svetlana Shabalina, Lev Yampolsky and I did in two laboratory populations of *Drosophila melanogaster* in 1995–1997. We called them middle class

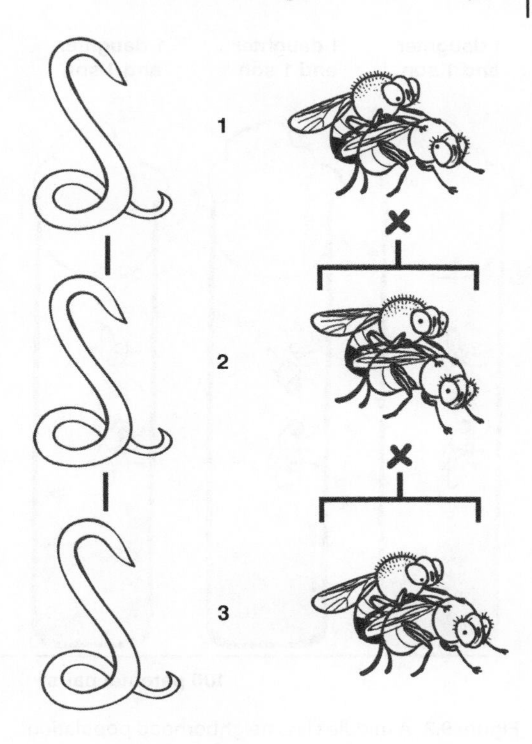

Figure 9.1 Three generations of life of an MA line of selfing worms and outcrossing flies.

neighborhood (MCN) populations, because they mimicked how we humans try to implement the American dream in middle class neighborhoods.

Each generation, 100 females and 100 males were paired randomly, and each pair was placed in its own vial. Parents were removed after a female laid 30–40 eggs – this way, the mortality of larvae was minimal, because they helped each other to consume the food but there was still more than enough for everyone. When the offspring flies emerged, a daughter and a son were picked randomly from each vial, before the siblings could interbreed. The resulting 100 females were randomly paired with 100 males, to start the next generation (Figure 9.2).

Despite all our efforts, there remained some residual selection in our MCN populations. Approximately 3% of pairs were sterile, and ~10% of eggs did not develop into adults. Still, we probably came closer than anybody else to a complete abolition of natural selection within a population of a natural outcrosser. To keep the population size constant, we replaced missing offspring of sterile pairs by randomly picking an extra daughter or son from offspring of fecund pairs.

For any measurement of fitness in the course of many generations, good controls are essential. We used three kinds of control flies, including those that emerged from cryopreserved embryos from the initial population. In the course of many months when the MCN populations went from generation to generation, these embryos sat in liquid nitrogen. Preserving *Drosophila* embryos is much more difficult than preserving sperm (see Chapter 15) but a cryobiologist, Peter L. Steponkus, taught us the process. After 10, 20, and 30 generations, we compared several properties of flies from the MCN populations with that of the controls, and estimated the relative

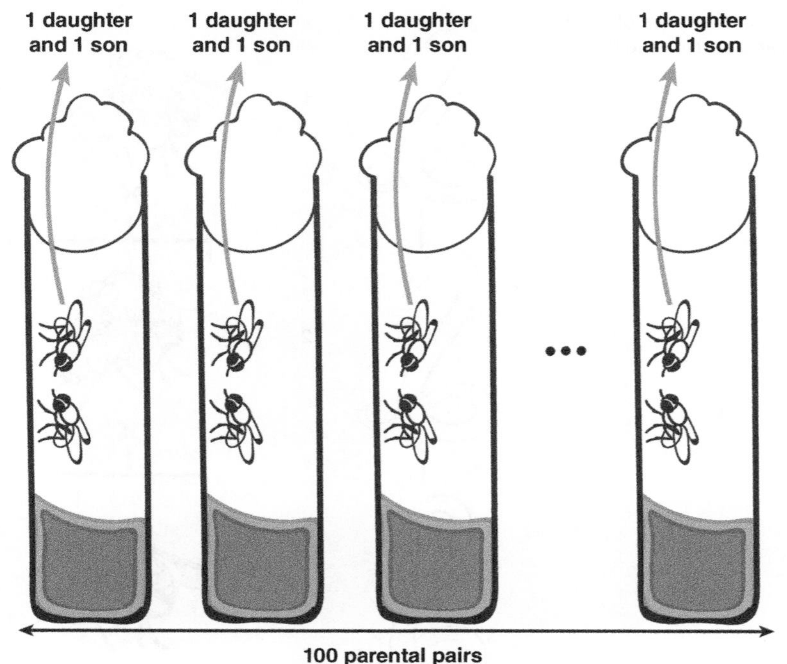

Figure 9.2 A middle class neighborhood population.

Table 9.1 Mutational pressures measured in the MCN experiment.

Trait	Per generation change
Competitive ability of larvae	−0.02
Motility of adults	−0.002
Fecundity of females	−0.003
Longevity	No change

(mean-normalized) rates of decline of fitness-related traits in the absence of selection (Table 9.1). The two MCN populations produced similar results.

With the exception of longevity, average values of all other traits that we measured declined in the MCN populations. By far the fastest was the decline of the ability of larvae to survive and eventually develop into adults under very crowded conditions, where the amount of resources was such that only ~10% of larvae could complete development. After 30 generations, the competitive ability of larvae was over two times below that of the controls. This key result is in good agreement with results of Mukai. In contrast, the speed with which adults could run, and the fecundity of females, both measured under benign conditions, declined ~10 times slower. The proportions of eggs which fail to develop into adults in a benign environment, and of couples that were sterile, also grew only slowly, by ~0.1% per generation, indicating that accumulation of mutations with drastic individual effects did not play a major role in our results.

The most plausible cause of the observed declines in fitness-related traits was unopposed accumulation of deleterious mutations. It is likely that if we could assay fitness comprehensively under a tough, complex, competitive environment, similar to what fruit flies experience in nature, its decline would be even faster than 2% per generation. Indeed, our key measurement of competitive ability did not take into account any traits of adult flies. Of course, it would be very desirable to repeat such measurements of fitness-related traits independently. This has not been done so far. However, in 2015 Katrina McGuigan and colleagues published an MCN experiment performed on another species of *Drosophila, D. serrata*, which showed that MA and MCN designs produce similar results.

9.2 Selection Against Deleterious Alleles

The only natural force that can check accumulation of deleterious mutations is negative selection. Let us first see how it operates against individual *de novo* mutations and pre-existing mutant alleles. Here, the key fact is that different mutations and alleles are deleterious to very different degrees. Even coefficients of selection against mutations that belong to the same functional class (see Chapter 2) vary widely. The same is, of course, true for pre-existing mutant alleles, although they are, on average, less deleterious than *de novo* mutations, because milder alleles persist in populations for longer (see below).

Among loss-of-function *de novo* mutations, ~20% are recessive lethals, but among such pre-existing alleles this proportion must be only ~1%, because there are only two recessive lethals (see Chapter 2) but over 100 loss-of-function alleles (see Chapter 11) per genotype. More generally, the coefficient of selection against a loss-of-function allele can be as high as 1.0, in the case of dominant lethals, and as low as ~0.001, in the case of those alleles that reach high frequencies. Still, the coefficient of selection against any loss-of-function allele of a gene must exceed $1/N_e$, because otherwise the gene that is no longer strongly needed would be on its way to becoming a pseudogene (see Chapter 8).

Among human missense *de novo* mutations, ~20% are strongly deleterious with s ~0.01 or higher, 60% are mildly deleterious with s ~0.001–0.0001, and ~20% are effectively neutral with s ~0.00001 or lower. For pre-existing missense alleles, the corresponding proportions are <1, ~10, and 90%, respectively. Coefficients of selection against deleterious mutations and alleles must be variable within other functional classes, too, although strong election ($s > 0.01$) appears to be rare for mutations that do not disturb a protein-coding gene. Sophisticated analyses suggest that a substantial proportion of deleterious alleles from all functional classes (except loss-of-function) are characterized by $s \sim 1/N_e$, which is $\sim 10^{-4}$ in the case of humans and $\sim 10^{-6}$ in the case of *D. melanogaster* (see Chapter 8), which may by explained by the Li–Akashi effect (see below).

Still, to ascertain precisely the distribution of coefficients of selection against *de novo* mutations and pre-existing mutant alleles is a difficult task. Strongly deleterious alleles are too rare within populations, and must be studied in MA experiments. In contrast, mildly deleterious mutations are hard to detect in such experiments, and selection against them needs to be inferred from their other properties (see Chapter 11). Thus, our knowledge of this distribution remains rather vague.

Of course, selection acts on complete genotypes, and not on individual deleterious alleles. Let us accept a simplifying assumption of one-dimensional epistasis (see Chapters 3 and 7), under which the fitness of a genotype is determined by the value of fitness potential, a single genotype-level quantitative trait. Here, this trait is genotype contamination, a measure of how badly a genotype is affected by all its deleterious alleles. As long as we ignore drastic alleles, it is natural to assume that genotype contamination is a sum of contributions of all deleterious alleles carried by the genotype, and that the contribution of an allele is equal to the coefficient of selection (see Chapter 7) against it (at the mutation–selection equilibrium, see below). For example, if I carry 10 deleterious alleles with $s = 0.1$, and another 100 deleterious alleles with $s = 0.001$, my genotype contamination is $(0.1 \times 10) + (0.001 \times 100) = 1.1$.

As long as mutations are unconditionally deleterious, selection on genotype contamination is directional (see Chapter 7): when it increases, fitness declines (at least, does not increase). Still, there could be many kinds of such selection (Figure 9.3). Exponential selection is a particularly simple option. In this case, a given increment of genotype contamination always produces the same relative decline of fitness, so that epistasis is absent (see Chapter 3). Suppose, for simplicity, that the coefficient of selection is the same for all deleterious alleles, being, say, 0.5. Then, under exponential selection, fitnesses of individuals with 0, 1, 2, 3, ... deleterious alleles (contaminations 0, 0.5, 1.0, 1,5, ...) constitute 100%, 50%, 25%, 12.5%, ... of the maximal fitness, because adding an extra deleterious allele always reduces fitness by one half.

In contrast, the remaining fitness landscapes shown in Figure 9.3 do not possess this simple property. In the case of three dark gray fitness landscapes, adding a deleterious allele to a more contaminated genotype causes a larger relative decline of fitness. This is synergistic epistasis (see Chapter 3): deleterious alleles reinforce the effects of each other. Truncation selection is the extreme form of synergistic epistasis: for a while, fitness is not affected by extra deleterious alleles, but, after the threshold genotype contamination is reached, it suddenly drops to zero. Selection against a completely recessive lethal (genotypes AA and Aa have a normal fitness, and genotype aa has zero fitness) is also truncation and represents the extreme case of within-locus synergistic epistasis. In the case of linear selection, absolute decline of fitness does not change when contamination of the genotype increases (as long as fitness remains above zero);

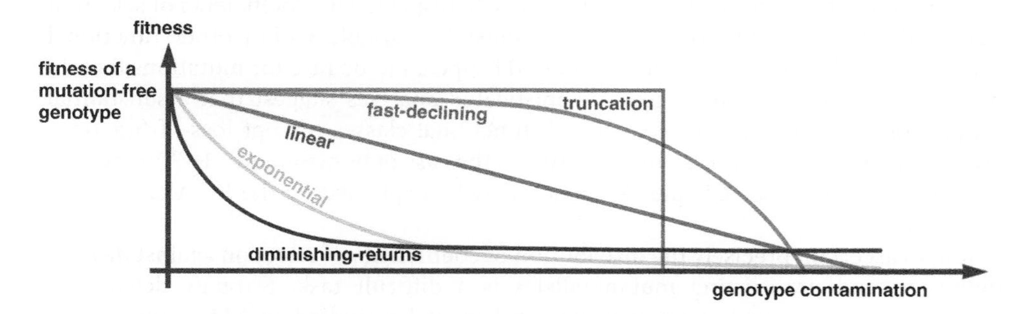

Figure 9.3 Various modes of selection against deleterious alleles. The light gray curve shows nonepistatic, exponential selection. Dark gray curves show different modes of selection with synergistic epistasis between deleterious alleles. The black curve shows a mode of selection with diminishing returns epistasis.

however, relative decline gets larger and this is all what matters here, because selection depends on relative fitnesses (see Chapter 7). In contrast, in the case of a black fitness landscape, the decrement of relative fitness due to a particular increment of the genotype contamination is smaller in more contaminated genotypes. This, rather weird, mode of epistasis is called diminishing returns: deleterious alleles attenuate effects of each other.

The ability of selection to counteract mutation depends on selection differential Δ (Figure 7.12). Because selection against deleterious alleles is directional, it reduces genotype contamination, so that Δ is always negative in this case. From the perspective of imperfection *I*, the most efficient mode of selections on a quantitative trait is truncation (Figure 9.3): it achieves a particular Δ by imposing the minimal *I*. For example, by culling 50% of individuals, which corresponds to $I = 1$, truncation selection shifts the mean of a trait with Gaussian density by ~80% of its standard deviation. However, from the perspective of evolvability of fitness *E*, the most efficient mode of selection is not truncation bur linear: it achieves a particular Δ by imposing the minimal *E*. Still, animal and plant breeders practice truncation selection: they want to minimize the proportion culled, and do not care about *E*. Thus, in the course of artificial selection for a high yield or any other desired trait a breeder ranks all the available individuals according to their values of the trait, and propagates only those with this value above some truncation point. Of course, direction of selection does not matter here: truncation selection for high yield and for low genotype contamination work in exactly the same way.

Diminishing returns epistasis in selection against deleterious mutations has never been observed and seems unlikely. In contrast, synergistic epistasis in the form of recessivity of severely deleterious alleles is very common. It seems plausible that deleterious alleles at different loci also should generally reinforce the effects of each other. Unfortunately, we still do not know how common synergistic epistasis is in selection against deleterious alleles. This is not surprising: neither genotype contamination nor fitness are easy to measure, and it is hard to determine subtle properties of a function when both its independent and dependent variables are poorly known! We will discuss this issue in Chapters 11 and 15.

9.3 Mutation–Selection Equilibrium

Beneficial mutations and positive selection favoring them are rare and, thus, are unlikely to result in a lot of genetic variation at any particular moment (see Chapter 7). The same is probably true for balancing selection (see Chapter 10). Acting alone, negative selection would eventually remove all substantially deleterious alleles, creating a population in which only the deleterious allele-free genotype is present. Does this mean that, not counting neutral and nearly-neutral variation, we expect populations to be almost uniform genetically? No, because ongoing mutation constantly supplies rare deleterious alleles and thus precludes purging all of them. Still, the resulting genetic variation is unconditionally deleterious and, thus, can never contribute to adaptive evolution.

If deleterious mutation and negative selection do not change for a long enough time, and the population is large enough so that random drift can be ignored, the population reaches a state of deterministic mutation–selection equilibrium. Let us first consider this equilibrium at an individual locus (perhaps, a nucleotide site), where deleterious mutations occur at a per generation rate μ. In 1923, Charles Danforth introduced a very useful concept of pervasiveness p, the expected total number of copies of an individual mutation which will appear in the population, likely in the course of many generations, until the mutant allele is lost. The equilibrium frequency f of a mutant allele is equal to the mutation rate times it pervasiveness: $f = \mu p$. For example, if one normal allele per 10^5 mutates every generation, and, on average, 1000 copies of a *de novo* mutant allele are eventually produced, its equilibrium frequency is 10^{-2}.

How do mutation and selection affect the frequency of a rare deleterious allele? Every generation, mutation increments it by μ, and selection decrements it by sf. For example, if the frequency of a mutant allele is 0.01, and the fitness of its carriers is reduced by 0.001, selection decrements this frequency by 10^{-5}. At equilibrium, the increment of f due to mutation and its decrement due to selection must be equal, so that $f = \mu/s$. This makes sense: the equilibrium frequency of a deleterious allele is directly proportional to the rate with which it appears by mutation and is inversely proportional to the strength of selection against it.

Comparing the above two results, we can see that $p = 1/s$: pervasiveness of a mutation is inversely proportional to the coefficient of selection against it. For example, if a mutation reduces fitness by 1%, 100 copies of it will be produced within the population, on average, before selection eliminates them all. Another way to arrive at this conclusion is to observe that p is the sum of a geometric series: the expected numbers of mutant alleles in successive generations are 1, $(1 - s)$, $(1 - s)^2$,....

How large is imperfection (see Chapter 7) caused in the equilibrium population by selection against mutations? A deleterious allele reduces fitness of its carrier by s, and the equilibrium frequency of this allele is μ/s. Thus, mean population fitness is $[(1 - \mu/s)1 + (\mu/s)(1 - s)] = 1 - \mu$. Because here the maximal fitness is 1, imperfection $I = [1 - (1 - \mu)]/(1 - \mu)$ (see Chapter 7) and, as long as μ is small, I is equal to μ, regardless of the strength of selection. This fact is known as the Haldane–Muller principle, which is traditionally formulated as "one mutation – one genetic death". Indeed, a severely deleterious mutation does not last long within the population and, thus, affects only a few (one, if it is a dominant lethal) individuals before being eliminated. In contrast, a mildly deleterious mutation has a high pervasiveness, affecting many individuals, albeit slightly, before being eliminated. The resulting imperfection is the same in both cases.

Imperfection at the mutation–selection equilibrium is called mutation imperfection (see Chapter 10), because in this case mutation is the cause of the mean population fitness being below the maximal possible fitness, that of the mutations-free genotype. Thus, the Haldane–Muller principle states that equilibrium mutation imperfection is equal to the mutation rate, in the simplest case of one locus and rare deleterious alleles. In contrast, the variance of relative fitness at the mutation–selection equilibrium at one locus depends on both μ and s: it is easy to show that $E = \mu s$, so that a stronger selection leads to a larger variance. Note, that mutation imperfection is different from mutational pressure (see Chapter 6). In particular, incomplete penetrance of a mutant allele reduces the mutational pressure but not the equilibrium mutation imperfection.

Of course, selection affects all variable loci of the genome simultaneously (see Chapter 7). Each generation, mutation increases the average number of deleterious alleles in a genotype by U (see Chapter 6). Thus, at equilibrium, negative selection must reduce this number also by U. This reduction occurs due to decrements of frequencies of individual alleles, and is equal to the sum of values of sf for all deleterious alleles present in the population (as long as these alleles are rare). Because the probability that an allele is present in a genotype is equal to the frequency of this allele, this implies that the (arithmetic) mean value of the sum of coefficients of selection against all substantially deleterious alleles carried by a genotype is U. In other words, at the mutation–selection equilibrium the average genotype contamination in the population is U.

This fundamental fact was discovered by Newton E. Morton, James F. Crow, and Hermann J. Muller and published in 1956 in a classical paper "An estimate of the mutational damage in man from data on consanguineous marriages". To better understand it, let us introduce the concept of a "muller", the unit of contamination of a genotype by its deleterious alleles. A set of deleterious alleles such that the sum of coefficients of selection against them is 1 contributes 1 muller of contamination to the genotype that carries them (Morton and colleagues used the term "expressed lethal equivalent"). Then, at the mutation–selection equilibrium an average genotype carries U mullers of contamination. In particular, assuming that our species is at this equilibrium, and that deleterious alleles are individually rare, contamination of an average human genotype must be ~10 mullers (see Chapter 8).

One generation of mutation introduces UMean[s] mullers of contamination into an average genotype, where Mean[s] is the arithmetic mean of coefficients of selection against a *de novo* mutation. Thus, without any opposition by selection, it will take 1/Mean[s] generations for mutation to introduce the average equilibrium amount of contamination into the population. In other words, if selection is shut down for 1/Mean[s] generations, the average contamination of a genotype will be doubled. For example, if every mutant allele reduces fitness by $s = 0.001$, it will take 1000 generations without selection to double the average per genotype number of such alleles in the population, because pervasiveness of every allele is 1000.

However, this example is oversimplified, and in reality different mutations are deleterious to different extents. Because pervasiveness of a mutant allele is inversely proportional to the coefficient of selection against it, less deleterious mutant alleles are overrepresented at equilibrium. Quantitatively, it is easy to show that the average (arithmetic mean) coefficient of selection against a deleterious allele extracted randomly from an equilibrium population is equal to the harmonic mean (reciprocal to the average of reciprocals) of coefficients of selection against *de novo* mutations. For example, if one mutation with $s = 0.1$ and two mutations with $s = 0.01$ appear on average in a genotype every generation, at equilibrium the average coefficient of selection against a deleterious allele is $3/(10 + 100 + 100) = 1/70$. Then, because the average genotype contamination is $U = 3$ in this case, an average genotype carries 210 deleterious alleles. Because Mean[s] $= (0.1 + 0.01 + 0.01)/3 = 0.04$, it would take $1/0.04 = 25$ generations of free accumulation of mutations to double the equilibrium genotype contamination of this population.

One can estimate the average coefficient of selection against a deleterious allele of a particular kind from the data on the rate of mutations that produces such alleles and their average number in a genotype. For example, the per genotype rate at which recessive lethals appear due to mutation is 2×10^{-2}, and there are on average ~2 recessive lethals in the genotype of a *Drosophila* (see Chapters 2 and 6). Thus, the average pervasiveness of a recessive lethal present in the population is 100, and the average coefficient of selection against it is 0.01. This analysis demonstrates that "recessive" lethals are, in fact, not completely recessive, and that selection eliminates them from the population mostly due to their mild effects on fitness of heterozygotes. Because harmonic mean is smaller than arithmetic mean, the average coefficient of selection against a *de novo* recessive lethal mutation must be above 0.01.

Properties of negative selection acting at the mutation–selection equilibrium on genotype contamination are of particular importance from the perspective of the Main Concern. Of course, the selection differential Δ must always be $-U$Mean[s]. In contrast, imperfection I and variance of relative fitness E depend on the mode of selection (Figure 9.3). It can be shown that in the case of exponential selection and, thus, of no epistasis, $I = \exp(U) - 1$ [it helps to think of $\exp(-U)$ as a probability of a newborn not getting any *de novo* mutations] and $E = \exp(sU) - 1$. Thus, even when U is large, E can be realistic, as long as individual mutations are not too deleterious. For example, with $U = 10$ and $s = 0.01$, $E = 0.11$, which is consistent with some data (see Chapter 7). In contrast, $U = 10$ implies $I > 10\,000$, which is clearly impossible, because a human population can survive such a high imperfection only if a mutation-free woman would have, on average, over 10 000 daughters.

If so, how can humans and other mammals survive $U >> 1$? A possible answer is synergistic epistasis. Even if we consider only one locus, under complete recessivity of deleterious alleles, which is the extreme form of synergistic epistasis in this case, one genetic death removes two mutations, instead of one. This halves the mutation imperfection, making it only 50% of the diploid mutation rate at the locus. Thus, the Haldane–Muller principle in its original form works only when selection removes mutations one-by-one, that is, as heterozygotes.

At the level of complete genotypes, the effect of synergistic epistasis may be even stronger. If, for example, distribution of genotype contamination is truncated at 10UMean[s], each genetic death eliminates more than 10UMean[s] mullers of contamination, which is 10 times more than what mutation introduces per genotype. With such an efficient selection, reduction of the mean population fitness and, thus, mutation imperfection, can be small at the mutation–selection equilibrium. In fact, by moving the truncation point rightward, one can achieve an arbitrarily small mutation imperfection at equilibrium, because one genetic death can eliminate any amount of genotype contamination.

Strict truncation is not needed for a profound reduction of the equilibrium mutation imperfection. Any synergistic epistasis produces this effect, as long as fitness does not decline much before genotype contamination becomes high enough. Efficiency of epistatic selection was first recognized by Muller in his classical paper "Our load of mutations", published in 1950, and was later studied in detail by Motoo Kimura, Takeo Maruyama, and James F. Crow. Unfortunately, we still do not have definite direct data on whether negative selection against deleterious mutations involves any substantial synergistic epistasis (see Chapter 11).

9.4 Inbreeding Depression

Although genotypes of individuals from natural populations carry a lot of mutant deleterious alleles (see Chapters 2 and 11), studying their phenotypic effects is difficult. Indeed, most of these alleles are only mild and cannot be detected individually by observing phenotypes (see Chapter 3). An important opportunity to investigate the cumulative effect of deleterious alleles is provided by inbreeding depression – perhaps, the most ubiquitous pattern in natural selection.

Inbreeding depression is a reduction of fitness and of values of fitness-related traits in inbred individuals, produced by closely related parents. This pattern is observed, almost without exception, every time when phenotypes of inbred and outbred offspring are compared. The magnitude of inbreeding depression varies between species and populations. Still, a 1% increase in the coefficient of inbreeding F of an individual (see Chapter 2) often reduces its fitness also by ~1%. Inbreeding depression is usually deeper under tough, competitive environments than under benign, protective environments (Figure 9.4, see Chapter 10).

The main cause of inbreeding depression is complete or, more likely, partial recessivity of deleterious alleles. In outbred individuals such alleles, which are individually rare, are mostly present as heterozygotes, due to the Hardy–Weinberg law (see Chapter 2). In contrast, in an inbred individual with coefficient of inbreeding F, the same proportion of deleterious alleles, $F/2$, are either lost or present as homozygotes. Thus, if an outbred individual carries, on average, U mullers of genotype contamination, an inbred individual carries, on average, $U(1 - F + kF/2)$ mullers, where k is the average ratio of coefficients of selection against homozygous over heterozygous deleterious allele. As a result, inbreeding increments genotype contamination by $UF(k/2 - 1)$, and this increment is positive and leads to inbreeding depression as long as $k > 2$, which means that a homozygous deleterious allele inflicts more than twice the harm of a heterozygous one.

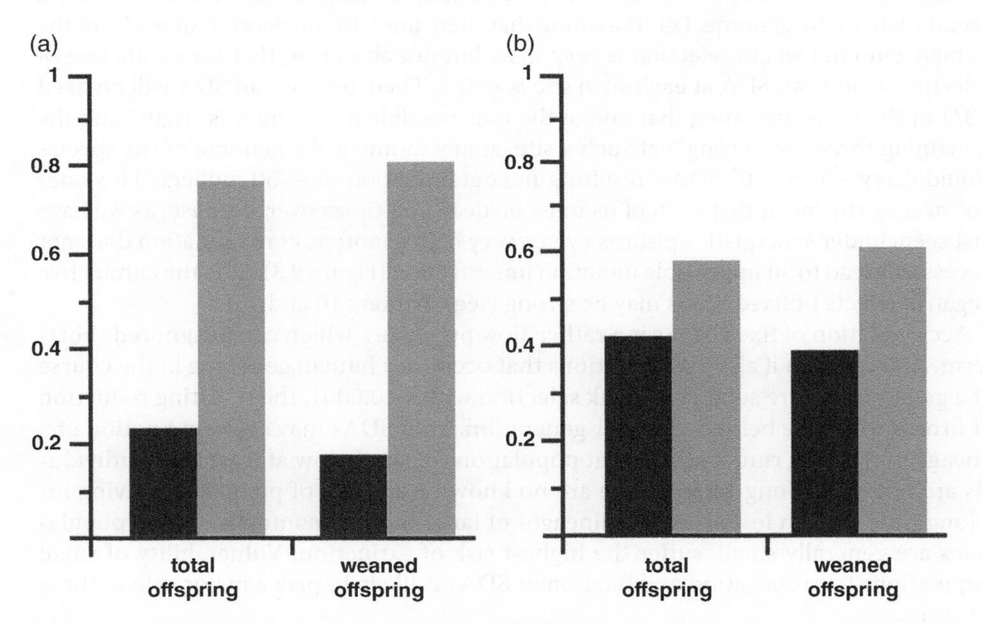

Figure 9.4 Proportions of offspring sired by inbred (black) and outbred (gray) male mice under a semi-wild environment where males were competing with each other (a) and in the laboratory where they were held separately, under benign conditions (b). In both cases, there were equal numbers of inbred and outbred males. Inbred males were produced by brother–sister mating ($F = 0.25$) (after Meagher *et al.*, *Proceedings of the National Academy of Sciences of the USA*, vol. 97, p. 3324, 2000).

It may be possible to use inbreeding depression to investigate some properties of the mutation–selection equilibrium (see Chapter 15).

9.5 Dangerous Slightly Deleterious Alleles

When the coefficient of selection s against a deleterious allele decreases, its expected pervasiveness $p = 1/s$ increases, and a selectively neutral allele has infinite pervasiveness. This is easy to understand: if a new mutation is not affected by selection, the expected number of copies of the mutant allele in every future generation is exactly 1. Thus, with passage of time, the total expected number of copies of such an allele grows without a limit.

Moreover, negative selection may be unable to prevent a slightly deleterious allele (SDA) from reaching a high frequency or even becoming fixed in the population. This is the case for alleles with $s < 1/N_e$ (see Chapter 8), because random drift plays a larger role in their dynamics than selection. Until the human population begun to multiply ~1000 generations ago (see Chapter 11), its effective size was only ~10 000, and dynamics of SDAs with $s < 0.0001$ were more like that of neutral alleles than of substantially deleterious ones. In the long run, at a locus that can harbor several such alleles each of them is fixed in the population for approximately the same proportion of time. Indeed, data on evolution of proteins show that slightly deleterious missense alleles reach fixation more often in populations with lower effective sizes.

Although almost harmless individually, SDAs may be dangerous when a lot of them accumulate in the genome. Let us assume that there are ~10^8 nucleotide sites (3% of the human genome) where selection is very weak but not absent, so that the coefficient of selection against an SDA at each such site is ~10^{-5}. Then, because an SDA will be fixed ~3/4 of the time (assuming that one of the four possible nucleotides is "right" and the remaining three are "wrong") at such a site, at any moment the genome of the species would carry ~0.75×10^8 SDAs, resulting in contamination of ~750 mullers. This does not necessarily mean that each of us must be dead 750 times over, because, as we have just seen, under synergistic epistasis even a very high genotype contamination does not necessarily lead to an impossible mutation imperfection (Figure 9.3). Still, the cumulative negative effects of fixed SDAs may be strong (see Chapters 10 and 15).

Accumulation of fixed SDAs is a rather slow processes, which can be ignored short-term. Indeed, even if all ~100 mutations that occur in a human genotype in the course of a generation were subject to weak selection with $s < 0.0001$, the resulting reduction of fitness would be below ~0.01 per generation. Still, SDAs may cause extinction of a lineage in the long run. It seems that populations of size below at least 1000 individuals are not viable long-term. There are no known examples of populations living for a long time at such low sizes, and lineages of large-bodied mammals, whose populations are generally small, suffer the highest risk of extinction. Vulnerability of small populations to propagation and fixation of SDAs is likely to play a major role in these patterns.

Obviously, the dynamics of an SDA in the population must be different from that of a substantially deleterious allele, which is always kept rare by negative selection and, thus, can never reach fixation. In 1963, Kimura, Maruyama, and Crow demonstrated that when selection against deleterious alleles is weak enough so that they can reach substantial frequencies, the equilibrium mutation imperfection increases. This happens because selection becomes less efficient when a deleterious allele is common: in this case, elimination of a copy of the deleterious allele benefits not only the beneficial allele but also other copies of the deleterious allele. Moreover, when the deleterious allele gets fixed, selection ceases, although imperfection remains (see Chapter 7).

Mutant alleles with $s \sim 1/N_e$ are particularly dangerous. Indeed, if negative selection is much stronger than that, it keeps mutant alleles rare (we ignore the possibility of mutation overwhelming selection without any help from random drift). In contrast, if negative selection is much weaker, a mutant allele cannot do much damage, even if it reaches fixation. Strikingly, a lot of data indicate that in many populations of rather diverse sizes, a high proportion of deleterious alleles are, indeed, opposed by selection with $s \sim 1/N_e$. Could nature conspire to always inflict the maximal harm on a population?

Perhaps, the answer is "yes", as Wen-Hsiung Li and Hiroshi Akashi were first to recognize. Let us assume that a functioning molecule (a transcription factor-binding segment of DNA, a protein, or something else) has some resilience, and can tolerate, with only negligible impairment to its function, several deviations from its optimal sequence. However, when deviations accumulate, eventually the function becomes strongly impaired – in other words, there is synergistic epistasis between deleterious alleles that affect the molecule. Then, negative selection would be unable to stop accumulation of mutations affecting the molecule before their number becomes so large that the coefficients of selection against them reach $\sim 1/N_e$, after which their further accumulation will be prevented. As a result, at equilibrium most selection coefficients against deleterious alleles at polymorphic loci will be within the dangerous range of $\sim 1/N_e$ (Figure 9.5). There is no definite proof that this is what actually happens, but this seems plausible, and could nicely explain the disproportional contribution of SDAs to genetic variation, discovered by Tomoko Ohta over 40 years ago.

Inefficiency of very weak selection is also responsible for a substantial proportion of large-scale mutations, which often occur due to illegitimate recombination between segments of similar non-allelic DNA (see Chapter 5). Because large-scale mutations are often strongly deleterious, selection must indirectly act against such segments.

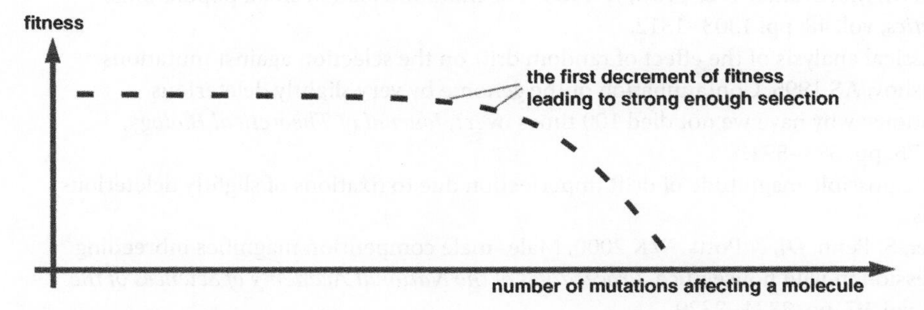

Figure 9.5 Li–Akashi effect. Mutations impairing the function of a molecule will be fixed, due to too weak opposition from negative selection, until enough of them accumulate to make the coefficient of selection against the next mutation $\sim 1/N_e$.

Indeed, every time when illegitimate recombination leads to a large-scale mutation which causes, say, severe hemophilia A (fatal until recently), both recombined DNA segments get eliminated. The coefficient of such indirect selection against a pair of similar segments is equal to the frequency of illegitimate recombination between them. Thus, in the human lineage selection is powerless to prevent fixation of a repetitive DNA segment if it engages in illegitimate recombination with probability below ~0.0001. In agreement with this reasoning, there are a lot of pairs of nearby genome segments with similar sequences which produce large-scale mutations with probabilities of ~0.00001, but not a single known example of a pair for which this probability is ~0.001.

Further Reading

Baer, CF, Joyner-Matos, J, & Ostrow, D 2010, Rapid decline in fitness of mutation accumulation lines of gonochoristic (outcrossing) *Caenorhabditis* nematodes, *Evolution*, vol. 64, pp. 3242–3253.
Mutation-accumulation lines of outcrossing nematodes deteriorate faster than that of self-fertilizing nematodes, in agreement with what the theory of evolution of mutation rates predicts.

Charlesworth, B 2012, The effects of deleterious mutations on evolution at linked sites, *Genetics*, vol. 190, pp. 5–22.
A lucid exposition of some basic properties of the mutation–selection equilibrium.

Crow, JF & Kimura, M 1979, Efficiency of truncation selection, *Proceedings of the National Academy of Sciences of the USA*, vol. 76, pp. 396–399.
One-dimensional synergistic epistasis can profoundly reduce mutational imperfection at the mutation–selection equilibrium.

Denisov, SV, Bazykin, GA, & Sutormin, R 2014, Weak negative and positive selection and the drift load at splice sites, *Genome Biology and Evolution*, vol. 6, pp. 1437–1447.
A discussion of the Li–Akashi effect.

Kimura, M & Maruyama, T 1966, The mutational load with epistatic gene interactions in fitness, *Genetics*, vol. 54, pp. 1337–1351.
A classical treatment of the mutation–selection equilibrium in large populations.

Kimura, M, Maruyama, T, & Crow, JF 1963, The mutation load in small populations, *Genetics*, vol. 48, pp. 1303–1312.
A classical analysis of the effect of random drift on the selection against mutations.

Kondrashov, AS 1995, Contamination of the genome by very slightly deleterious mutations: why have we not died 100 times over?, *Journal of Theoretical Biology*, vol. 175, pp. 583–594.
On the possible magnitude of drift imperfection due to fixations of slightly deleterious alleles.

Meagher, S, Penn, DJ, & Potts, WK 2000, Male–male competition magnifies inbreeding depression in wild house mice, *Proceedings of the National Academy of Sciences of the USA*, vol. 97, pp. 3324–3329.
Inbreeding depression in mice is deeper under a harsh, competitive environment.

Shabalina, SA, Yampolsky, LY, & Kondrashov, AS 1997, Rapid decline of fitness in panmictic populations of *Drosophila melanogaster* maintained under relaxed natural selection, *Proceedings of the National Academy of Sciences of the USA*, vol. 94, pp. 13034–13039.

Fitness deteriorates rapidly in Middle Class Neighborhood populations.

Yampolsky, LY, Kondrashov, FA, & Kondrashov, AS 2005, Distribution of the strength of selection against amino acid replacements in human proteins, *Human Molecular Genetics*, vol. 14, pp. 3191–3201.

Estimates of coefficients of selection against missense nucleotide substitutions in humans.

10

Phenomenon of Imperfection

It is clearly impossible to say what is the "best" phenotype unless one knows the range of possibilities. If there were no constraints on what is possible, the best phenotype would live for ever, would be impregnable to predators, would lay eggs at an infinite rate, and so on.

John Maynard Smith, 1978.

We may be interested in imperfection of phenotypes and of genotypes, in terms of either fitness or wellness. From the evolutionary perspective, a genotype can be imperfect because reaching perfection would require (i) a radical redesign, impossible due to low fitnesses of intermediate genotypes, (ii) introducing substantially beneficial alleles, which will be acquired eventually only after a long wait, (iii) shutting down Mendelian segregation that keeps producing maladapted homozygotes when heterozygotes have the highest fitness, (iv) getting rid of individually rare substantially deleterious alleles, constantly supplied by mutation, and (v) reverting fixations of multiple slightly deleterious alleles (SDAs), which became a part of the genome due to random drift. We concentrate on the mutation imperfection and regard any genotype that is free of rare deleterious alleles as weakly perfect. Because the average number of deleterious alleles in a genotype is very high, weakly perfect genotypes are too improbable to appear in actual populations. In the course of their history, humans have experienced Pleistocene, Preindustrial, and Industrialized environments, none of which is optimal. Under a suboptimal environment, imperfection is likely to be higher than under the optimal environment.

10.1 Phenotypic and Genotypic Imperfection

The Main Concern is about "undesirable effects" of deleterious mutations. Thus, we need a concept of "quality" – otherwise, how can we tell desirable from undesirable and deleterious from beneficial? The quality of an individual is a number which tells us how "good" is the phenotype of this individual. This quality can depend on genotype, controllable environment, and chance (see Chapter 3). For example, a human can have a low-quality phenotype due to a drastic mutation in their genotype, lead-based paint in the house where they grew up, or an accidental trauma – or to any combination of adverse factors.

Crumbling Genome: The Impact of Deleterious Mutations on Humans, First Edition. Alexey S. Kondrashov.

Any useful definition of quality should provide a comprehensive assessment of the phenotype. Thus, quality is a phenotypic quantitative trait which is secondary in the sense that its value can be deduced from values of other traits, which together describe the phenotype fully. If we take into account only partial phenotypes, individuals of the same phenotype may be of different quality, due to effects of neglected traits. For example, people with the same blood pressure can still produce different numbers of offspring and die at different ages. In this case, quality of some partial phenotype must be understood as the expected quality of an individual with this phenotype.

Although quality assays the phenotype, and not the genotype, of an individual, one can also indirectly ascribe quality to genotypes, because they serve as blueprints for phenotypes. This is essential for our purposes, because only genotypes are directly affected by mutation. Under a particular environment, the quality of a genotype is an inherited (expected) value of quality of individuals with this genotype (see Chapter 3).

From the evolutionary perspective, quality is fitness, the efficiency of self-propagation (see Chapter 7). Fitness provides the most comprehensive assessment of the phenotype, because every complex adaptation must contribute to fitness: if one does not, it will soon be lost. However, with all due respect to evolution, the Main Concern is about modern humans, where efficiency of self-propagation *per se* does not matter very much. Instead of the number of children of a human, we mostly care about their wellness.

Admittedly, wellness is hard to quantify. It may be helpful to think of wellness as what a sensible individual wishes for their future child. Among two phenotypes, the one which a would-be parent prefers confers a higher wellness ("parent's test"). The relationship between wellness and fitness is discussed in Chapters 12, 13, and 15.

As it is the case for any phenotypic trait (Figure 3.13 and Figure 7.2), it is helpful to think about quality in terms of its landscape. A quality landscape that flies over the space of genotypes or of phenotypes shows the quality of each of them and may depend on two kinds of factors. One is the controllable environment: a polar bear is less fit (and less well) than a brown bear in a forest, but more fit and well on drifting ice. The other is the composition of the population, that is, frequencies of different genotypes and phenotypes in it. Frequency dependence of a fitness landscape can naturally lead to selective advantage of rare genotypes (Figure 10.1). Imagine a population of organisms capable of utilizing multiple resources – apples, oranges, plums, and so on. If different genotypes are better at utilizing different resources, a rare genotype gets a boost in fitness, because there is less competition for its preferred resource. As a result, multiple genotypes can coexist indefinitely, instead of the best one replacing all others, which is the outcome of frequency-independent selection. If this is the case, we say that the population is under balancing (or negative frequency-dependent) selection and at the state of balanced polymorphism. At equilibrium, when frequencies of different genotypes correspond to the abundance of their preferred resources, the frequency-dependent fitness landscape may become flat. Frequency-dependent wellness landscapes will be discussed in Chapter 15.

From the perspective of genetics of quantitative traits, fitness is just one of them, although the most complex one. However, from the evolutionary perspective, fitness is unique. The effect of an allele on fitness has consequences for the dynamics of this allele in the population. As a result, data on genetic composition of the population carry information on whether a particular allele is deleterious, neutral, or beneficial, fitness-wise (see Chapter 11). In contrast, the effect of an allele on any other phenotypic

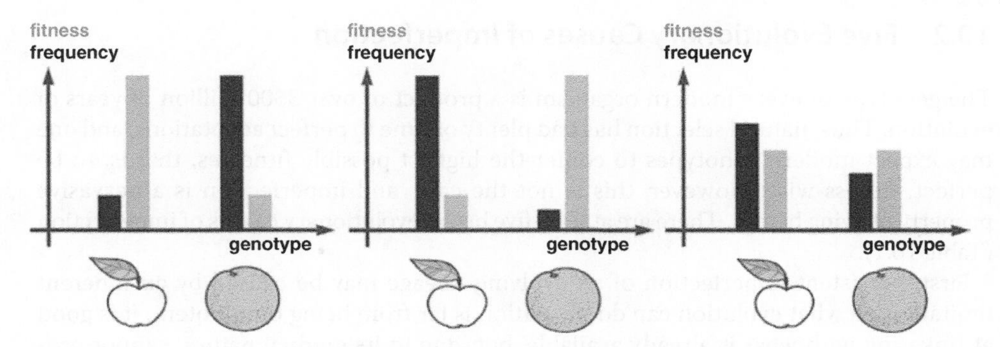

Figure 10.1 Frequency-dependent fitness landscape. When apple (orange) eaters are rare, they are more fit. At equilibrium, a population that depends on 2 apple trees and 1 orange tree consists of 2/3 of apple-eaters and 1/3 of orange-eaters, both having the same fitness.

trait (not counting such specific traits as rates of mutation or recombination) does not directly entail any population-level consequences. Thus, effects of alleles on wellness are much more difficult to study than their effects on fitness (see Chapter 12).

How can we determine whether the quality of a phenotype or a genotype is "high" or "low"? This can be done only by comparing it to some paradigm, the perfect phenotype or genotype, which has the highest possible quality. Obviously, this paradigm corresponds to the highest peak on the quality landscape, at least within its portion that we are willing to consider. Once we have identified the perfect quality, imperfection can be understood as any deviation from it. The concept of imperfection was introduced formally, for the case of genotypes and fitness, in Chapter 7, and applied to the mutation–selection equilibrium in Chapter 9. Let us now examine it in a little more detail.

As Maynard Smith noticed, defining perfection may be a problem. Indeed, what is the highest possible fitness? Having 1000, or even 10 000 children? Or what is the best possible wellness? Living for 1000, or even for 1 000 000 years?

If we consider phenotypes only, the problem can be solved if the perfect phenotype can be defined *a priori* (seeing one quantum of light is the best an eye can do), or if our measure of quality can accept only a fixed set of values (nothing can be better than "excellent" health), or if the quality landscape is stabilizing (Figure 7.11) (the perfect pH of blood is 7.40, and not higher or lower). In such cases, we can talk about phenotypic imperfection. In contrast, if our measure of quality is open-ended, it is necessary, in order to identify the perfect phenotype and to quantify phenotypic imperfection, to impose some additional restrictions on what phenotypes are possible.

The situation is simpler, at least conceptually, in the case of genotypic imperfection. Evolutionary considerations, discussed below, can help to identify the genotype which is perfect in terms of fitness. Then, we can define imperfection of a genotype as a deviation of the expected quality of an individual with this genotype from the expected quality of an individual with the perfect genotype. Unfortunately, this conceptual clarity does not help us to determine the actual magnitude of genotypic imperfection, because we know nothing about phenotypes of individuals with deleterious alleles-free genotypes. This issue, as well as genotypic imperfection of wellness, will be considered in Chapters 12 and 15.

10.2 Five Evolutionary Causes of Imperfection

The genotype of every modern organism is a product of over 3500 million of years of evolution. Thus, natural selection has had plenty of time to perfect adaptations, and one may expect modern genotypes to confer the highest possible fitnesses, that is, to be perfect, fitness-wise. However, this is not the case, and imperfection is a pervasive property of living beings. There are at least five broad evolutionary causes of imperfection (Table 10.1).

First, persistent imperfection of an evolving lineage may be caused by an inherent limitation on what evolution can do. Evolution is far from being omnipotent: it is good at tinkering with what is already available, but, due to its gradual nature, cannot produce instant, radical changes. A human can erase *Hamlet* from a file and type *Ulysses* in its stead. However, to gradually transform, by replacing letters or words one-by-one, *Hamlet* into *Ulysses*, in such a way that all intermediate texts are sensible, to say nothing of being masterpieces, is obviously impossible. Evolution cannot do anything like erasing and typing from scratch, because mutation does not produce novel DNA sequences

Table 10.1 Five evolutionary causes of imperfection.

Kind of imperfection	Cause	What would it take to get rid of it	Persistence time in evolution	Relevance to the Main Concern
Imperfect fundamental designs of adaptations	Inherent limitation of gradual evolution	Coordinated, individually deleterious allele replacements in very many genes	Permanent	Irrelevant
Adaptations, imperfect under the current environment, after its recent change	Inherent limitation of slow evolution	Individually beneficial allele replacements in some number of genes	Short-term, after each change of the environment	Imperfect adaptation to Industrialized environments may exacerbate deleterious effects of mutations
Presence of homozygotes, when heterozygotes have the highest fitness	Mendelian segregation and, more generally, sexual reproduction	Switching to asexual reproduction	Permanent, as long as the advantage of heterozygotes persists	Irrelevant
Presence of low-frequency substantially deleterious alleles	Incessant influx of deleterious mutations	Reverting multiple individually rare deleterious alleles	Permanent, as long as mutation keeps acting	Directly relevant
Fixations of slightly deleterious alleles	Inefficiency of weak selection in finite populations	Reverting very numerous, fixed slightly deleterious alleles	Permanent, as long as the effective population size remains not very high	Perhaps, directly relevant, but poorly understood

that are long and sensible and because selection does not have a foresight. A deleterious mutation will be removed by negative selection, even if it represents the necessary first step along a long path that could eventually lead to a higher fitness.

Thus, evolution has to work through gradual accumulation of many individually beneficial (or, at the very least, approximately neutral) mutations, which limits its reach. As a result, complex adaptations often retain features that betray their evolutionary origin, and make them suboptimal. When shared by multiple species, such suboptimal adaptations provide evidence for their common ancestry.

A well-known example of a fundamental imperfection which persists because reaching perfection would require a radical redesign of the genotype and the phenotype are inverted eyes of vertebrates, in which blood vessels and nerves approach the retina not from behind but from the front, intercepting most of the light that enters the eye. We experience the place where the optical nerve penetrates the retina as a blind spot. In contrast, in cephalopods the retina is supplied from behind, so that their eyes have a better overall design that does not involve a blind spot. Another example of persistent imperfection is a suite of generic mammalian traits that became maladaptive after our ancestors evolved bipedal locomotion some 10 Ma. These traits are still present in humans, and do not show any tendency to disappear. As a result, humans have difficult childbirth, and are prone to hernias and spine problems.

Secondly, the genome of an evolving lineage can become imperfect temporarily, due to a change in its environment, after which the current environment becomes different from the one to which the lineage became adapted in the course of its past evolution. This phenomenon is known as lag imperfection. Eventually, positive selection of new, beneficial mutations will lead to allele replacements (see Chapter 7) which will adapt the lineage to the new environment, thus abolishing this temporary imperfection. However, this requires a considerable amount of time, because adaptive evolution is slow: a population may have to wait for many generations for a necessary beneficial mutation to appear, a fresh beneficial mutation may be accidentally lost, and replacement of the old allele with the new one, effected by positive selection, takes time.

Thirdly, even when the environment remains stable for a long time, imperfection can persist because imperfect genotypes keep being produced by Mendelian segregation. A famous example is the advantage of heterozygotes at the locus that encodes human beta hemoglobin. In populations living where malaria is common, the highest fitness is conferred by heterozygotes Aa, where A is the common allele of beta hemoglobin, and a is its mutant allele, which provides some protection against malaria. Homozygotes AA, which are the most fit in the absence of malaria, are more vulnerable to malaria than Aa individuals, and homozygotes aa suffer from sickle-cell anemia, a severe Mendelian disease.

In an asexual population, such an advantage of heterozygotes would lead to fixation of a clone of Aa individuals in the population. However, with sexual reproduction less fit homozygotes keep being produced every generation by Mendelian segregation. Still, the polymorphism is maintained because, according to the Hardy–Weinberg law (see Chapter 2), a rare allele is mostly exposed to selection in heterozygotes and, thus, gets an advantage when heterozygotes have the highest fitness. As a result, in a sexual population the advantage of heterozygotes leads to balancing selection and stable polymorphism. In contrast to a situation where multiple resources are available (Figure 10.1), here balancing selection is involved with persisting segregation imperfection.

Fourthly, a population can be imperfect due to the presence of substantially deleterious alleles that never reach fixation in the population and, instead, always remain individually rare. Rarity of a deleterious allele is contingent on selection acting against it consistently and being strong enough. In the case of humans, strong enough means that coefficients of selection against mutant alleles exceed 10^{-4}, because the effective size of the human population was ~10 000 until recently (see Chapters 8 and 9). Because such rare, derived, and substantially deleterious alleles are an easy target for negative selection, they would be soon eliminated from the population, if there was not a constant influx of new deleterious mutations. Thus, here we are dealing with mutation imperfection (see Chapter 9).

Fifthly, a population can be imperfect due to fixation of SDAs (see Chapter 9). Such drift imperfection is caused by joint action of mutation and drift and can be substantial if drift made negative selection inefficient for a long time, due to a small effective population size (see Chapter 8). In particular, species of mammals, with their large genomes and low long-term effective population sizes, are likely to carry very many fixed SDAs.

Figure 10.2 presents these five evolutionary causes of imperfection from the perspective of fitness landscapes. How important is each of them, in particular, for humans? Currently, we do not know for sure, but here are some thoughts.

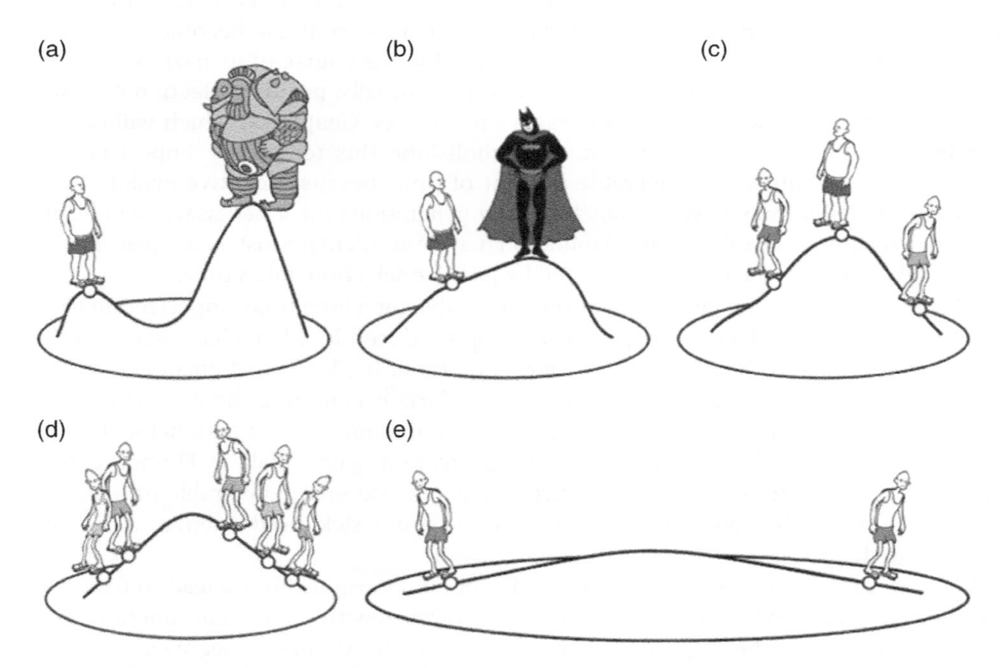

Figure 10.2 Fundamental, lag, segregation, mutation, and drift imperfection of the present genotype(s), shown by open circle(s). (a) The perfect genotype corresponds to a different fitness peak, which can be reached only after crossing a fitness valley. (b) The perfect (under the current environment) genotype corresponds to the nearest fitness peak, but the current genotypes did not have time to reach it. (c) The perfect genotype is heterozygous, and Mendelian segregation keeps producing imperfect genotypes. (d) Rare substantially deleterious alleles make all the present genotypes imperfect. (e) Fixed slightly deleterious alleles make all the present genotypes imperfect.

We have no clue what kind of superhumans (or superfruitflies) could be produced by deliberate, coordinated multiple changes in the *Homo sapiens* (or *Drosophila melanogaster*) genome. It is not yet possible to study distant high fitness peaks, separated from existing genotypes by fitness valleys. Thus, fundamental imperfection of the designs of adaptations, although fascinating from the point of view of evolutionary theory, is currently mute for all practical purposes.

Changes of the environment definitely occurred in the course of recent human history, and we are not perfectly adapted to Industrialized environments (see below). An allele replacement which would lead to sugar aversion or cause the human immune system to work properly under a low pathogen load is likely to benefit those living in developed countries, by preventing obesity or autoimmune diseases. Unfortunately, we do not have even a vague idea about the number of allele replacements that could be favored by positive selection in the current human populations living under Industrialized environments, and about the magnitude of the lag imperfection in humans.

Only very few definite cases of heterozygote advantage and, thus, of segregation imperfection have been described in humans. Every such case must leave a footprint on within-population genetic variation (see Chapter 15). Such footprints are apparently rare, which makes it plausible that segregation imperfection is not that important.

Mutation imperfection due to presence of rare, substantially and unconditionally deleterious alleles is the key element of the Main Concern, and is likely to be high (see Chapter 15). Relaxation of negative selection leads to a rapid increase of mutation imperfection due to accumulation of such alleles.

Drift imperfection can also be quite high, and should also be regarded as a part of the Main Concern. However, we know very little about it, because fixed SDAs do not result in any ongoing selection and, thus, are especially hard to study. Indeed, we can compare effects of a rare deleterious allele to that of its normal, common counterpart, but cannot do the same for a fixed SDA. Thus, we will concentrate on mutation imperfection.

10.3 Weakly Perfect Human Genotypes and Phenotypes

Ignoring all but the fourth cause of imperfection, let us regard as perfect any human genotype that does not carry any rare substantially deleterious alleles. To emphasize that perfect, in this sense, genotype can still suffer from imperfection due to other causes, I will call it "weakly perfect". In fact, there could be many weakly perfect genotypes, which differ from each other due to neutral variation and variation under balancing selection (see Chapter 15).

Here, I need to make explicit a key assumption about negative selection that has not been spelled out so far (see Chapter 9). I assume that the issue of perfection can be addressed separately for each locus, in the sense that at every locus the ancestral allele is unconditionally beneficial, no matter which alleles are present at other loci of the genotype and, thus, rare derived allele(s) is unconditionally deleterious. In other words, we assume that sign epistasis (see Chapter 3) is rare, at the scale of within-population variation. The validity of this assumption will be discussed in Chapter 15.

The weakly perfect genotype is tantalizingly close, because the building blocks from which it can be assembled are readily available. Indeed, by combining ancestral alleles from just two genotypes one can assemble a genotype which is weakly perfect at a vast majority of its loci, excluding only those where both the original genotypes carry a deleterious allele (such loci are very rare, because deleterious alleles are individually rare). One can easily generate the weakly perfect genotype *in silico*, as a consensus of multiple individual genotypes, or the genome of a species (see Chapter 1). However, there is a long way from an *in silico* to an *in vivo* genotype. Current technologies make it possible to create a living organism with the artificially produced genotype only in the case of bacteria and yeast, and even this requires a lot of labor (see Chapter 15).

Because polymorphic loci which harbor deleterious alleles are very numerous, the weakly perfect genotype is extremely improbable, and we are not going to find it in the human, or any other, population. To illustrate this point, let us assume that there are 100 000 polymorphic loci where alleles which are deleterious enough to be taken into account are present, and that the frequency of a deleterious allele is 0.01 at each of these loci. Then, the expected number of deleterious alleles in a haploid genotype is 1000, and the formula for Poisson distribution informs us that, as long as alleles at different loci are distributed independently, the probability that a genotype is deleterious alleles-free is only $e^{-1000} \sim 10^{-400}$. Perhaps more intuitively, a diploid individual that carries 1000 deleterious alleles, each in heterozygous state, will produce a deleterious alleles-free gamete, assuming independent transmission of alleles at each locus, with the same probability as that of getting 1000 heads (or tails), after flipping a coin 1000 times, which is $(1/2) \times (1/2) \times \ldots \times (1/2)$ (1000 terms) $= 2^{-1000} \sim 10^{-300}$ (Figure 10.3). Thus, there would be not a single individual with the weakly perfect genotype even in a population of the size equal to the number of electrons in the observable Universe ("only" $\sim 10^{80}$).

Although we do know reasonably well which genotype would be weakly perfect – the one that lacks rare, derived alleles – we paradoxically cannot tell which of the two actual genotypes is closer to perfection. Indeed, all deleterious alleles are not created equally deleterious. It is currently impossible to say what is worse – my 120 nonsense alleles or your 110 nonsense alleles – because functions of some genes are more important than that of others and also because some nonsense alleles may not lead to total loss of function, due to a number of molecular mechanisms. In other words, we are still rather far from being able to infer the quality of a genotype from its sequence (see Chapter 12).

Let us call weakly perfect the expected phenotype of an individual of the weakly perfect genotype. Because the correspondence between genotypes and phenotypes is known rather poorly (see Chapter 3), having the weakly perfect genotype only *in silico* does not help to infer the properties of the weakly perfect phenotype. What would be an expected intelligence quotient or longevity of a weakly perfect human? This is discussed in Chapter 15.

Still, we can be reasonably sure that the weakly perfect phenotype is not the best possible one, because random events can adversely affect even individuals with weakly

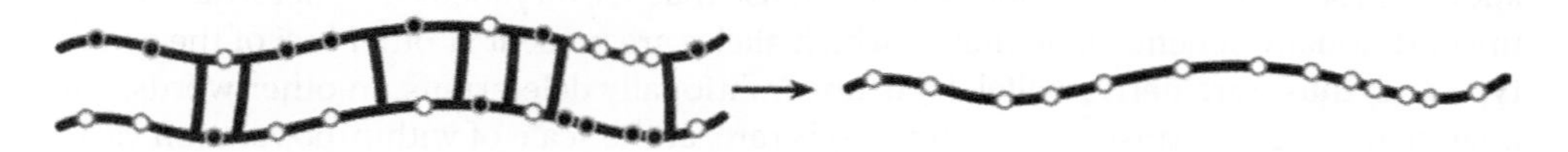

Figure 10.3 Production of a deleterious alleles-free gamete by a diploid genotype carrying many heterozygous deleterious alleles is extremely improbable, as it requires multiple, independent recombination events (thick lines) to occur only at exactly prescribed locations. Deleterious alleles are shown by filled circles and normal alleles by open circles.

perfect genotypes. Even a weakly perfect zygote, carried by a young, weakly perfect mother, is still likely to occasionally develop, even under the optimal environment, into a child with a serious birth defect. Of course, the probability of this is unknown.

10.4 Native, Novel, and Optimal Environments

Imperfection of genotypes and phenotypes depends on their controllable environment. If we ignore the evolutionary history of a genotype, its environment can be either optimal, under which quality of the genotype is higher than under any other environment, or suboptimal. From the perspective of past evolution of the lineage to which a genotype belongs, its environment can be native, if the lineage lived and evolved under it for a long time, or novel.

Let us relate these general concepts to the environments of modern humans and our not-too-distant ancestors. Broadly speaking, humans can live under any of the following three environments (Figure 10.4):

1) Pleistocene (named after a geological epoch which lasted from 2 588 000 to 11 700 years ago), which was the environment of all our hunter–gatherer ancestors. Very few contemporary human populations, such as !Kung, Hadza, and Ache (see Chapter 11), are still living under a mostly Pleistocene environment.
2) Preindustrial, which ancestors of the majority of modern humans experienced since the transition from hunting and gathering to agriculture. This transition, known as the Agricultural, or Neolithic, Revolution began ~10 000 years ago, at the beginning of the current geological epoch, Holocene, and eventually affected most of the range of our species.
3) Industrialized, which gradually appeared in Europe ~200 years ago, and is now rapidly spreading across the globe. The Industrial Revolution created a supportive environment, with reduced exposure to pathogens, easy availability of food, limited need for tough physical work, and advanced health care.

Of course, there is a lot of variation within these three broad categories. After out-of-Africa expansion of *Homo sapiens*, which began ~100 000 years ago (and was at least the third such expansion of humans), Pleistocene environments of our ancestors became as

Figure 10.4 Four human environments – (a) Pleistocene, (b) Preindustrial, (c) Industrialized, and (d) optimal.

diverse as tropical rain forests and savannahs, temperate forests, tundras, and mountains. This led to evolution of rather different local adaptations in different populations. A lot of changes also occurred between the Agricultural and Industrial Revolutions, so that Preindustrial environments encompass small pre-historic villages, ancient cities, and medieval ghettoes. Currently, the Industrialized environments in Finland, Italy, and Japan are also far from being the same; in particular, due to the rather different diets.

The Agricultural Revolution led to a substantial increase in the density of human populations and of sizes of local groups. However, this apparently was not accompanied by reduction of early mortality or increased longevity (see Chapter 11). Indeed, increased carrying capacity of the environment does not necessarily make it better, as long as the per capita amount of resources does not change. Infectious diseases may be even a bigger problem within a denser population. In contrast, the Industrial Revolution led to the demographic transition, which involved a dramatic reduction of early mortality and increased longevity (see Chapter 11).

Because evolution occurs slowly, the Pleistocene environment, as a first approximation, remains native even for modern humans. Conversely, the Preindustrial and, especially, Industrialized environments are novel for us. Still, there was enough time for some human populations to evolve some adaptations to their particular varieties of the Preindustrial environment. The ability of adults to digest lactose (see Chapter 7) is just one of many examples of such adaptations. On top of the inherent slowness of response to positive selection, adaptation of humans to Industrialized environments may be impeded by relaxation of selection: these days less-than-perfect health does not necessarily result in fewer children. Nevertheless, some adaptation to the Industrialized environments appears to be ongoing (see Chapter 11).

Obviously, a novel environment is generally not optimal for a genotype. For example, easy availability of high-calorie foods under the Industrialized environments is deleterious, and may cause obesity and other diseases of civilization, because we are adapted to living under a threat of starvation. However, the native environment, to which a genotype is maximally adapted in the sense that all the possible beneficial allele replacements have already occurred, is also not necessarily optimal. Although we are adapted, probably to the best of our ability (save a hypothetical radical redesign), to deadly pathogens and mortal physical danger, we are better off without them. Apparently, the optimal human environment is a mixture of Pleistocene and Industrialized features: a loving family, no overeating, no pollution, reasonable physical and intellectual challenges and, perhaps, not too much hygiene and comfort – but no famine, smallpox, or sabretooth tigers. According to Hippocrates, "If we could give every individual the right amount of nourishment and exercise, not too little and not too much, we would have found the safest way to health". Although nobody yet lives under the exactly optimal environment, it is definitely easier to attain than the perfect genotype.

We will consider imperfection of humans under Pleistocene and Preindustrial environments mostly in terms of fitness, primarily because population genetic data shed light on past effects of alleles on fitness only (see Chapter 11), and phenotypic data are available almost exclusively for modern populations. Still, the Main Concern is primarily about wellness, which we will consider mostly for populations living under Industrialized environments. Hopefully, the proportion of humanity living under such environments will keep growing. The relationship between Pleistocene fitness and Industrialized wellness is discussed in Chapter 15.

10.5 Factors, Exacerbating Mutation Imperfection

By definition, a genotype performs worse under a worse controllable environment. However, weak genotypes may suffer disproportionately under a poor environment. This effect, known as genotype–environment (G × E) interaction (Figure 10.5), makes intuitive sense: while under a benign environment performance of weaker individuals can be only moderately inferior, a poor environment may present challenges that they are unable to meet. For example, I can walk 10 miles, although slower than a trained athlete, but would surely drown attempting to swim across the English Channel.

Fortunately, environments are much easier to manipulate than genotypes, and it is possible to study experimentally effects of rare, deleterious alleles under a variety of them. In 1935, Timofeev-Ressovsky observed that under a harsh, competitive environment the effect of mutations induced by X-rays on the viability of larvae of *D. melanogaster* was relatively larger. In 2001, we confirmed this effect for ethyl methanesulfonate (EMS)-induced mutations (see Chapter 6). In particular, treatment of both grandfathers of a male fly by EMS increased its average development time from 228.5 to 283.0 h (2%) at 25°C (optimal temperature), and from 358.9 to 375.5 h (4.6%) at 20°C. In 1994 David Houle and I observed the same pattern for deleterious effects of spontaneous mutations in *D. melanogaster*. However, we also observed that under an extremely poor environment the differences between fitnesses of genotypes became smaller, and not larger. In this case, only a very small proportion of larvae, perhaps those who hatched from the very first eggs laid in a vial, were able to complete development. As a result, mortality became mostly random. The G × E effect has also been observed for inbreeding depression: fitness of inbred individuals declines deeper under a semi-natural environment than under a protective environment (Figure 9.4).

G × E interactions with respect to fitness and wellness imply that one must take the environment into account when imperfection is considered. Being the most benign of

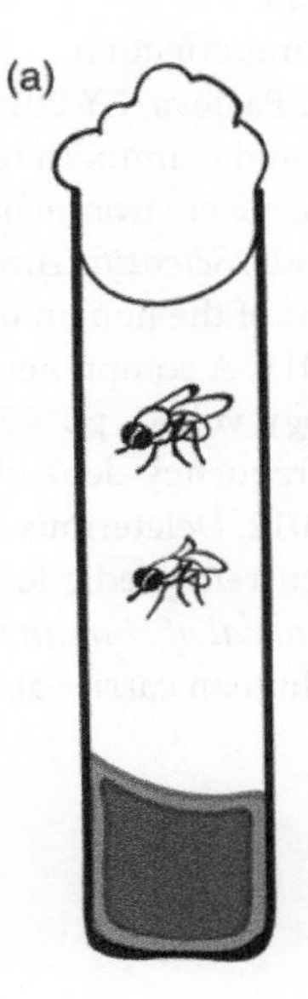

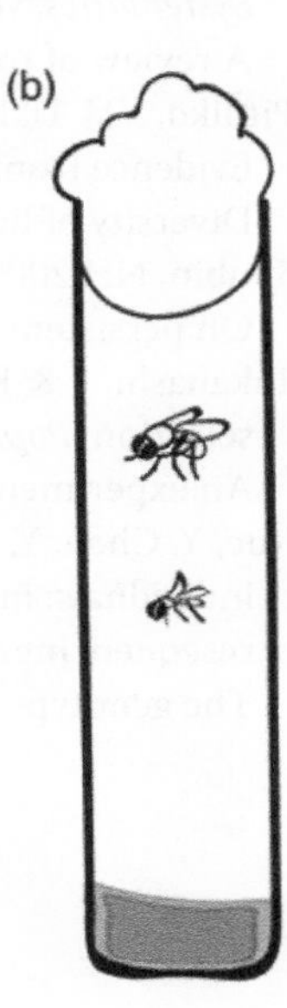

Figure 10.5 The phenomenon of G×E: on a good food (a), a genetically inferior fly (bottom) has only a slightly inferior phenotype, in comparison with a genetically superior fly. However, on a poor food (b) the difference between phenotypes produced by the same two genotypes is much larger.

the available human environments, the Industrial environments likely lead to the lowest imperfection. Thus, data on imperfection under harsher environments may overestimate imperfection in many modern human populations (see Chapter 11).

On top of a suboptimal environment, mutation imperfection may be also exacerbated by other kinds of imperfection of genotypes. Indeed, an already-impaired genotype may be more sensitive to the harm of substantially deleterious alleles. We will briefly consider possible synergism of mutation imperfection with lag and drift imperfection in Chapters 12 and 15.

Further Reading

Bocquet-Appel, J-P 2011, When the world's population took off: the springboard of the Neolithic Demographic Transition, *Science*, vol. 333, pp. 560–561.
Comparison of the Neolithic and the contemporary demographic transitions.

Crespi, BJ 2000, The evolution of maladaptation, *Heredity*, vol. 84, pp. 623–629.
A review of evolutionary causes of imperfection.

DeBusk, R 2010, The role of nutritional genomics in developing an optimal diet for humans, *Nutrition in Clinical Practice*, vol. 25, pp. 627–633.
On the optimal human environment.

Diamond, J 2002, Evolution, consequences and future of plant and animal domestication, *Nature*, vol. 418, pp. 700–707.
On the Agricultural Revolution.

Kondrashov, AS & Houle, D 1994, Genotype–environment interactions and the estimation of the genomic mutation rate in *Drosophila melanogaster*, *Proceedings of the Royal Society B*, vol. 258, pp. 221–227.
The relative effect of deleterious mutations increases under poor environments.

Maynard Smith, J 1978, Optimization theory in evolution, *Annual Reviews of Ecology and Systematics*, vol. 9, pp. 31–56.
A review of the concept of imperfection.

Pitulko, VV, Tikhonov, AN, & Pavlova, EY 2016, Early human presence in the Arctic: Evidence from 45,000-year-old mammoth remains, *Science*, vol. 351, pp. 260–263.
Diversity of human Pleistocene environments.

Shubin, NH 2009, This old body, *Scientific American*, vol. 300, pp. 64–67.
On persistent imperfections of the human body.

Takahashi, Y & Kawata, M 2013, A comprehensive test for negative frequency-dependent selection, *Population Ecology*, vol. 55. pp. 499–509.
An experimental study of frequency-dependent fitness landscapes.

Xue, Y, Chen, Y, & Ayub, Q 2012, Deleterious- and disease-allele prevalence in healthy individuals: insights from current predictions, mutation databases, and population-scale resequencing, *American Journal of Human Genetics*, vol. 91, pp. 1022–1032.
The genotype of a healthy human carries at least ~1000 substantially deleterious alleles.

11

Our Imperfect Fitness

With savages, the weak in body or mind are soon eliminated; and those that survive commonly exhibit a vigorous state of health. We civilised men, on the other hand, do our utmost to check the process of elimination ... Thus the weak members of civilised societies propagate their kind. No one who has attended to the breeding of domestic animals will doubt that this must be highly injurious to the race of man.

Charles Darwin, *Descent of Man*, 1871.

A deleterious allele can be characterized by its alteration of the DNA sequence, effect on molecular function, frequency in the population, age, and effect on fitness and/or wellness. The first three properties can be assayed directly. In contrast, ages and selection coefficients of alleles can only be estimated indirectly. Within a sample of 10 000 diploid human genotypes, ~90 million loci will be polymorphic. However, at a vast majority of these loci the derived allele is very rare. The genotype of an individual carries, on average, ~4 million derived alleles, thousands of which are substantially deleterious. The average contamination of a human genotype is ~10 mullers. Different human populations carry mostly the same benign or slightly deleterious alleles, but mostly different substantially deleterious alleles. Deleterious alleles at different loci are distributed approximately independently within human populations. Still, underdispersed distributions of genotype contamination may suggest that selection against them involves synergistic epistasis. Opportunity for selection in human populations declined after the demographic transition. However, some selection is ongoing even in the current populations living under Industrialized environments, although its targets are obscure.

11.1 Properties of an Allele

Finally, we are ready to directly deal with the Main Concern, which is the effects of deleterious mutant alleles on humans. In this chapter, we will approach this problem from the evolutionary perspective by considering fitness. Then, in Chapters 12 and 13 we will focus on wellness. Chapter 12 treats effects of mutations that occurred in the past generations but are still present in the population, whereas Chapter 13 deals with effects of *de novo* mutations.

Crumbling Genome: The Impact of Deleterious Mutations on Humans, First Edition. Alexey S. Kondrashov.

Let us start from the elementary units of weak imperfection, individual deleterious derived alleles. Each derived allele has five key properties (see Chapter 2):

1) nature of its alteration of the DNA sequence;
2) effect on molecular function;
3) frequency in the population (derived allele frequency, DAF);
4) age;
5) effect on fitness and/or on other traits we care about.

For example, an allele may (1) be a single-nucleotide substitution of G with A, (2) produce a premature stop codon (a TGG > TGA nonsense substitution) within a coding exon, (3) have frequency 0.005, (4) be 10 000 years (~500 generations) old, and (5) reduce the fitness of its heterozygous carriers by 0.001 (and increase the body mass index, under Industrialized environments, by 0.01). Or it may (1) be a long deletion of 100 kb of DNA, (2) result in the removal of two successive protein-coding genes, (3) have frequency 0.0001, be 1000 years (~50 generations) old, and (5) reduce the fitness of its heterozygous carriers by 0.01 (and increase the risk of schizophrenia by 0.05).

The first three properties of an allele are easy to ascertain. The nature of a DNA sequence alteration follows directly from comparison of sequences of the ancestral and the derived alleles at the locus. After this, one can deduce the impact of the derived allele on molecular function by relating this alteration to the genome annotation (see Chapter 2). True, current genome annotations are incomplete, and we may erroneously conclude that a mutation that disturbs a yet unknown function is functionally mute. Still, it seems that even the already available annotations of some well-studied genomes, including human, are reasonably good, at least as far as protein-coding genes are concerned.

The frequency of an allele within the population can be determined by sequencing a large enough sample of genotypes. Currently, the sizes of such samples are in the thousands, making it possible to reliably measure DAFs above ~0.01. However, samples of millions of genotypes will soon become available. Of course, a sample must be unbiased, in the sense that the chances of an individual being included must not depend on whether it carries the allele.

In contrast to the first three properties, the age of an allele cannot be ascertained directly and instead has to be estimated indirectly, on the basis of several clues. First, young alleles tend to have low frequencies, because it takes a lot of time for a new mutation to reach high frequency, unless it is strongly beneficial. Secondly, young alleles are usually present only in one local population, in contrast to old alleles which often have wide and even global geographical ranges. Thirdly, young alleles tend to reside within long uniform stretches of genotypes, marked by common alleles that are identical by descent (see Chapter 2), because recombination has not had enough time to chisel away the edges of the genotype in which the original mutation occurred (Figure 11.1). The best methods of estimating the age of an allele are rather sophisticated and require inferring the complete history of genealogical relationships and recombination events among all the sequenced genotypes, which can be represented by the "ancestral recombination graph".

No matter how hard we try, the age of an individual derived allele can be estimated only with a substantial uncertainty. In contrast, it is often possible to estimate with good confidence the distribution of ages of alleles at multiple polymorphic loci with similar properties. This is sufficient to address many questions.

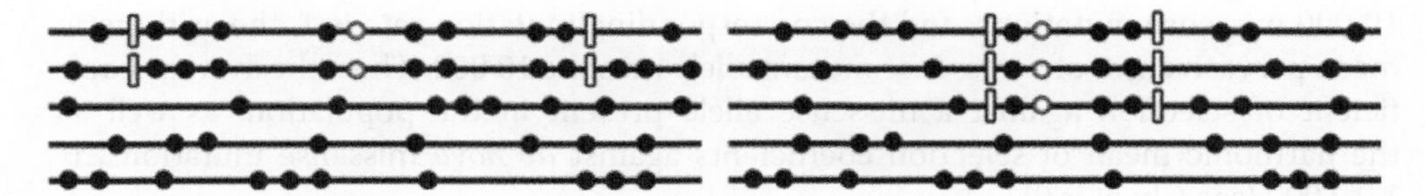

Figure 11.1 Estimating the age of a derived allele (shown by open circles) at a locus. (Left) In different genotypes, copies of a young derived allele are still embedded into a long segment of the genotype (bounded by bars) in which the mutation that produced this allele first appeared. (Right) Copies of an old derived allele are embedded only into a short segment of the original genotype. Derived alleles at other loci are shown by black circles.

The effect of a derived allele on fitness, relative to that of the ancestral allele, is characterized by its selection coefficient s (see Chapter 7), and an analogous coefficient can be used to characterize the effect of the derived allele on any other phenotypic trait (see Chapter 3). For our purposes, these coefficients are the most important properties of an allele; however, they are also the most difficult to ascertain. One option is to measure selection coefficients by comparing fitnesses of many individuals carrying the ancestral and the derived alleles (see Chapter 3). Surprisingly, such a direct approach has so far never been applied to human fitness, although data on the number of children of individuals are readily available. Unfortunately, this approach would not work when one of the alleles is rare, or if the selection coefficient is small. Moreover, selection in modern human populations is not necessarily identical to selection that operated until recently (see below).

Fortunately, the effect of an allele on fitness, in contrast to any other phenotypic trait, influences its dynamics in the population. If an allele reduces body size by, say, 1%, this fact *per se* tells us nothing about how long this allele will persist in the population and what frequency it is likely to reach. As a result, direct measurements are the only feasible way of studying the effects of alleles on all traits other than fitness (see Chapter 12). In contrast, if an allele reduces fitness by 1%, this greatly reduces its chances of being old and common, which opens a number of possibilities for estimating selection coefficients indirectly.

First, evolutionary conservatism of a genome segment or site implies that most derived alleles that affect it must be deleterious. The trouble is that any negative selection with a selection coefficient well above $1/N_e$ (where N_e is effective population size) is sufficient to prevent evolution (see Chapter 8). Thus, in mammals, where $N_e \sim 10^5$, lack of evolution at a site only tells us that coefficients of selection against derived alleles at this site are above $\sim 10^{-5}$, but does not help to distinguish, say, $s = 10^{-3}$ from $s = 10^{-2}$. Still, there are semi-rigorous ways to work around this problem, by using data on multiple species.

Secondly, the selection coefficient of an allele can be inferred from its frequency and age. The expected frequency of an allele is μ/s, where μ is the rate of mutations that can create this allele (see Chapter 9). The expected selection coefficient of an allele of a particular age can also be calculated. Unfortunately, these inferences are only probabilistic: due to random drift (see Chapter 8) different alleles generally have different frequencies and ages, even if their values of μ and s are the same.

Thirdly, we can estimate the typical strength of selection against a particular class of alleles by using data on *de novo* mutations. For example, if an average genotype carries

10 000 missense mutations, and the corrresponding mutation rate is 1, the arithmetic mean pervasiveness of a missense mutant allele must be 10 000. Thus, the average coefficient of selection against a missense allele present in the population, as well as the harmonic mean of selection coefficients against *de novo* missense mutations, is 1/10 000 (see Chapter 9).

Some information on the coefficient of selection against a derived allele is also provided by its effect on the molecular function. In particular, loss-of-function alleles of protein-coding genes are usually the most deleterious. Finer details of the molecular function may be helpful in the case of missense alleles, where coefficients of selection can be inferred from comparison of the properties of the ancestral and the derived amino acids, from the analysis of protein structure, and from data on similar proteins in other species. Still, such deductions are never precise. In particular, there is a lot of variation in the strength of selection against derived alleles within each functional class. Depending on the gene, its loss-of-function alleles can be dominant lethals, recessive lethals, or cause only a mild decrease of fitness even when homozygous (see Chapter 2).

Taken together, these clues provide some idea about coefficients of selection against deleterious alleles that are currently present in the population. As in the case of allele ages, the distribution of selection coefficients for a class of derived alleles can be estimated with a much higher confidence than individual selection coefficients. Also, straightforward methods shed light only on the average coefficient of selection against an allele. However, an allele can reside in a variety of genetic backgrounds in different individuals, which in the presence of epistasis can lead to variation in the strength or even the very direction of selection (see Chapter 3). Detecting epistasis is discussed in Chapter 15.

11.2 Human Derived Alleles

What are properties of pre-existing derived alleles, produced by mutation in the past, which are present in contemporary human populations? Attempts to address this fundamental question begun in the early 20th century, when phenotypically drastic alleles that can cause Mendelian diseases were discovered (see Chapters 2 and 6). Late in the 20th century, sequencing of genotype fragments made it possible to detect phenotypically mild and mute alleles (see Chapter 1). In the last decade, next-generation sequencing (NGS) produced a lot of data on (almost) complete human genotypes. As a result, we now possess a rather good knowledge of the sequence alterations, effects on the molecular function, and frequencies of human derived alleles. In contrast, their ages and selection coefficients are known with much less confidence.

The description of the set of all derived alleles in any population is provided by their total number and a five-dimensional joint distribution of allele properties. Indeed, different properties of an allele are not independent of each other (Table 11.1), so that five separate one-dimensional distributions, one for each property, do not tell the whole story. The causes of correlations between some allele properties were discussed above, and the causes of others should now be obvious for a careful reader.

On top of pairwise correlations between allele properties, there are also multiway correlations. This makes the exact description of the set of derived alleles that are present in a population rather complex. However, we can afford to ignore sophisticated details and, instead, concentrate on the basic properties of this set.

1.1 Correlations between the five properties of derived alleles.

	Effect on molecular function	Frequency	Age	Selection coefficient
f the ed DNA ent	Small-scale with mute (weak): a large-scale allele can be functionally mute (if it affects only junk DNA), and a loss-of-function allele can be small-scale	Small-scale with common (weak): a large-scale allele can be common (if it is selectively neutral or only mildly deleterious), and a rare allele can be small-scale	Small-scale with old (weak): a large-scale allele can be old (if it is selectively neutral or only mildly deleterious), and a young allele can be small-scale	Small-scale with neutral (weak): a large-scale allele can be neutral if it is functionally mute, and a deleterious allele can be small-scale
on ular on	—	Mute with common (moderate): a loss-of-function allele can be common if the corresponding function is not very important for fitness, and a rare allele can be mute	Mute with old (moderate): a loss-of-function allele can be old if the corresponding function is not very important for fitness, and a young allele can be mute	Mute with neutral (strong): a loss-of-function allele can be only slightly deleterious if the corresponding function is not very important for fitness, but a deleterious allele cannot be mute
ency		—	Rare with young (strong): a common allele cannot be young (unless it is favored by positive selection), although an old allele can be rare	Rare with deleterious (strong): a common allele cannot be deleterious, although a neutral allele can be rare
			—	Young with deleterious (strong): an old allele cannot be deleterious, although a neutral allele can be young

First, we need to ask how many polymorphic loci are present in the population. Surprisingly, this straightforward question does not admit a definite answer, because it depends on the minimal DAF at a locus that we are still willing to take into account. When this minimal DAF approaches 0, the number of polymorphic loci keeps increasing.

The key reason for this fundamental property of genetic variation is not hard to grasp. A polymorphism appears due to a mutation, and, therefore, the initial frequency of the derived allele is 1/2N, where N is the size of a diploid population. After this, random drift (chapter 8) usually eliminates this allele rather soon, before it has a chance to reach a high frequency, which is a lot of only few, lucky ones. Sewall Wright showed that, if we imagine an infinitely large population of organisms with infinitely large genotypes at a stochastic equilibrium between rare mutation and random drift, at any given moment of time the number of selectively neutral polymorphisms with DAF = x is proportional to 1/x. Thus, in such a population the number of polymorphic loci approaches infinity when the minimal DAF approaches 0.

There are also two "extra" reasons. One of them is negative selection. In every species, many derived alleles are neutral, but many others are deleterious, and only very few are substantially beneficial. A strong enough negative selection eventually eliminates a deleterious derived allele from the population, but before this is accomplished it keeps this allele rare (see Chapter 9), creating an excess of rare alleles even over the $\sim 1/x$ neutral expectation.

The other "extra" reason for the excess of rare alleles, specific to humans, is the recent dramatic increase of the human population size. The Agricultural Revolution ~10 000 years ago (see Chapter 10) definitely was a factor, although it now seems that this increase began before it. In a growing population, mutation produces more new alleles every generation; however, even a lucky neutral allele, spared by random drift, cannot reach a high frequency fast. Thus, like negative selection, population expansion creates an excess of rare alleles.

Within the current human population of $N = 7$ billion, every nucleotide site in the genome is, at the same time, a single-nucleotide polymorphism (SNP; except those very few where all the three possible single-nucleotide substitutions are dominant lethals). Indeed, because the per nucleotide mutation rate in humans is $\sim 10^{-8}$ (see Chapter 6), among $\sim 10^{8}$ children that are born every year one, on average, carries a *de novo* mutation at every particular site. Thus, unless we impose some minimal DAF, there would be 3 billion SNPs in the global human set of polymorphic loci, most of them with extremely low DAFs. It does not make much sense to consider such a set.

A convenient way to deal with this situation is to take into account only those derived alleles that are present, perhaps only as a heterozygote in just one genotype (singletons), within some sample of genotypes from the population. Indeed, all the data we currently have are on such samples, because not every person on Earth has yet have their genotype sequenced. Most extremely rare derived alleles will be missing even in a large sample, which makes the resulting set of polymorphic loci manageable. Of course, some information is lost this way.

Recently, results of four large-scale whole-genotype sequencing projects were published – of 2504 humans from 26 populations from all continents, of 2636 Icelanders, of almost 10 000 residents of the UK, and of 2120 Sardinians. On the basis of these data, I deduced what appear to be typical characteristics of a set of loci that are polymorphic within a sample of ~10 000 humans.

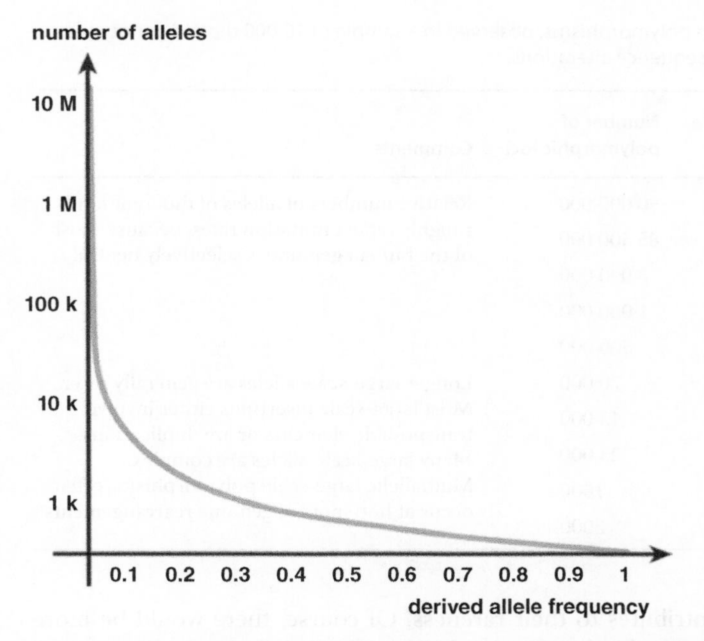

Figure 11.2 Spectrum of DAFs in genotypes of 10 000 humans (after 1000 Genomes Project Consortium, *Nature*, vol. 526, p. 68, 2015).

A vast majority of human polymorphic loci are effectively diallelic, because only one derived allele has a noticeable frequency (see Chapter 2). Figure 11.2 shows the density of DAFs. When a heterozygous derived allele appears as a singleton, its frequency is recorded as 1/20 000 which is the minimal possible DAF within a sample of 10 000 diploids. For most such singletons their true DAFs are below 1/20 000, and most polymorphic loci with true DAFs below 1/20 000 are missed altogether. Only loci with DAF > 0.0005 are ascertained (almost) completely, and these DAFs are measured with a reasonable precision, because at least 10 copies of the derived allele are expected to appear within the sample. Despite this underascertainment of very rare derived alleles, most loci that are polymorphic within the sample have very low DAFs, close to the minimal possible value of 1/20 000.

Table 11.2 presents numbers of polymorphic loci, subdivided according to the kind of DNA sequence alterations. These numbers should be treated with caution, because current methods of NGS cannot detect every allele in every genotype with absolute confidence, so that some rare alleles are likely to be routinely missed. Also, African human populations are significantly more variable than non-African populations, which lost some proportion of ancient polymorphisms in the course of the third out-of-Africa expansion (see Chapter 10), which involved population bottlenecks. Still, it is likely that, as a rough approximation, the numbers presented in Table 11.2 are applicable to a sample of genotypes of 10 000 individuals from any human population.

Relative numbers of derived alleles of different kinds reflect, first of all, relative rates of mutations that produce them. In particular, large-scale alleles are much rarer because large-scale mutations are much rarer. However, a stronger negative selection against

Table 11.2 Numbers of human polymorphisms, observed in a sample of 10 000 diploid genotypes, classified by the kinds of DNA sequence alterations.

Scale of an allele	Sequence nature of an allele	Number of polymorphic loci	Comments
Small-scale (<50)	All	90 000 000	Relative numbers of alleles of different kinds roughly reflect mutation rates, because most of the human genome is selectively neutral
	SNP	85 500 000	
	Deletions	3 000 000	
	Insertions	1 000 000	
	Complex	500 000	
Large-scale (>50)	All	70 000	Longer large-scale alleles are generally rarer. Most large-scale insertions either involve transposable elements or are duplications. Many large-scale alleles are complex. Multiallelic large-scale polymorphisms often occur at hot-spots of genome rearrangements
	Deletions	43 000	
	Insertions	23 000	
	Inversions	1000	
	Multiallelic	3000	

large-scale alleles also contributes to their rareness. Of course, there would be more (less) polymorphic loci of every kind present within a larger (smaller) sample.

Table 11.3 presents numbers of polymorphic loci from the same set, subdivided according to the effect of the derived allele on molecular function. Crude descriptions of distributions of DAFs are also included in this table.

The remaining two properties of a derived allele, its age and selection coefficient, are known with much less confidence than the former three. Figure 11.3 presents estimates, subject to a considerable uncertainty, of average ages of derived alleles with different effects on molecular function. Not surprisingly, polymorphisms that affect function more are younger, due to stronger negative selection. The average age of a polymorphism of a particular functional kind depends on the size of the sample of genotypes: the larger the sample, the higher is the proportion of rare alleles, and, thus, the lower the average age. Alleles of all kinds are, on average, younger in non-African populations, because some old alleles were accidentally lost in the course of the third out-of-Africa expansion.

The selection coefficient is the most important property of a derived allele, from the perspective of the Main Concern – and it would be really nice if it were directly observable. Unfortunately, this is not the case. Still, we can be reasonably certain regarding two key general features of selection acting on derived alleles. First, a majority of derived alleles (perhaps, ~90% of them in mammals) are selectively neutral for all practical purposes – simply because they affect only junk DNA. Secondly, non-neutral derived alleles are mostly deleterious, thus being subject to negative selection, and positively selected derived alleles are very few. This claim follows from general considerations (see Chapter 7) and from one of the most salient and universal patterns in genetic variation, observed in all species studied so far, including humans: those derived alleles that affect the molecular function more are also more likely to be rare (Table 11.3).

Also, there is definitely a wide variation in coefficients of selection against derived alleles even within a functional class. Indeed, on the one hand even among loss-of-function

Table 11.3 Numbers of human polymorphisms, observed in a sample of 10 000 diploid genotypes, classified by their effects on molecular function.

Effect on molecular function	Number of polymorphic loci	Distribution of DAFs (singletons, <0.5%, 0.5–5.0%, >5.0%)	Comment
Intergenic mute	43 000 000	0.38, 0.35, 0.16, 0.11	Genome segments where selection is not strong enough to slow evolution down
Introns mute	36 000 000	0.40, 0.35, 0.15, 0.10	Excluding edges of introns and some other important sites
Non-coding important	9 000 000	0.40, 0.35, 0.15, 0.10	Boundaries of important non-coding segments are determined mostly through evolutionary conservation
UTRs	900 000	0.42, 0.35, 0.15, 0.08	Selection on UTR alleles is mostly weak, but non-trivial
Synonymous	300 000	0.41, 0.35, 0.15, 0.09	Selection on synonymous alleles is mostly weak, but non-trivial
Missense	550 000	0.48, 0.35, 0.08, 0.05	Missense alleles can have widely different impacts on protein function, including total loss, and on fitness
Loss of function of a protein-coding gene	18 000	0.63, 0.33, 0.03, 0.01	Only definite loss-of-function alleles are included, such as those disturbing splicing consensuses on intron tips, start-codon abolishing, nonsenses, frameshifts, and large-scale affecting a gene

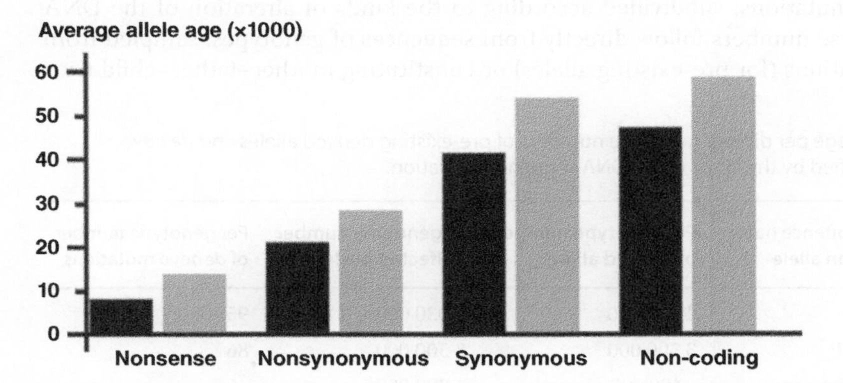

Figure 11.3 Average ages of derived alleles from different functional classes in Europeans (shown in black) and Africans (shown in gray) (after Fu *et al.*, *Nature*, vol. 493, p. 216, 2013).

derived alleles there are some with frequencies >5%, which testifies to only a rather weak selection against them. On the other hand, we know from direct data on phenotypes (see Chapter 2) that many alleles are individually strongly deleterious. Obviously, a strongly deleterious allele must be rare, although the converse statement is not true: a rare allele can be neutral, and most of them probably are. We will continue this consideration of negative selection below, in the context of properties of complete genotypes.

11.3 Average Imperfection of a Genotype

Alleles are building blocks, from which genotypes are made. Thus, it is impossible to understand genotypes without first understanding individual polymorphic loci. Still, the Main Concern is, first of all, about genotypes, and now it is time to consider their properties. The most important of them is the average complement of derived alleles in a genotype.

By definition, an allele is carried by a haploid genotype with probability that is equal to its frequency within the population. This obvious statement has three important implications. First, every genotype carries only a small proportion of the available derived alleles, because most of them are rare. Still, because polymorphic loci are very numerous, the average number of derived alleles in a genotype is high. Secondly, the average frequency of a derived allele that is randomly drawn from a genotype is much higher than the average DAF of a polymorphic locus, because more frequent derived alleles are present in a higher proportion of genotypes. As a result, a derived allele sampled from a genotype is also older and less deleterious than a derived allele sampled from the set of all polymorphic loci.

Thirdly, the average per genotype number of derived alleles can be immediately obtained from the complete data on polymorphic loci, on the basis of their DAFs. For example, in a population that harbors 1000 polymorphic loci with DAF = 0.005 and 30 polymorphic loci with DAF = 0.05, an average genotype carries $(1000 \times 0.005) + (30 \times 0.05) = 6.5$ derived alleles. In practice, however, because at most polymorphic loci DAFs are very low, genotypes are investigated directly – it is easier to characterize a genotype than a population.

Table 11.4 presents average per genotype numbers of pre-existing derived alleles and *de novo* mutations, subdivided according to the kinds of alteration of the DNA sequence. These numbers follow directly from sequences of genotypes, sampled from human populations (for pre-existing alleles) or constituting mother–father–child trios

Table 11.4 Average per diploid genotype numbers of pre-existing derived alleles and *de novo* mutations, classified by the kinds of the DNA sequence alteration.

Scale of an allele	Sequence nature of an allele	Per genotype number of derived alleles	Per genotype number of affected nucleotides	Per genotype number of *de novo* mutations
Small-scale (<50)	All	4 060 000	5 030 000	95
	SNP	3 500 000	3 500 000	86
	Deletions	400 000	1 200 000	6
	Insertions	150 000	300 000	2
	Complex	50 000	100 000	1
Large-scale (>50)	All	4300	18 200 000	0.55
	Deletions	2800	5 600 000	0.23
	Insertions	1200	1 200 000	0.12
	Inversions	40	80 000	0.004
	Complex	300	11 300 000	0.20

Table 11.5 Average per diploid genotype numbers of pre-existing derived alleles and *de novo* mutations, classified by their effects on molecular function.

Effect on molecular function	Per genotype number of derived alleles	Distribution of DAFs (<0.5%, 0.5–5.0%, >5.0%)	Per genotype number of *de novo* mutations
Intergenic mute	2 000 000	0.02, 0.10, 0.88	50
Introns mute	1 560 000	0.03, 0.12, 0.85	33
Non-coding important	430 000	0.03, 0.13, 0.84	7
UTRs	50 000	0.05, 0.18, 0.77	3
Synonymous	11 200	0.04, 0.15, 0.81	0.5
Missense	10 200	0.07, 0.20, 0.73	1
Loss of function of a protein-coding gene	100	0.16, 0.24, 0.60	0.1

(for *de novo* mutations, see Chapter 6), and, thus, must be reasonably precise. Perhaps, only rates of large-scale mutations might need a substantial revision in the future, because the current figures are based on small numbers of observed events. Not surprisingly, an average genotype carries much fewer derived alleles than a sample of 10 000 genotypes (Table 11.2).

Table 11.5 presents data on average per genotype numbers of pre-existing derived alleles and *de novo* mutations, subdivided according to their effects on molecular function. Genotype-level rates of mutations of different functional classes are mostly proportional to the sizes of the corresponding mutational targets (see Chapter 3). An important exception are rates of nonsynonymous and synonymous mutations, which are higher than the total length of coding exons would imply. This discrepancy appears because coding exons are enriched with mutagenic CG contexts, which elevates their per nucleotide mutation rate by about a factor of two (see Chapter 8).

Similarly to Table 11.3, Table 11.5 also presents the distribution of DAFs for each functional class of polymorphic loci. As expected, the corresponding distributions in these tables are very different. While at most polymorphic loci which are present within a population DAF < 0.005, frequencies of most derived alleles sampled from a genotype exceed 0.05.

If, according to Wright's theoretical prediction, the number of polymorphic loci in a population with DAF x is proportional to $1/x$, a derived allele randomly sampled from a genotype has equal chances to have any frequency. In particular, 95% of polymorphisms found in a genotype should have DAF > 0.05. However, the real proportion of such common derived alleles is lower (Table 11.5), even for alleles that affect segments of the genome which appear to be functionally mute. A slight excess of rare alleles among apparently selectively neutral polymorphisms must be primarily due to the recent expansion of the human population, and a much larger excess in functionally important polymorphisms is a signature of pervasive negative selection.

As long as the population is at equilibrium and negative selection is strong enough, the average coefficient of selection against a pre-existing allele, as well as the harmonic mean of coefficients of selection against *de novo* mutations, must be equal to the ratio of the per genotype numbers of *de novo* mutations over pre-existing alleles (see Chapter 9).

Of course, this statement remains true if we consider only mutations and alleles of a particular kind. For example, on the basis of data presented in Table 11.5, we may conclude that the average coefficient of selection against a pre-existing missense allele is 1/10 200 = 0.0001. However, even within a functional class of alleles and mutations selection coefficients vary widely (see Chapter 9), and there must be some loci where selection against derived alleles is very weak or even absent (except, perhaps, for the loss-of-function class), undermining such estimates.

Variation of coefficients of selection against all deleterious alleles is, of course, even wider than within their functional classes (see Chapter 9). Table 11.6 presents my guesses regarding coefficients of selection against pre-existing rare deleterious alleles and *de novo* mutations present in an average human genotype. This table should be primarily viewed as a prediction, to be compared with estimates that will hopefully appear in the not-too-distant future.

From the perspective of the Main Concern, we are primarily interested in just one, integral characteristic of the presence of deleterious alleles in an average genotype – its overall contamination C, that is, the sum of coefficients of selection against all of them (see Chapter 9). Because at the mutation–selection deterministic equilibrium $C = U$ (where U is the genomic deleterious mutation rate), we can expect an average human genotype to carry ~10 mullers of contamination (see Chapter 9). In fact, this figure can be even higher, because many mildly deleterious alleles are present at high frequencies, which reduces the efficiency of selection against them.

In 2015, Brenna Henn and colleagues estimated the average contamination of human genotypes using a different approach. They subdivided the nucleotide sites in the human genome into classes of different functional importance, using data on rates of evolution, instead of the genome annotation. For each such conservatism class, there is an estimate of the average coefficient of selection against a derived allele, based on the excess of rare alleles in human populations (apparent in Table 11.5, in particular, for loss-of-function and missense alleles). This excess makes it possible to quantify negative selection. Assuming that derived alleles are intermediately dominant (which makes sense), they concluded that $C = 15$ in a variety of human populations.

Their analysis ignored the theoretical prediction that $C = U = 10$. Thus, estimates $C = 15$ and $C = 10$ are independent, and their agreement is remarkable. If true, such a high value of C has important implications for the mode of negative selection and properties of the mutation–selection equilibrium (see Chapter 15).

Table 11.6 Selection against human deleterious pre-existing alleles and *de novo* mutations.

Range of coefficients of selection	Per genotype number of pre-existing alleles	Per genotype number of *de novo* mutations
0.1–1.0	0.3	0.03
0.01–0.1	30	0.3
0.001–0.01	1000	1
0.0001–0.001	100000	9

Deleterious alleles are not distributed randomly along the human genome. Not surprisingly, they are more common in regions where the mutation rate is higher. They are also more common in regions where the rate of crossing-over is lower. Indeed, there are several reasons to believe that crossing-over facilitates removal of deleterious alleles by negative selection.

So far, we have ignored differences between human populations. In fact, Africans are more genetically variable than non-Africans, and an average African genotype carries more derived alleles. Thus, one may think that the average genotype contamination of Africans is also higher. However, this logic is flawed: a higher overall genetic diversity indicated a higher effective population size and less intense random drift in the history of African populations, which must result in a more efficient selection (see Chapter 8). Thus, it could be natural to assume that the average genotype contamination of Africans is lower. Data indicate, however, this is also not true, and that the average numbers of derived deleterious alleles, and average genotype contaminations, are pretty much the same in all human populations. Perhaps, effects of population bottlenecks and random drift in the history of non-African populations were compensated by increased efficiency of selection against partially recessive deleterious alleles, which become exposed to selection as homozygotes in small populations.

Despite this quantitative uniformity, different populations mostly carry different substantially deleterious alleles. This is not surprising: an allele that reduces fitness by, say, 0.001, rarely persists in the population for over 1000 generations (~20 000 years), and ancestral lineages of many human populations separated much earlier. In contrast, all human populations share high-frequency, effectively neutral alleles, which were already present in the common ancestral population and which represent the majority of derived alleles in every genotype.

Among all species of animals investigated so far, *Homo sapiens* is at the lower end of the spectra of nucleotide diversities and effective population sizes (see Chapters 2 and 8). Thus, selection against mildly deleterious alleles is less efficient in humans than in many other species. Data on genome evolution indicate that the burden of slightly deleterious alleles is exceptionally high in humans and great apes, due to the decline of the effective population size in our ancestral lineage ~20 Ma. This decline was likely caused by an increased body size which evolved at that time – it is hardly surprising that a particular territory cannot support as many gorillas as mice.

However, even in humans random drift does not affect strongly the removal of substantially deleterious alleles with $s > 0.001$. Thus, in this case the mutation rate and strengths of negative selection are the only factors determining equilibrium genotype contamination. We do not know much about relative strengths of negative selection in different species. One may think that selection should be the strongest in species with very high fertility – only 1 newborn in millions will make it in cod or sea urchins, in contrast to more than 1 in 10 in humans, elephants, or blue whales. But this is, apparently, not the case – data on loss-of-function alleles of important, ubiquitous genes indicate that they tend to be equally common in genotypes of more and less fecund species. Perhaps, to win the struggle for existence against 10 well-provisioned baby elephants is not easier than against millions of cod larvae, most of which will perish purely by chance.

11.4 Variation Among Genotypes

Properties of an average genotype is not the whole story. Every genotype carries a unique complement of derived alleles, and we need to consider variation between genotypes. The basic properties of this variation are surprisingly simple, thanks to sexual reproduction. Indeed, free interbreeding between individuals and meiotic recombination keep randomizing the gene pool of a sexual population, making distribution of alleles at different loci approximately independent (see Chapter 2).

Let us start from considering the number of derived deleterious alleles in a genotype, without taking into account their individual effects. If rare alleles are distributed independently of each other, their total number in a genotype has a Poisson distribution. When its mean Q is large, Poisson distribution becomes very close to Gaussian distribution (Figure 3.3). The key property of a Poisson distribution is that its mean and variance are equal. Thus, the standard deviation of a Poisson distribution is $Q^{1/2}$, so that its mean is separated from 0 by $Q^{1/2}$ standard deviations. When Q is large the number of deleterious alleles in almost all genotypes is confined within a relatively narrow range, centered at the mean.

Consider, for example, deleterious missense alleles. The average genotype carries ~900 of such alleles, out of all 10 200 derived missense alleles (Table 11.5). Thus, assuming Poisson (and, thus, approximately Gaussian) distribution of the number of deleterious missense alleles in a genotype, its standard deviation is ~30. Because 99.7% of a Gaussian distribution is confined within 3 standard deviations from the mean, only 0.3% of individuals carry below 810 or above 990 deleterious missense alleles, and essentially nobody has less than 700 or more than 1100 (Figure 11.4).

Now let us take into account coefficients of selection against independently distributed deleterious alleles and consider, instead of their raw number, genotype contamination. When different alleles are deleterious to different degrees, the distribution of genotype contamination varies more widely, relative to the mean, than the Poisson distribution of their overall number. For example, if 99% of deleterious missense alleles have only negligible effects, the mean per genotype number of alleles that make substantial contributions to contamination is only 9, instead of 900. The number of such alleles still has a Poisson distribution, but its standard deviation is only 3 (instead of 30) times smaller than its mean. Generally, it is easy to show that the ratio of the variance of genotype contamination over its mean is equal to Mean[s] + Variance[s]/Mean[s], where s is the coefficient of selection against a deleterious allele sampled from the population.

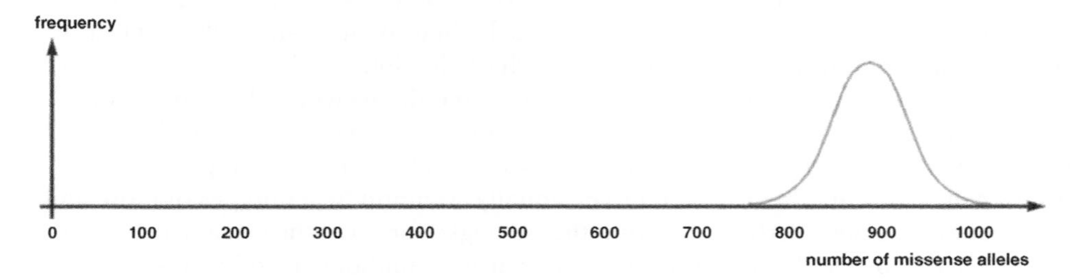

Figure 11.4 Distribution of the per genotype number of missense alleles predicted to be deleterious in human populations (after Fu *et al.*, *American Journal of Human Genetics*, vol. 95, p. 421, 2014).

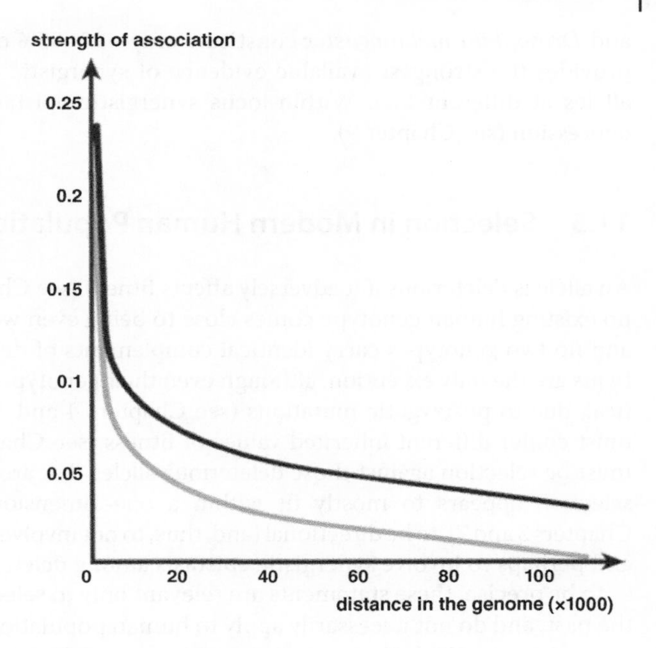

Figure 11.5 Strengths of associations between alleles at different loci decline with the distance between them more rapidly in African (shown in gray) than in non-African (shown in black) populations (after 1000 Genomes Project Consortium, *Nature*, vol. 526, p. 68, 2015).

Still, independence of distributions of different derived alleles within the population is not exact, because there are several processes that can all produce associations between alleles at different loci. Such associations are often denoted by a rather unfortunate term "linkage disequilibrium" (neither linkage nor deviations from any equilibrium are directly relevant here). Three of these processes need to be mentioned. The first is random drift. Of course, the signs of associations produced by random drift are also random: two derived alleles have equal chances to preferably occur together or to avoid each other. Because random drift is a rather slow process, it can produce strong associations only between tightly linked loci. Remarkably, in non-African populations these associations involve less tightly linked loci, because random drift was slower in larger African populations (Figure 11.5).

Secondly, if a population is the product of a recent admixture of two or more distinct populations, alleles of the same origin will be associated with each other. For example, in the USA alleles of African and non-African origin are still mostly found in the same genotypes. However, unless the barriers to interbreeding within the population are extremely strong, eventually its gene pool will be randomized.

Thirdly, associations between deleterious alleles may be created if selection against them involves synergistic epistasis (see Chapter 9). In this case, different deleterious alleles tend to avoid each other, which results in underdispersed, relative to Poisson, distribution of the number of such alleles in a genotype. Thus, synergistic epistasis between deleterious alleles (which is equivalent to diminishing returns epistasis between beneficial alleles) can also be called narrowing epistasis. Associations produced by epistatic selection between alleles at unlinked loci cannot be very strong, because sexual reproduction with random mating and free recombination halves them every generation. Still, recently Mashaal Sohail, Olga Vakhrusheva and Shamil Sunyaev demonstrated that the variances of the per genotype numbers of loss-of-function alleles in humans and *Drosophila melanogaster* constitute only ~90–95% of the mean. This observation provides the strongest available evidence of synergistic epistasis between deleterious alleles at different loci. Within-locus synergistic epistasis is revealed by inbreeding depression (see Chapter 9).

11.5 Selection in Modern Human Populations

An allele is deleterious if it adversely affects fitness (see Chapter 7). As we have just seen, no existing human genotype comes close to being even weakly perfect (see Chapter 10) and no two genotypes carry identical complements of deleterious alleles. Monozygotic twins are the only exception, although even their genotypes may not be completely identical, due to postzygotic mutations (see Chapters 4 and 13). Thus, different genotypes must confer different inherited values of fitness (see Chapter 3). In other words, there must be selection against those deleterious alleles that are not fixed in our species. This selection appears to mostly fit within a one-dimensional epistasis framework (see Chapters 3 and 7), to be directional (and, thus, to not involve sign epistasis, see Chapter 15), and perhaps to involve synergistic epistasis among deleterious alleles (see Chapter 9).

To be precise, these statements are relevant only to selection that acted on humans in the past, and do not necessarily apply to human populations living under Industrialized environments. With the exception of the most deleterious alleles that cannot persist for a long time, genetic composition of the contemporary human populations has mostly been shaped by mutation and selection that occurred many generations ago, under Preindustrial and even Pleistocene environments. Thus, pervasive population genetic signatures of selection against mildly deleterious alleles do not necessarily mean that this selection is ongoing, and are consistent with its partial or even complete recent relaxation. Darwin, the discoverer of selection, was sure that such relaxation indeed occurred (see epigraph).

Today we know that drastic alleles responsible for severe Mendelian and complex diseases are still selected against, because such diseases are inconsistent with life or reproduction despite the best available care. But could Darwin be right, as far as selection against mild alleles is concerned? Sequencing of genotypes of our closest extinct relatives, Neanderthals and Denisovans, demonstrated that genotype contamination of Neanderthals was very similar to ours. In contrast, Denisovans carried a slightly larger number of deleterious alleles, apparently because of their lower effective population size which resulted in a less efficient selection. Moreover, genotypes of modern humans from populations that switched to agriculture ~10 000 years ago and from populations that live as hunters–gatherers today or until very recently also carry essentially the same numbers of deleterious alleles. Thus, any relaxation of selection, if any, could have happened only very recently, and could not, in particular, be caused by the Agricultural Revolution (see Chapter 10).

Still, such comparisons cannot rule out relaxation of selection after the Industrial Revolution, because even a free accumulation of *de novo* mutations in the course of a few generations would be hard to detect by studying genotypes alone. Thus, to study selection under Industrialized environments, one needs to explicitly take phenotypes into account. In particular, we need direct data on fitness, the most important and complex among phenotypic quantitative traits.

Selection acts through differential survival and reproduction of individuals (see Chapter 7). Apparently, the Agricultural Revolution did not cause a dramatic change in the life history of humans. Data on skeletons of people who lived in southern Levant as hunter–gatherers 12 500–10 300 years ago and as Neolithic food-producers 10 300–7,500 years ago revealed very similar probabilities of dying before the age of 5 (13.9 and 16.5%, respectively) and life expectancies at birth (24.6 and 25.5 years, respectively). Of course, under these two environments selection could have favored rather different phenotypes; however, its overall strength apparently did not change.

In contrast, the Industrial Revolution caused dramatic changes in human life histories, known as the demographic transition. In Preindustrial Europe, ~20% of newborns died during the first year of their lives, and life expectancy at birth was 25–30 years. In the course of the demographic transition mortality, especially early in life, declined dramatically due to better sanitation, improved availability of food and, later, advances in medicine. After that, fertility also declined, due to gains in education and social status of women and the introduction of contraception and family planning. In Western Europe, the demographic transition began in the late 18th century, so that Darwin witnessed the beginning of this process.

An almost complete elimination of pre-reproductive mortality abolished the opportunity for selection through differential viability and, thus, definitely reduced its strength. In contrast, a lower average fertility does not necessarily imply a lower variation in relative reproductive successes of individuals. Moreover, the strength of selection is determined not by this variation *per se* but by its proportion that is due to influences of genotypes (see Chapters 3 and 7). We cannot say *a priori* how the demographic transition affected evolvability of inherited values of human fitness.

On the one hand, data on some populations indicate that under benign environments selection may be stronger, due to weaker random influences on fitnesses of individuals. Comparison of different species also suggests that low fecundity does not make strong negative selection impossible. On the other hand, there are data showing that fitnesses of different genotypes become closer to each other under benign environments, so that selection becomes weaker (G × E interactions, see Chapter 10). Also, when effective family planning is possible, the number of children may depend on cultural and economic factors more than on genotypes. Finally, strong selection in a population with low maximal fertility inevitably leads to the decline of its size and its eventual elimination (see Chapter 7), although this fact does not exclude the possibility of strong selection for a limited time. Clearly, only direct data can settle this issue.

Opportunity for selection (or Crow's index, see Chapter 7) was estimated for a number of human populations. Very important information is provided by a few populations, such as !Kung, Hadza, and Ache, which still live as hunter–gatherers. These populations are characterized by high pre-reproductive mortality (0.40 in !Kung, 0.46 in Hadza, 0.40 in Ache) and only moderate fertility (the average life-time number of children is 4.5 in !Kung, 4.6 in Hadza, and 8.2 in Ache women). Variation in this number is also moderate; for example, over 50% of Ache women have 7, 8, or 9 children. Values of Crow's index are 1.2, 1.5, and 0.8 for !Kung, Hadza, and Ache women, respectively, and most opportunity for selection comes from pre-reproductive mortality. These values may be typical for humans under Pleistocene environments. Opportunity for selection may be higher among men, due to a larger variance in the number of offspring, but probably not by very much.

In a number of European countries, high-quality records of births, marriages, and deaths in the course of the last several centuries are available. These records show that in several late Preindustrial populations, Crow's index was as high as in hunter–gatherers, or even higher: 2.3 in Finns, 2.0 in Italians, and 1.6 in Swedes. However, demographic transition reduced Crow's index substantially, and its current value is 0.5–0.6 in these three populations.

This pattern is not confined to developed countries. A rather recent demographic transition in Gambia also drastically reduced Crow's index there; and its current value for a number of contemporary populations in India is within the range of 0.4–0.6. In agreement with what Darwin said, this recent reduction of the opportunity for selection is primarily due to a much lower early mortality.

Such values of Crow's index must be typical for populations of stable size with low pre-reproductive mortality. Indeed, in a population where a woman produces, on average, two children, and their number has a Poisson distribution (so that its variance is equal to the mean), the value of Crow's index is 0.5. Usually, the number of children of women who had at least one child is underdispersed, relative to the Poisson density, which is, however, compensated by some excess of childless women. Still, even in China, where the government limits the number of children in a family, Crow's index is ~0.4, so that its further reduction is unlikely.

Thus, opportunity for selection in modern human populations that have already completed the demographic transition is definitely reduced, but still far from absent. A Crow's index of "only" 0.5 is sufficient for a rather strong selection. But does this opportunity for selection translate into real selection? To what extent is the number of children of a contemporary human determined by their genotype?

To answer this question, we need to estimate evolvability of inherited (instead of actual) values of human fitness (see Chapters 3 and 7). Such estimates, performed for a number of populations, involve two problems. On the one hand, due to the methods used, what gets estimated is usually the variance of transmittable, instead of inherited, values of fitness. This is likely to lead to an underestimation of the strength of selection; however, in the absence of sign epistasis this underestimation should not be large (see Chapter 3). On the other hand, there is a danger of overestimating the strength of selection, because many methods are based on resemblance between close relatives, which can be partially due to shared cultural and other environmental influences. However, recently NGS made it possible to use data on resemblance between genotypes and phenotypes of socially unrelated individuals (see Chapter 3), and such estimates must be free from this problem.

The data show that variation of fitnesses of contemporary humans is, to a substantial extent, due to their genetic differences. In the populations of the UK and the Netherlands, 10% of the variance in the total number of children of an individual is due to genetic factors that are associated with common alleles. Perhaps, the real proportion that the variance of inherited values of the number of children makes from the variance of their actual values is ~20% in this case (see Chapter 12). Similar estimates were obtained in several other studies. Moreover, an analysis of 300 years of genealogical data from Finland showed that the variance of the inherited number of children was not affected by the demographic transition, and was essentially the same before and after 1880. However, because the average number of children after 1880 was two times smaller than before, this implies that evolvability of fitness increased by a factor of 4.

Thus, some selection definitely keeps operating even in modern human populations. Unfortunately, data on the overall strength of selection tell us nothing about its specific properties (see Chapter 7). For our goals, genotype-level properties of selection are of particular importance. Indeed, only negative selection against rare, derived, deleterious alleles (see Chapter 7) is directly relevant to the Main Concern. In the course of history of the human (or any other) population, negative selection was much more common than positive selection. Thus, so far in this chapter I have ignored positive selection altogether, although human genetic variations possess a number of signatures of recent positive selection at individual loci. Often, these signatures are population-specific, reflecting human adaptation to local environments. However, data on genotypes alone tell us almost nothing about what happens in the last few generations and, thus, are mute on the strengths of negative and positive selection under Industrialized environments.

At the level of phenotypes, direct data on the properties of ongoing selection are already available. Apparently, directional selection was and remains rather common in Preindustrial and Industrialized populations. For example, in forager–horticulturalist Tsimane (Amazonian Bolivia) several basic features of personality were associated with higher fertility. Selection in a late Preindustrial French-Canadian population favored earlier onset of reproduction. Selection in Dutch people, already the tallest people on Earth, keeps favoring the larger height. Such directional selection can be due to adaptation to a recently changed environment (see Chapter 10). However, data on a large number of quantitative traits in many species, analyzed by Sarah Diamond and Joel Kingsolver, indicate that fitness landscapes for most of them are directional, instead of stabilizing, even under apparently stable environments, perhaps because selection acts against the mutational pressures. Still, stabilizing selection is also common in modern humans. For example, children born with weight deviating in either direction from the optimum (3500 g), which is close to the population mean, are at higher risk of early death. In all these cases, we do not know whether selection on a trait is real or only apparent (see Chapter 7).

It seems, however, that Industrialization did produce some important changes in the direction of selection. A number of studies detected ongoing selection for lower general intelligence. Similarly, a recent study demonstrated that alleles which are associated with more years of education currently reduce fitness. Such selection must be a very recent phenomenon, because high intelligence is a fragile adaptation that must have been protected by selection until very recently. Unfortunately, studies of selection at the level of phenotypes depend on direct data on phenotypes and, thus, cannot be extended deep into the past. Selection on traits that can be inferred from records of births, marriages, and deaths can be studied since late Preindustrial times, and selection on intelligence and many other traits only since the 20th century. In this respect, our knowledge of genotype- and phenotype-level properties of selection is complementary. Thus, we cannot rule out a possibility that contemporary selection is very different from that before the Industrial Revolution. If so, many alleles identified as mildly deleterious by the population genetic data may, in fact, increase fitness currently.

One way to gauge the strength of selection against deleterious alleles in modern humans, as well as in any other population, is to measure the depth of inbreeding depression, which is primarily due to recessive deleterious alleles (see Chapter 9). Marriages between first cousins are accepted in many cultures, and, worldwide, ~10% of children are produced in such marriages. Inbreeding depression definitely persists in modern humans. A higher frequency of autosomal recessive diseases in the offspring of consanguineous marriages directly follows from Mendelian mechanisms of inheritance, and most of these diseases still reduce fitness and wellness despite the best available health care. However, inbreeding depression in contemporary humans is not confined to Mendelian diseases. Inbred women tend to have fewer children, and inbreeding increases the risk of schizophrenia and a number of other diseases which reduce fitness, as well as wellness. Also, inbreeding moves average values of traits such as blood pressure, height, and intelligence quotient in offspring into the undesired directions. Still, according to some data inbreeding depression has declined in the last decades.

So far, we have ignored one major source of opportunity for selection – prenatal mortality. In industrialized countries only ~30% of conceptions by healthy young couples lead to birth, and the rest result in spontaneous abortion, mostly in the first weeks of pregnancy, when they usually go unnoticed. There are no data on prenatal mortality in humans living under pre-industrial environments, but it could not be much higher, as this would interfere with reproduction. However, it is also hard to imagine that a hunter–gatherer woman has lower chances of losing her pregnancy that a woman living in an American suburb.

The very phenomenon of high prenatal mortality may look like an evolutionary puzzle. For example, in healthy, outbred fishes over 99% of fertilized eggs successfully develop into larvae. Why cannot humans, who invest a lot more resources into each offspring and keep them, during 40 long weeks after conception, under an allegedly protective environment inside the womb, do at least equally well? Apparently, the difference is that, in the case of external fertilization without parental care, a mother already commits all the resources she is going to invest into the offspring before the egg is even fertilized. In contrast, in the case of internal fertilization, most resources are invested into an offspring after conception. Thus, it is in the evolutionary interest of the mother to enforce post-conception selection of offspring early in the pregnancy. This way, she can abort a weak fetus before investing too much time and other resources into it, and to conceive again. The metaphor "evolutionary interest" simply means that an allele which makes pregnant women to abort weak fetuses would spread in the population. Indeed, some seed plants, which, like mammals, invest a lot of resources into an offspring after fertilization, also abort 50% or more of the developing seeds.

Thus, we can expect prenatal mortality to lead to some selection against deleterious alleles. Indeed, we know that grave genetic defects such as aneuploidy, large copy number variations, or dominant early-acting lethal point mutations can cause early pregnancy loss, although their exact contribution remains unknown. However, we also know that many even fatal autosomal recessive Mendelian diseases appear in ~25% of children of heterozygous carriers, and, therefore, do not lead to a substantial rise of prenatal mortality. Still, a large-scale analysis of transmission of loss-of-function alleles in Iceland showed that the chances of a homozygote to be born are reduced by ~1%.

To summarize, selection against mutant alleles definitely keeps operating even in those contemporary human populations that enjoy the best available environments. Nevertheless, I believe that Darwin was, as usual, essentially right, and that the Industrial Revolution and the demographic transition resulted in a substantial relaxation of this selection. However, we do not really know. This issue can be resolved only by direct data relating fitnesses of modern humans to the properties of their genotypes. At the mutation–selection equilibrium, selection differential of the number of deleterious alleles in a human genotype must have been ~10 (see Chapter 9) – how large is it now?

How does the number of children of a modern human depend on contamination of their genotype? Hopefully, rapid progress in NGS methods will soon make it possible to produce very large data sets that are needed to answer such questions.

Further Reading

Adzhubei IA, Schmidt, S, & Peshkin, L 2010, A method and server for predicting damaging missense mutations, *Nature Methods*, vol. 7, pp. 248–249.

A sophisticated method for inferring the strength of selection against a missense allele.

Beauchamp, JP 2016, Genetic evidence for natural selection in humans in the contemporary United States, *Proceedings of the National Academy of Sciences of the USA*, vol. 113, pp. 7774–7779.

Currently, selection favors lower educational attainment in the USA population.

Courtiola, A, Pettayd, JE, & Jokelae, M 2012, Natural and sexual selection in a monogamous historical human population, *Proceedings of the National Academy of Sciences of the USA*, vol. 109, pp. 8044–8049.

Opportunity for selection in pre-Industrial Finns.

Fu, W, O'Connor, TD, & Jun, G 2013, Analysis of 6,515 exomes reveals the recent origin of most human protein-coding variants, *Nature*, vol. 493, pp. 216–220.

Derived alleles tend to be older in African than in non-African human populations.

Henn, BM, Botigué, LR, & Bustamante, CD 2015, Estimating the mutation load in human genomes, *Nature Reviews Genetics*, vol. 16, pp. 333–343.

An average human genotype carries ~10 mullers of contamination by rare deleterious alleles.

Larsen, EC, Christiansen, OB, & Kolte, AM 2013, New insights into mechanisms behind miscarriage, *BMC Medicine*, vol. 11: 154.

Prenatal mortality in modern humans is ~70%.

Li, AH, Morrison, AC, & Kovar, C 2015, Analysis of loss-of-function variants and 20 risk factor phenotypes in 8,554 individuals identifies loci influencing chronic disease, *Nature Genetics*, vol. 47, pp. 640–642.

An average human genotype carries ~100 loss-of-function alleles of protein-coding genes.

Rasmussen, MD, Hubisz, MJ, & Gronau, I 2014, Genome-wide inference of ancestral recombination graphs, Retrieved January 9, 2017, *PLoS Genetics*, http://dx.doi.org/10.1371/journal.pgen.1004342.

A sophisticated method for inferring the age of an allele.

Sohail, M, Vakhrusheva, OA, & Sul, JH 2017, Natural selection narrows the distribution of the genomic number of rare deleterious alleles, *Science* (accepted).

Evidence for synergistic epistasis between deleterious alleles.

The UK10K Consortium 2015, The UK10K project identifies rare variants in health and disease, *Nature*, vol. 526, pp. 82–90.

Analysis of genotypes of nearly 10 000 people from the UK.

12

Our Imperfect Wellness

If all Trees were Writers or Clerks, and all Branches were Pens, and all Hills were Books, and all Waters were Ink, yet they could not sufficiently describe the lamentable Misery which Lucifer ... has brought into ... that World wherein he was created.

Jacob Boehme, *Aurora*, 1612.

Under the best Industrialized environments, ~99% of those who are born now will reach the age of 30. Before this age, a substantial proportion of people experience at least one chronic disease. Still, only ~5% of individuals under the age of 30 are disabled, and over two-thirds are in excellent or very good health. More detailed descriptions of wellness by quantitative traits reveal its substantial imperfection and variation, which is to a large extent caused by genotypes. Wellness-impairing alleles that can cause Mendelian diseases are almost always derived and deleterious. The same is likely true for wellness-impairing alleles that affect complex traits, although this is still only a hypothesis. Many alleles that impair wellness only through their effects on complex traits affect molecular function only mildly and are not strongly deleterious, and, therefore, can reach substantial frequencies. Every human genotype harbors a large number of wellness-impairing alleles, and the genetic architectures of the Industrialized wellness and Preindustrial fitness are probably similar.

12.1 Qualitative Characteristics of Wellness

Boehme, a Christian mystic, blamed Lucifer for the misery that pervades our world (see epigraph). Natural sciences are mute on transcendental roots of things, and instead strive to reveal their proximal causes. Biologically, the key cause of suffering is vulnerability of phenotypes of organisms to insults of all kinds, including those due to deleterious alleles in their genotypes.

In the previous chapter, we considered only one, albeit fundamental, phenotypic trait – fitness. Here we concentrate on wellness and its particular facets (see Chapter 10). This requires a change in perspective. Because genetic variation in contemporary populations has been shaped by selection, fitness landscapes that existed over a long period of time can be studied using data on the current frequencies of genotypes (see Chapter 11). In contrast, such data do not directly shed any light on wellness

Crumbling Genome: The Impact of Deleterious Mutations on Humans, First Edition. Alexey S. Kondrashov.

landscapes. Thus, in this chapter I proceed not from genotypes to phenotypes, as in Chapter 11, but in the opposite direction. I start from data on human wellness-related phenotypic traits, mostly ignoring underlying genotypes. After this, I consider overall effects of heredity on these traits, and, finally, discuss specific genetic mechanisms that are behind these effects. This approach has already been used in Chapter 3.

My consideration of human wellness is limited in three ways. First, I ignore a surprisingly high prenatal mortality, of which very little is known (see Chapter 11), and consider only wellness of those who are born. Secondly, I ignore effects of poor environments, and concentrate on wellness under Industrialized environments. Thirdly, I ignore deterioration of wellness due to aging, and only deal with phenotypes of young people. Rare deleterious alleles likely play a particularly prominent role in imperfection of wellness of young people who live under good conditions, which makes it especially relevant to the Main Concern.

The basic facets of wellness can be qualitatively characterized by "coarse-grained" phenotypic traits that can accept only a small number of states, one of which, such as "alive" or "excellent health" is the best possible option. Phenotypic imperfection of the population in terms of such a trait can be directly assessed (see Chapter 10). Qualitative characteristics of wellness, or rather of its imperfection, are concerned with four different but related phenomena – death, diseases, disability, and less-than-excellent health. Let us consider them in this order.

The most fundamental facet of wellness, life versus death, is inherently two-state. After a dramatic decline of infant and childhood mortality brought by the Industrial Revolution (see Chapter 11), today over 99% of those who are born in the most advanced industrialized countries will reach the age of 30 (Figure 12.1).

Some diseases, such as retinal detachment or stroke, are naturally discrete, although their cases can be of different severity. Other conditions, however, such as intellectual disability or hypertension, correspond to deviating values of quantitative traits (see Chapter 3), which makes gray areas, where diagnosis is ambiguous, unavoidable. Nevertheless, as a rough but still useful approximation, any disease can be described by a two-state trait – no disease or disease.

Diseases can be subdivided into congenital and others. A congenital disease (CD) already "exists" at birth, and other diseases appear only later in life. Admittedly, this concept is fuzzy. Of course, a disease that is already unambiguously manifested at birth, such as an anatomical birth anomaly (also known as a birth defect or a congenital malformation or anomaly or abnormality), exists at this point. It also makes sense to say that a fully penetrant (see Chapter 3) Mendelian disease (see Chapter 2) exists at birth, and even at conception, although some of them, such as many cancer predisposition syndromes (see Chapter 5), may remain clinically mute for decades. But what about genetically complex diseases? Although an autistic spectrum disorder obviously cannot be diagnosed in a newborn, it seems that in many cases its eventual manifestation is essentially preordained very early. Still, CDs are traditionally understood restrictively, to encompass only Mendelian diseases and birth anomalies.

Many CDs can be diagnosed prenatally. The frequency of induced abortions due to prenatal diagnosis of a CD varies widely between countries, reaching ~0.5% of all registered pregnancies in some industrialized countries (see Chapter 14). I count pregnancies terminated due to a CD towards the birth frequency (prevalence among the newborns) of the CD. Among ~1% of newborns who do not reach the age of 30, about a half die of a CD.

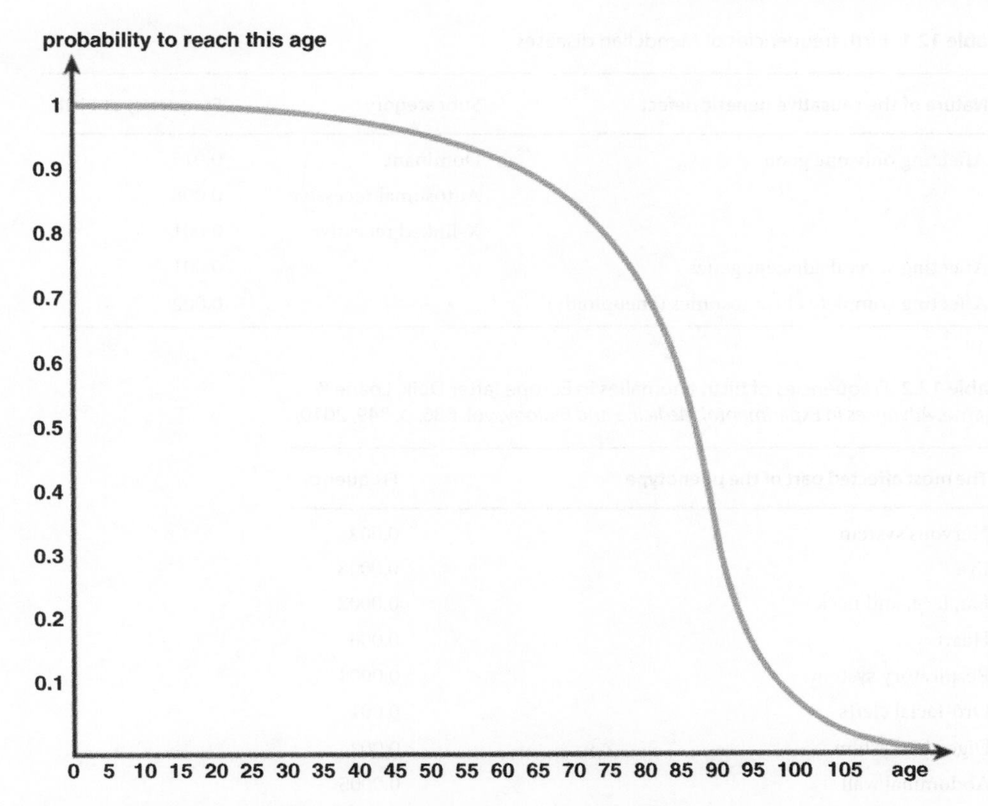

Figure 12.1 Estimated probability of a person born in Japan in 2014 to live until a particular age (data from http://www.ipss.go.jp/p-toukei/JMD/00/index-en.html).

Table 12.1 presents estimates of overall birth frequencies of the main categories of Mendelian diseases, including those that are manifested only later in life. We can see that ~3% of humans are born with a Mendelian disease. This estimate, however, is subject to a substantial uncertainty, due to two reasons.

First, severity of Mendelian phenotypes varies continuously. Such a phenotype can cause prenatal death (which is not counted as a disease here), death at early or later age, life-long disability, a treatable condition, a mild inconvenience (at least under Industrialized environments), or only a barely perceptible alteration. Borderline cases appear due to inherently mild diseases (colorblindness), mild cases of potentially severe diseases that are caused by alleles that disturb the function of the gene only modestly (mild hemophilias), and low expressivity (see Chapter 3) of an allele that sometimes causes a severe disease (some forms of Charcot–Marie–Tooth neuropathy). Also, penetrance of deleterious alleles varies from 100% to very low. Thus, there is no "right" way to draw the lines separating Mendelian diseases from complex diseases and from normal phenotypic variation.

Secondly, frequencies of a number of autosomal recessive diseases vary widely among populations, due to different prevalences of inbreeding and to local overdominance of some disease-causing alleles (see Chapter 14). In contrast, frequencies of dominant and X-linked recessive Mendelian diseases are rather uniform, because they are determined

Table 12.1 Birth frequencies of Mendelian diseases.

Nature of the causative genetic defect	Subcategory	Frequency at birth
Affecting only one gene	Dominant	0.015
	Autosomal recessive	0.008
	X-linked recessive	0.001
Affecting several adjacent genes		0.001
Affecting complete chromosomes (aneuploidy)		0.002

Table 12.2 Frequencies of birth anomalies in Europe (after Dolk, Loane & Garne, *Advances in Experimental Medicine and Biology*, vol. 686, p. 349, 2010).

The most affected part of the phenotype	Frequency
Nervous system	0.002
Eye	0.0003
Ear, face, and neck	0.0002
Heart	0.006
Respiratory system	0.0004
Oro-facial clefts	0.001
Digestive system	0.002
Abdominal wall	0.0005
Urinary	0.003
Genital	0.002
Limb	0.0004
Musculoskeletal	0.0008
Other	0.0006
Teratogenic	0.0001

by the equilibrium between mutation and negative selection (see Chapter 9). Both these factors acted about the same in all populations, at least until effective treatments of some Mendelian diseases became possible in countries with advanced health care.

Table 12.2 presents data on birth frequencies of salient birth anomalies that are typical for industrialized countries. About 1/5 of birth anomalies are due to known Mendelian diseases, and such cases are not included in these data. Some of the included cases are likely due to undiagnosed or yet undescribed Mendelian diseases. However, most birth anomalies lack a simple genetic cause and, instead, should be viewed as disease states of two-state complex traits (see Chapter 3). As it was the case with Mendelian diseases, severity and treatability of birth anomalies varies widely.

Table 12.2 suggests that a salient birth anomaly is present in ~2% of newborns. However, this is only the tip of the iceberg. Mild and borderline cases of birth anomalies, as well as of any complex disease, are more frequent than severe cases. For example, recent data show that at least 5% of newborns have a cardiovascular anatomical anomaly, detectable by modern diagnostic tools. Many of these mild anomalies will remain clinically mute, but they all confer increased risk of an overt disease later in life (and likely reduce wellness). Due to this continuum, the line between disease and no disease in the case of birth anomalies is perhaps even more arbitrary than in the case of Mendelian diseases. Still, at least ~10% (and, perhaps, even a majority) of newborns are likely to carry an individually rare anatomical anomaly that is not totally harmless.

Let us now consider diseases of young people. I ignore infectious diseases, diseases associated with pregnancy and childbirth, and acute diseases that leave no lasting damages. Only chronic diseases, defined as those lasting for at least 3 months, are taken into account. Thus, if an individual has occasional colds, or broke a leg once, this is ignored, while being prone to recurrent respiratory infections or having brittle bones are symptoms of chronic diseases.

Table 12.3 presents frequencies of broad classes of chronic diseases, according to the WHO classification, that may occur during the first three decades of life in industrialized countries. Only mental, behavioral, and neurodevelopmental diseases are subdivided further. Many chronic diseases, such as schizophrenia, are life-long, so that after the onset of symptoms a person will (almost) never become disease-free. However, other chronic diseases can be either cured (many cases of cancer) or eventually go away mostly on their own (childhood asthma). Thus, not all cases of diseases on which figures presented in Table 12.3 are based are actually present in 30-year-olds.

Disease frequencies shown in Table 12.3 are subject to a considerable uncertainty, due to at least two reasons. First, there are some real differences between frequencies of a disease between populations living under different Industrialized environments. Secondly, borderline cases are common. For example, it is often not clear whether an adolescent with behavioral problems should be diagnosed with ADHD (attention deficit hyperactivity disorder), and diagnostic criteria differ widely between countries. Understanding diseases too broadly is involved with the risk of "medicalization" of the society and overtreatment, and understanding them too narrowly with the risk of neglecting treatable conditions and cases.

Broad classes of diseases, listed in Table 12.3, hide the fact that many individual diseases are rare. A chronic disease is regarded as rare if its frequency is below 0.0005 (1:2000). Most CDs fit this definition. A majority of people with a CD now live past 30 years, and a substantial proportion of rare diseases in 30-year-olds are CDs. However, there are also many rare diseases that are not included in the CD category. The total number of rare diseases is ~7000, and their total frequency is ~6%. Still, very little is known about most of them. A high proportion of rare diseases have various degrees of intellectual disability among the symptoms.

Many diseases mostly affect older people. Thus, life-long chances of having, for example, a cardiovascular disease are much higher than what is shown in Table 12.3, although anatomical warning signs of vascular damage can often be already detected in asymptomatic young adults. In contrast, some other non-congenital diseases are usually manifested early. In particular, ~80% of all cases of neurodevelopmental disorders are diagnosed before the age of 30. Some 30-year-old people are (or were) affected by more than one chronic disease, so that the total frequency of those affected by at least one disease is less than the sum of all frequencies.

Table 12.3 Frequencies of diseases and impairments that are manifested before the age of 30.

Disease	Frequency	Comment
Neoplasms (benign and cancerous tumors)	0.01	
Diseases of the blood and blood-forming organs	0.03	
Endocrine, nutritional, and metabolic diseases	0.05	This frequency would be ~0.2, if obesity were included
Mental, behavioral, and neurodevelopmental disorders	0.4	Estimates of frequencies of some of these disorders vary particularly widely
Intellectual disability, severe	0.005	
Intellectual disability, moderate	0.01	
Schizophrenia	0.005	
Autism spectrum disorders	0.01	
Bipolar disorder	0.01	
Major depression	0.10	
Anxiety disorders	0.15	
Behavioral disorders	0.10	
Diseases of the nervous system	0.15	Frequency of migraine is ~0.1
Diseases of the eye	0.01	This frequency would be ~0.3, if myopia were included
Diseases of the ear	0.02	
Diseases of the circulatory system	0.01	
Diseases of the respiratory system	0.10	Frequency of asthma in children is 0.08
Diseases of the digestive system	0.04	
Diseases of the skin and subcutaneous tissue	0.12	Frequency of eczema is ~0.1
Diseases of the musculoskeletal system and connective tissue	0.01	
Diseases of the genitourinary system	0.02	

Despite all uncertainties, it is safe to say that a majority of 30-year-old people suffer, or suffered, from at least one chronic disease. This conclusion may seem to be weird, because we are not used to associating young age with widespread illness. However, there is no contradiction here: many of these diseases are mild, and do not severely restrict the ability of the affected people to function and enjoy life.

Thus, the overall state of wellness may be better characterized by data on disability and health. Table 12.4 shows self-reported statistics of disability in Canada. The definition of disability used in this survey "includes anyone who reported being 'sometimes', 'often' or 'always' limited in their daily activities due to a long-term condition or health problem, as well as anyone who reported being 'rarely' limited if they were also unable to do certain tasks or could only do them with a lot of difficulty".

Table 12.4 Self-reported disability in Canada in 2012 (data from http://www.statcan.gc.ca/pub/89-654-x/89-654-x2013002-eng.htm).

Age group (years)	Percent disabled
15–24	4.4
25–44	6.5
45–64	16.1
65–74	26.3
75 and over	42.5

Table 12.5 Self-reported health in Canada in 2012 (data from http://www4.hrsdc.gc.ca/.3ndic.1t.4r@-eng.jsp?iid=10#M_3).

Age group (years)	Excellent or very good	Good	Fair or poor
12–19	69.6	26.2	4.2
20–34	68.4	25.8	5.8
35–44	64.8	27.6	7.6
45–64	56.1	31.2	12.7
65 and over	44.2	33.6	22.2

Institutionalized individuals (~1% of the total population) were not included. Over a quarter of all cases of disability were "very severe". Comparison of Tables 12.3 and 12.4 clearly indicates that not all chronic diseases, past or even ongoing, lead to self-reported disability. Direct studies also showed that less than 30% of cases of chronic diseases are associated with disability, because people often manage to cope well even with a serious disease.

Finally, wellness can be characterized by health. Table 12.5 presents data on self-reported health in Canada. Such data partially depend on culture. For example, in Japan, despite its exceptionally high life expectancy, self-reported health is, on average, much poorer than in Canada and many other industrialized countries.

Nevertheless, comparison of Tables 12.3 and 12.5 clearly shows that a chronic disease, or history of a chronic disease, does not always prevent a person from regarding their health as at least good. This is in line with the current understanding of the concept of health, which emphasizes not a complete well-being and freedom from illness (as did the definition that was formulated in 1948 by the WHO) but the ability to adapt and self-manage. Thus, it is not surprising that fair or poor self-reported health is about as common as at least some degree of disability. Of course, different people with "the same" disability, such as loud tinnitus, can adapt, self-manage, and function at very different levels.

Thus, the overall qualitative picture of human wellness is rather complex. On the positive side, the current human gene pool is good enough to ensure survival of a vast majority of people until the beginning of aging and well beyond that, at least under relatively benign Industrialized environments. On the negative side, a high proportion of even young people is affected by at least one recognizable chronic disease. However, not all these diseases are severe, and only ~5% of young people are strongly impaired, being at least to some extent disabled and having only fair or poor health. Over 2/3 of young people in industrialized countries are generally doing rather well. From this perspective, the colloquial meaning of "normal" as "good" is currently justified.

12.2 Quantitative Traits

Qualitative characteristics of human wellness by coarse-grained traits are important, but do not tell the whole story. There are a lot of important differences even between disease-free, disability-free and excellent-health people. One can have intelligence quotient, IQ = 85 and still belong to this category, although this IQ is one standard deviation below the average. Such a person is not going to be able to obtain, say, a PhD in physics, and in the USA their expected income is ~30% below the average. Thus, we need to characterize wellness quantitatively.

Unfortunately, this is easier said than done. Wellness is much more difficult to quantify than fitness. In the case of fitness, we know exactly what needs to be measured – the efficiency of reproduction, which could be characterized crudely by the number of offspring (see Chapter 7) and more precisely by a quantity known as the Malthusian parameter. In contrast, there is no obvious quantity that can describe wellness of a person. Using fitness as a proxy clearly would not work in a modern, industrialized environment, where a healthy, happy, and productive individual often decides to have only one child, or even to remain childless. Relying on subjective self-assessment, acceptable in the case of coarse-grained traits such as disability and health, also would not work, because we do not have a clear idea of the maximal possible wellness. I may feel perfectly well, but still fail miserably as a triathlete or a violinist.

To quantify wellness objectively, we need some kind of a comprehensive phenotypic score, a quantitative trait that takes into account all the wellness-related facets of a phenotype. Perhaps, these facets should be parts of the phenotype ("archetypes") which perform distinct, wellness-related functions. Decomposing the phenotype into quantitative traits may not be suitable to deal with such facets, which may be better characterized by more complex traits that directly correspond to shapes, dependencies, and so on (see Chapter 3). However, our current understanding of human, or any other, phenotypes is not enough for this approach. We even do not know what set of quantitative traits could provide an adequate description of a human phenotype. Clearly, this set must be large. It would take a lot of measurements to fully characterize Michelangelo's *David* – and it is just a shape. In a recent paper, Krapohl and colleagues attempted to comprehensively characterize human behavior ("the behavioral phenome"), and used 50 traits for this purpose. Of course, even this number of quantitative traits could not fully describe a human. Still the overall number of truly independent traits cannot be astronomical – after all, there are only ~20 000 protein-coding genes in the human genome (see Chapter 1).

Thus, let us set a more modest goal, and consider wellness-related quantitative traits individually, together with the corresponding wellness landscapes (see Chapters 7 and 10). Despite their biological heterogeneity, almost all quantitative traits (or their logarithms) have population distributions that are close to Gaussian (see Chapter 3).

Together, the distribution of a quantitative trait and the corresponding quality (fitness or wellness) landscape produce the distribution of the quality (Figure 7.13). We are particularly interested in imperfection that characterizes these distributions.

Almost all quantitative traits are variable. Unless the wellness landscape is flat, variation of a trait causes imperfection, because some individuals must possess imperfect values of the trait. As long as there are very many more-or-less independently varying quantitative traits, the phenotype of every individual must deviate substantially from the values that confers the highest wellness with respect to many traits. If so, why is not imperfection of every phenotype prohibitive? Perhaps, the number of "real" traits is not very high, and effects on wellness of many easily-measurable quantitative traits are mostly apparent, instead of real (see Chapter 7)? We do not really know.

Some variation is ubiquitous, but evolvabilities of different quantitative traits are strikingly different (see Chapter 3). On the one hand, traits such as blood pH are rather uniform: its average value is 7.4, and deviations of more than 0.1 occur only when an individual is gravely ill. On the other hand, many apparently important traits, nevertheless, vary widely. For example, blood concentration of von Willebrand factor, a crucial component of the blood coagulation system, varies within a factor of 6 in healthy people, from 40 to 240% of the population average. We do not understand the reasons for such contrasts.

Wellness landscape associated with a quantitative trait can be either stabilizing or directional (see Chapter 7). Purely directional wellness (and fitness) landscapes probably do not exist, even for traits that seem to be directly related to wellness, such as longevity or general intelligence. When the value of such a trait increases over a certain limit, it is likely that other facets of the phenotype, and, therefore, its wellness would suffer, due to trade-offs, imposed by the overall design of the phenotype (see Chapter 10). For example, there are data that worms selected for extremely long lifespan are inept. Still, if the fitness or wellness maximum is so far away from the population average that essentially all individuals are on the same side of it, the landscape experienced by the population is effectively directional.

Under a stabilizing wellness landscape, imperfection exists solely due to variation of the trait. Ascertaining imperfection caused by deviations from the average, which in this case is also perfect, value of the trait is easy, because perfection is known. In contrast, under an effectively directional wellness landscape imperfection is mostly due to discrepancy between the average and the perfect values of the trait, and variation does not affect it much (Figure 12.2). When the perfect value of the trait is way outside the range of values that are actually present in the population, it may be impossible to directly estimate imperfection (see Chapter 15). Estimating evolvability of wellness is always easier, as it depends only on its mean and variance, which can be measured. However, in contrast to the case of fitness, where knowing evolvability is crucial as it characterizes the strength of selection, evolvability of wellness is much less important.

Purely stabilizing fitness and wellness landscapes are definitely possible. For example, the average blood pH is very close to perfect. In fact, one could expect stabilizing selection (and, perhaps, stabilizing wellness landscapes, too) to be the most common case: when the average value of the trait in the population deviates from the perfect one, selection works to reduce, and eventually eliminate, the discrepancy (see Chapter 10). Nevertheless, it seems that in many species effectively directional fitness landscapes are

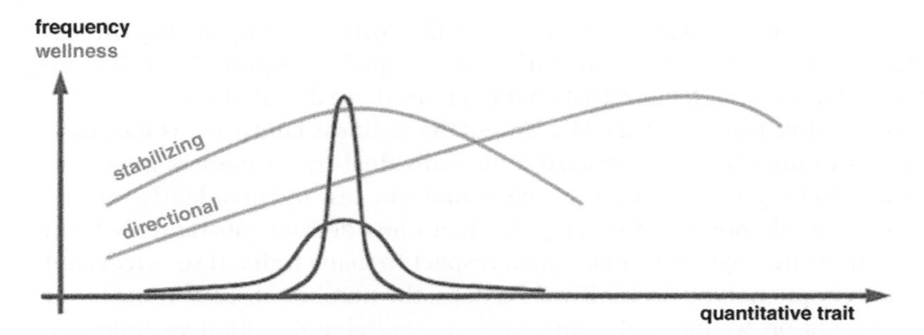

Figure 12.2 Under a stabilizing wellness landscape, imperfection appears only because values of the quantitative trait in some individuals deviate from the average. Thus, a wider distribution of the trait leads to a larger imperfection. In contrast, under (effectively) a directional wellness landscape, imperfection appears because all individuals deviate from the perfect value of the trait in the same direction and, thus, does not depend much on variation of the trait. As a result, the two distributions of the trait shown in the figure produce essentially the same imperfection.

more common than stabilizing (see Chapter 11). The same is apparently true for human wellness landscapes under Industrialized environments. Discrepancies between average and perfect values of quantitative traits are the rule – think of blood pressure, height, BMI, general intelligence, and longevity. Deleterious mutations is one of the possible mechanisms of this discrepancy and resulting imperfection (see Chapter 10).

Among straightforward human quantitative traits, general intelligence, a quantity behind correlation between various measures of human intelligence, is likely to be a close proxy for wellness. Perhaps, academic achievement, which reflects many genetically influenced traits, on top of intelligence, would be even a better proxy, but this trait is currently less well-studied. General intelligence is measured as IQ, which is standardized in such a way that its average is 100 and standard deviation is 15. High IQ is correlated with better overall health and increased longevity. Of course, IQ is not synonymous to wellness – one can be very smart and depressed, for example – but it appears to be closer than other obvious traits.

Phenotypic imperfection of the human population in terms of IQ is huge. There are individuals with IQ > 170 (0.03% of all individuals), and they do not generally suffer from any apparent problems. Thus, the perfect IQ, although likely not infinite, appears to be at least 170, implying that the average constitutes only ~60% of it and is almost 5 standard deviations below. Directional wellness landscapes of a number of other quantitative traits whose average values deviate substantially from those that confer maximal wellness must also lead to imperfection. Still, we do not know how to quantitatively translate these deviations into impairment of wellness. After all, one does not need a stratospheric IQ to be happy.

12.3 Contributions of Heredity and Environment

Qualitative and quantitative imperfection of human wellness is relevant to the Main Concern only to the extent to which it is caused by weak imperfection of genotypes (see Chapter 10). Obviously, genotypes are the root cause of Mendelian diseases: an individual with a weakly perfect genotype would never have hemophilia. But what is the contribution of genotypes into impairments of wellness due to imperfect values of complex traits?

Let us first consider a two-state trait that is so closely related to wellness that its states can be simply called "well" and "unwell". If so, phenotypic imperfection of wellness caused by this trait is equal to the frequency of the "unwell" state P, and genotypic imperfection I_{gen} is equal to the value of imperfection (see Chapter 7) that characterizes the distribution of inherited values of the trait. The distribution of inherited (as well as of transmittable) values of even a two-state trait is generally continuous (Figure 3.7). Still, I_{gen} is simply the normalized difference between the frequency of the "well" state among individuals with the perfect genotype and P. For example, if the "unwell" state occurs in 0.01 of all individuals and in 0.003 of individuals with the perfect genotype, $I_{gen} = (0.997 - 0.990)/0.990 \sim 0.007$ (see Chapter 7), because the denominator can be ignored, as long as the "unwell" state is rare. Thus, when even the perfect genotype does not always produce the "well" state, I_{gen} is below P.

P is directly observable. In contrast, for want of individuals with perfect genotypes (see Chapter 10), we cannot observe frequency of the "unwell" state among them. Fortunately, it is easy to show that this frequency (or, in other words, the minimal inherited value of the trait) cannot exceed $P(1 - h^2_i)$, where h^2_i is heritability of inherited values of the trait (Figure 12.3). Thus, I_{gen} must be at least Ph^2_i.

Heritabilities can be estimated from data on the available genotypes (see Chapter 3) and are definitely substantial. For example, a woman who had a child with a birth anomaly has an increased risk that the next child by the same father will have an anomaly of the same kind, but this risk is reduced by a factor of 3 or 4 if the two children have

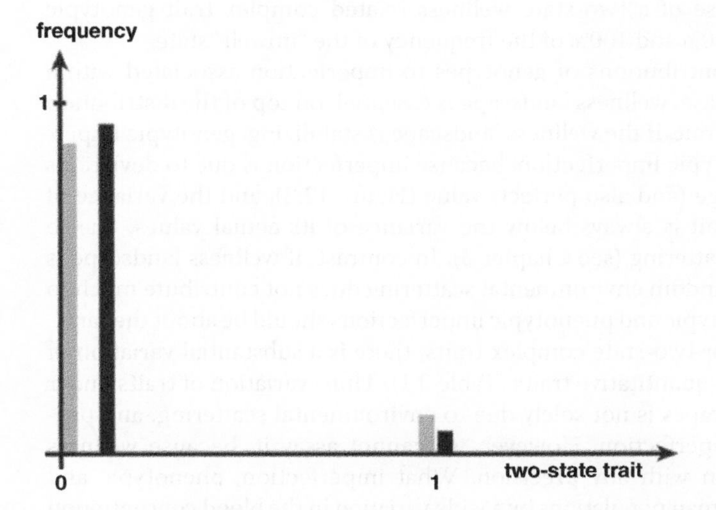

Figure 12.3 Distribution of actual values of a two-state trait with mean P is always concentrated in two points, 0 ("well") with probability $1 - P$, and 1 ("unwell") with probability P (gray bars). Assuming that P is small, the variance of this distribution is $\sim P$. I_{gen} is less than P only if the minimal inherited value of the trait is above zero. This minimal value is the highest, under the condition that the variance of the distribution of inherited values of the trait is Ph^2_i, when the distribution is concentrated at just two points: $P(1 - h^2_i)$ with probability $1 - Ph^2_i$, and 1 with probability Ph^2_i (black bars).

different fathers. Quantitative estimates of heritabilities of transmittable values of a number of two-state traits are high (Table 3.1), and heritabilities of inherited values must be even higher. Thus, it is fair to assume that I_{gen} constitutes at least 50% of the frequency of the "unwell" state of a typical wellness-related two-state trait.

On top of estimates of heritability, there is other evidence of a large contribution of genetic factors into the "unwell" states of two-state complex traits, provided by a recently discovered role of *de novo* mutation in the origin of diseases. For example, ~15% of cases of schizophrenia are triggered by *de novo* mutations in genotypes of affected people (see Chapter 13). Thus, non-genetic factors alone cannot be responsible for more than 85% of cases of schizophrenia. Very likely, genotypes contribute to much more than 15% of cases, because pre-existing deleterious alleles likely play a role even when the disease was not provoked by a *de novo* mutation. I believe that if the "unwell" state of a complex two-state trait is triggered by *de novo* mutations in more than, say, 10% of cases, this implies that the perfect, mutation-free genotype would (almost) never produce this state, because such mutations would be phenotypically mute in the absence of pre-existing deleterious alleles. If so, genotypic imperfection is not much below phenotypic imperfection.

The contribution of definitely non-genetic factors to the incidence of the "unwell" state of a two-state trait also sheds light on the role of genotypes in imperfection. For example, in young people ~30% of cases of disability can be attributed to environmental insults (accidents, consequences of infections, etc.). Even individuals with weakly perfect genotypes must be prone to such insults, perhaps to a somewhat smaller extent (e.g., they would get sick less often even if infected, and would recover sooner). Thus, genotypic imperfection due to disability likely constitutes no less than ~70% of the phenotypic imperfection, which is known to be 0.05 (Table 12.4). To summarize, it is likely that in the case of a two-state wellness-related complex trait genotypic imperfection is between 50% and 100% of the frequency of the "unwell" state.

Let us now consider contributions of genotypes to imperfection associated with a quantitative trait. In this case, wellness landscape is essential, on top of the distribution of inherited values of the trait. If the wellness landscape is stabilizing, genotypic imperfection is less than phenotypic imperfection, because imperfection is due to deviations of the trait from its average (and also perfect) value (Figure 12.2), and the variance of inherited values of the trait is always below the variance of its actual values, due to random environmental scattering (see Chapter 3). In contrast, if wellness landscape is (effectively) directional, random environmental scattering does not contribute much to imperfection, so that genotypic and phenotypic imperfections should be about the same.

As it was for the case for two-state complex traits, there is a substantial variation of inherited values of human quantitative traits (Table 3.1). Thus, variation of traits under stabilizing wellness landscapes is not solely due to environmental scattering, and produces some genotypic imperfection. However, we cannot assay it, because wellness landscapes are not known with any precision. What imperfection, phenotypic and genotypic, is induced in human populations by a wide variation in the blood concentration of von Willebrand factor? We have no idea.

Even less is known about the magnitude of genotypic imperfection associated with quantitative traits under directional wellness landscapes. In this case, genotypic imperfection is determined primarily by the inherited value of the trait of individuals with the weakly perfect genotype, which can be far outside the range of values that are present in the population. If so, treating the best observable value of the trait as the perfect one will cause underestimation of genotypic imperfection (see Chapter 15).

So far, we have considered human imperfection under Industrialized environments, which are relatively benign but definitely not optimal. What difference would the optimal environment make? By definition, the phenotype produced by a particular genotype under the optimal environment must be the best. Thus, phenotypic imperfection must be the lowest under the optimal environment.

The same is likely to be true for genotypic imperfection, due to genotype–environment (G × E) interactions (see Chapter 10). Of course, we have no direct data on this, as we had no chance to study weakly perfect genotypes under any environment, to say nothing about the optimal one. Still, some undesirable phenotypes, such as tuberculosis or obesity, may disappear altogether under the optimal environment, which would eliminate the corresponding imperfections. In contrast, it is possible that the optimal environment would not lead to a large reduction of genotypic imperfection due to diseases such as schizophrenia.

12.4 Wellness-impairing Alleles

As we have just seen, genotypes are, to a substantial extent, responsible for imperfection and variation of human wellness and its facets. What do we know about alleles that are behind this responsibility? There is a striking contrast between the current states of knowledge about alleles that affect simple versus complex phenotypic traits.

A simple phenotypic trait is, by definition, primarily affected by alleles of one gene or by large-scale alleles that disturb a particular group of consecutive genes. In almost all simple human traits, one state is rare and pathological and corresponds to a Mendelian disease (see Chapter 2). In the course of the last decades, responsible genes were identified for over 3000 Mendelian diseases, and the total number of known pathogenic alleles is now ~200 000. Currently, a causative allele is found in almost every case of a Mendelian disease that has been carefully investigated. Usually, such an allele produces a major disturbance of some molecular function.

In contrast, a complex phenotypic trait can be affected by alleles of multiple unlinked genes. Even identifying all the genes that can harbor these alleles is challenging. We already know that complex traits are, indeed, complex – alleles of dozens of unlinked genes can contribute to conditions such as asthma or Crohn's disease, and alleles of hundreds of genes can affect quantitative traits such as height or blood pressure. Both two-state and quantitative traits that describe functioning of the brain, such as schizophrenia and general intelligence, are particularly complex, which is not surprising, because the brain is by far the most transcriptionally complex human organ. However, the current state of knowledge is far from being final: hundreds of genes whose mutations can contribute to intellectual disability are still discovered every year.

Identifying complex trait loci (CTLs; see Chapter 3) within genes that affect a complex trait is even more difficult, and only the very first steps have been made so far towards this goal. In the case of a complex disease, alleles of each particular gene and at each particular CTL contribute to only a minority of cases. Conversely, each case appears due to effects of an essentially unique set of many causative alleles, and identifying this set has so far been impossible.

Still, some general patterns are already clear. In terms of changes in the DNA sequence, wellness-impairing alleles as a whole are not too different from the overall set of human polymorphic loci (see Chapter 11). All kinds of changes – from single-nucleotide substitutions to large-scale deletions and insertions – can affect the phenotype. Large-scale alleles are disproportionately common among wellness-impairing alleles, because a small-scale allele has a higher chance of being phenotypically mute. Nevertheless, single-nucleotide polymorphisms are the most common class of even Mendelian alleles, which impair wellness most drastically. There is no doubt that the same is true for alleles at CTLs.

In contrast, in terms of their effects on the molecular function wellness-impairing alleles are radically different from other polymorphisms. While ~90% of all polymorphic loci reside within junk DNA and are therefore functionally mute, every wellness-impairing allele must affect some molecular function. A substantial majority of alleles that cause all kinds of Mendelian diseases affect either coding exons or the edges of introns, in which case they disrupt splicing. Apparently, individual effects of alleles that affect only non-coding DNA are usually not strong enough to change the phenotype drastically. Instead, a lot of such alleles are phenotypically mild, and non-coding DNA harbors a high proportion of all CTLs. Many alleles that directly alter sequences of proteins also fail to produce Mendelian phenotypes but affect complex traits.

Because a lot of Mendelian wellness-impairing alleles have already been identified, their general properties are known with confidence. The majority of Mendelian diseases are caused by alleles that are deleterious, both when homo- and heterozygous. This is true not only for dominant but also for a majority of recessive diseases, because even heterozygotes are mildly impaired (see Chapter 9). Most cases of severe dominant diseases are due to *de novo* mutations, but cases of dominant diseases that impair fitness only mildly, as well as of recessive diseases, are mostly due to pre-existing alleles. Because a lot of different mutations can destroy or impair the function of a gene, a population usually harbors a large number of very rare alleles that can cause a particular Mendelian disease. However, some Mendelian diseases are due to a specific change or gain of the molecular function, and there may be just one or very few possible causative alleles for such a disease. For example, all cases of achondroplasia occur due to exactly the same recurrent single-nucleotide substitution at a mutation hotspot.

There are also several recessive Mendelian diseases that are caused by alleles that confer selective advantage to heterozygous carriers in some populations (see Chapter 10). One or several alleles that cause such a disease are often old and relatively common locally (see Chapter 14). Still, the advantage of heterozygotes, which is an example of sign epistasis (see Chapters 3 and 15), is a relatively rare phenomenon, and a vast majority of Mendelian diseases are caused by unconditionally deleterious alleles. Crucially, these alleles are deleterious under any environment, including Pleistocene and Preindustrial environments. That is why most alleles responsible for Mendelian diseases are derived, young (produced by mutation not-too-many generations ago), and rare.

In contrast, our knowledge of alleles that affect complex phenotypic traits is still rather fuzzy. Let us first assume one-dimensional epistasis (see Chapter 3) and consider two-state ("well" versus "unwell") traits and quantitative traits under directional wellness landscapes. In both these cases, at each CTL one allele always impairs wellness, by increasing susceptibility to the "unwell" state or moving the expected value of the trait into an undesired direction, and the other allele increases wellness. In contrast to drastic Mendelian alleles, effects of mild wellness-impairing alleles on fitness can be only small, if any. Currently, such alleles may both reduce and increase fitness (according to some data, lower intelligence is now associated with more children, see Chapter 11). However, the crucial question is: how did alleles that currently impair wellness affect fitness before the demographic transition, when the genetic composition of modern human populations evolved? If these alleles mostly were, like Mendelian alleles, deleterious, they must also be mostly derived, young, and rare.

Because not very many human CTLs have so far been precisely identified, we do not have a firm answer to this question. Industrialized environments are very different from Pleistocene and Preindustrial ones, and an allele that currently impairs wellness in the USA did not necessarily reduce fitness 10 000 or even 1000 years ago. In particular, there are "diseases of civilization", such as asthma, obesity, type 2 diabetes, and colorectal cancer, which are common only under Industrialized environments. An allele that currently increases susceptibility to such diseases could be mute or even beneficial, both wellness- and fitness-wise, under earlier environments. Nevertheless, I believe that most alleles that currently impair wellness were deleterious, at least until very recently, so that in this respect Mendelian and mild wellness-impairing alleles are not different. Five arguments support this hypothesis.

First, there is no clear-cut boundary between alleles that cause Mendelian diseases, which are and were deleterious almost universally, and alleles that affect complex traits. In fact, alleles that cause recessive Mendelian diseases when homozygous often exert mild effects on complex traits when heterozygous. Of course, an average wellness-impairing allele that only affects complex traits must be less deleterious than an average allele that causes a Mendelian disease.

Secondly, direct data show that CTLs that affect a number of complex traits, including height and BMI, disproportionately harbor rare alleles. Very rare protein-altering alleles, both small- and large-scale, play a major role in schizophrenia and macular degeneration. CTLs with derived allele frequencies (DAFs) between 0.1% and 1% account for a striking 60% of all genetic contributions to prostate cancer, although only ~3% of polymorphisms found in an average genotype have such DAFs (see Chapter 11). However, CTLs with DAF > 5% also appear to be important for some complex diseases. Mildly deleterious, derived susceptibility alleles can occasionally reach high frequencies as a result of hitch-hiking, if positive selection occurred at a locus nearby (see Chapter 7). Also, those alleles that reached higher frequencies by chance will affect the trait more. In contrast, causative Mendelian alleles rarely reach frequencies above ~0.1%, except when they provide an advantage for heterozygotes.

Thirdly, for several complex traits sophisticated statistical analyses revealed a negative relationship between the effect of an allele on the trait and its frequency: large-effect wellness-impairing alleles are rarer in the case of prostate cancer, height, and BMI (Figure 12.4). This pattern can be naturally explained by negative selection: the more an allele affects the trait, the more deleterious it is or was, until very recently.

Fourthly, a majority of causative alleles of several complex diseases were shown to be population specific. An African and a non-African are predisposed to hypertension, asthma, and rheumatoid arthritis mostly due to different susceptibility alleles. This pattern implies that these susceptibility alleles are young and derived and, thus, were deleterious.

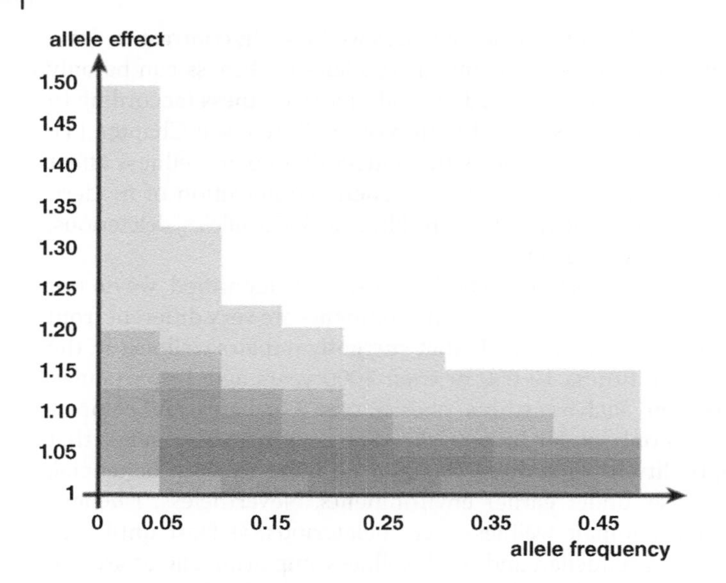

Figure 12.4 Current understanding of the joint distribution of frequencies and effects of human disease-causing alleles: alleles with large effects tend to be rarer (after Stahl *et al.*, *Nature Genetics*, vol. 44, p. 483, 2012).

Fifthly, complex phenotypic traits are subject to inbreeding depression (see Chapter 9). Inbreeding increases frequencies of complex diseases, such as schizophrenia, and shifts distributions of quantitative traits under directional quality landscapes in the undesired direction, as it is the case with blood pressure. Thus, there is no difference between the effects of inbreeding on fitness and on other complex traits, which suggests that the same alleles likely impair them all.

Still, ancestral susceptibility (and, thus, derived protective) alleles do exist, in particular, among those responsible for diseases of civilization. For example, at a locus CDHR3 that makes a substantial contribution to early childhood asthma, the protective allele is derived. At one CTL affecting ankylosing spondylitis, the protective allele is maintained by balancing selection (see Chapter 10). However, susceptibility derived alleles are also known for such diseases. For example, one derived allele which increases risk of type 2 diabetes creates a spurious binding site for a regulatory protein, and many of them simply disrupt pre-existing binding sites.

The general patterns just described suggest that derived protective alleles must be exceptions. Also, it is likely that derived alleles that protect against diseases of civilization are currently beneficial, both wellness- and fitness-wise and, therefore, are subject to positive selection. However, such selection must be of recent origin, because otherwise these alleles would be already fixed. Thus, if we assume that derived protective alleles exist at a large proportion of very numerous CTLs that affect diseases of civilization, this implies ongoing positive selection-driven allele replacements at very many loci, at least in human populations living under Industrialized environments. This is unlikely, because data on genetic variation within humans and all other species studied so far show a pervasive signature only of negative selection: the more functionally important genome segments and sites are, the larger is the excess of low-frequency alleles at them (see Chapter 11). In contrast, although many individual cases of positive selection are known, at any given moment of time derived alleles are advantageous only at a small proportion of polymorphic loci.

If even in the cases of diseases of civilization most susceptibility alleles were always deleterious (and, thus, are derived, rare and young) such diseases appear due to synergistic interaction between lag and mutation imperfection of modern humans. Indeed, their root cause is imperfect adaptation of humans to Industrialized environments. Still, this lag imperfection leads to clear-cut diseases only when it is exacerbated by derived alleles that are deleterious under all environments. Thus, humans that are free from alleles that were deleterious even in the Pleistocene epoch would be unlikely to suffer from diseases of civilization. If, as it happened many times in the course of evolution, some number of allele replacements eventually adapts us to life among fast food restaurants and computers, we will become more or less immune to asthma and diabetes even if mutation imperfection does not change.

Selection against wellness-impairing alleles does not need to be strong. If, for example, there is a disease with frequency 1% which reduces fitness by 10%, an allele that increases the risk of this disease by 10% would directly reduce fitness by only 0.0001. Of course, the actual selection coefficients can be larger, due to pleiotropy.

If my hypothesis is correct, and most alleles that impair wellness under Industrialized environments were subject to negative selection until very recently, such alleles are directly relevant to the Main Concern. Indeed, in this case their current presence must be solely due to deleterious mutations and, moreover, mutation will keep producing them in the future. However, the matter will be settled only when many individual CTLs and their causative wellness-impairing alleles are precisely identified and carefully studied.

Even less is known about alleles that affect traits under stabilizing wellness (and fitness) landscapes. If a stabilizing landscape is real (Figure 7.14), this implies sign epistasis, because every allele can both increase and decrease wellness, depending on the rest of the genotype. Stabilizing selection is discussed in Chapter 15.

12.5 Genetic Architecture of Wellness

Properties of individual wellness-impairing alleles are important, but our "Holy Grail" is genetic architecture of wellness as a whole. Genetic architecture of any trait is defined by its target (see Chapter 3), that is, the set of genes whose alleles can affect it, individual properties of these alleles, and modes of their interactions. In Chapter 11 we considered genetic architecture of fitness.

How large is the target of human wellness under Industrialized environments? It is definitely very large. Even those genes whose mutations can cause Mendelian diseases and, thus, can impair wellness severely are very numerous. Currently, ~2000 dominant, ~1200 autosomal recessive, and ~150 X-linked recessive (single-gene) Mendelian diseases are known. New Mendelian diseases are constantly being described, and their total number may be above 5000. Almost all individual Mendelian diseases are rare, because of strong negative selection against disease-causing alleles. Still, some of them are less rare, due to weak negative selection (several cancer predisposition syndromes that lead to tumors mostly after the age of 40), very high mutation rates (neurofibromatosis I or Down's syndrome), or selective advantage of heterozygotes carrying a recessive disease-causing allele (sickle-cell anemia, thalassemia; see Chapter 10). In this latest case, a disease may be relatively common locally (see Chapter 14).

In contrast, many other Mendelian diseases are extremely rare, sometimes with only a single case known. An autosomal recessive disease can be extremely rare if disease-causing alleles reduce fitness of even heterozygous carriers, who do not have a clear-cut abnormal phenotype (see Chapter 2). A dominant disease that is due to haploinsufficiency can hardly be extremely rare, because the total rate of mutations producing loss-of-function alleles of a typical protein-coding gene is above 10^{-6} (see Chapter 6), so that several affected children must be born every year in the USA alone. Only those dominant diseases that are due to specific gains or changes of the molecular function can be extremely rare, due to very low rates of causative mutations. Thus, most single-gene Mendelian diseases that are still awaiting discovery are likely to be either autosomal recessive or dominant with exotic molecular mechanisms. Mendelian diseases caused by large-scale alleles affecting several consecutive genes can also be extremely rare, due to low mutation rates and strong negative selection. The overall birth frequency of all Mendelian diseases, known and yet unknown, is probably not much higher than ~3% (Table 12.1).

Genes that are behind Mendelian disease can also affect complex phenotypic traits, because their alleles that impair molecular function only mildly, as well as heterozygous loss-of-function recessive alleles, often have only mild phenotypic effects. On top of these, thousands of loci whose mutations apparently never cause Mendelian diseases, are already known to affect complex wellness-related traits, and their number keeps increasing. The real target of wellness under Industrialized environments must consist of at least ~10 000 genes, and likely includes (almost) all 20 000 human protein-coding genes. Those, presumably few, genes whose functions are no longer relevant under Industrialized environments are at the very beginning of the path that may eventually convert them into pseudogenes (see Chapter 7).

How many wellness-impairing alleles are present in a human genotype? A typical individual carries ~50 mostly heterozygous alleles that were reported to cause Mendelian diseases. Some of these reports may be false, and most of these alleles cause diseases only when homozygous. Still, most of them are likely to always impair wellness. The same is true for ~150 loss-of-function alleles, present in an average genotype (Table 11.5), unless the target of Industrialized wellness is radically smaller than that of Pleistocene fitness. These two sets of alleles partially overlap. Also, an average human carries ~900 missense alleles that PolyPhen predicts to be deleterious (see Chapter 11). All these alleles are likely to impair wellness, as well as fitness, substantially. On top of these alleles, every genotype also harbors wellness-impairing alleles, mostly affecting non-coding genome segments, with slight individual effects, and their number is likely to be in the many thousands.

How do wellness-impairing alleles interact with each other? The answer to this question is determined by the wellness landscape (see Chapter 7). Currently, we do not know its properties with any confidence. Here, the key question is: what is the relationship between wellness landscape under Industrialized environments and fitness landscapes?

I believe that genotypes determine Industrialized wellness and Pleistocene fitness in rather similar ways (see Chapter 9). If so, genotypes that are weakly perfect in terms of fitness would also have the maximal wellness. Paradoxically, Industrialized wellness is likely to be more genetically similar to Pleistocene fitness than to Industrialized fitness.

Indeed, Industrialized fitness, but not wellness, declines with general intelligence (see Chapter 11). Of course, such a complex adaptation as high intelligence must have been a product of selection and, thus, must increase Pleistocene fitness.

Still, Industrialized wellness and Pleistocene fitness are not identical: the ability to slew a mammoth is not directly important for wellness in, say, London, and proficiency in using email is not directly important in a cave. Nevertheless, it is likely that, due to pleiotropy, functioning of mostly the same genes contribute to both Pleistocene fitness and Industrialized wellness. If so, wellness provides a rather comprehensive assessment of the whole genotype and phenotype, and most deleterious alleles, recognized as such by their negative effects on fitness under Pleistocene and Preindustrial environments, are, at the same time, detrimental to wellness under an Industrialized, or even optimal, environment (see Chapter 8). Then, weak imperfection of wellness must be substantial (see Chapter 15).

Obviously, these claims are currently supported only by the preponderance of evidence, and should not be regarded as settled facts. Still, if they are correct, the Industrialized wellness landscape has properties described in Chapter 11: it fits into the framework of one-dimensional epistasis (see Chapter 3), wellness declines monotonically with genotype contamination, and this decline likely involves synergistic epistasis.

Within-locus synergistic epistasis is revealed by inbreeding depression which affects many wellness-related traits (see Chapter 11). In contrast, little is known about possible synergistic epistasis between alleles of different loci. Still, such epistasis is suggested by pervasive incomplete penetrance of wellness-impairing alleles. For example, an allele which, as a *de novo* mutation, can lead to a complex disease such as schizophrenia may, nevertheless, be present in genotypes of some healthy people (see Chapter 13). Apparently, a susceptibility allele leads to a disease only on a genetic background that includes many other susceptibility alleles (Figure 3.14).

Again, this conclusion is only tentative, in particular, because we cannot even approximately estimate contamination of a genotype (see Chapter 11) or, more generally, its score that corresponds to any complex trait. To illustrate this point, let us again consider general intelligence. Heritability of this trait is high, and an inherited value of IQ of an individual can be several standard deviations above the average. And yet very little is currently known about how genotypes determine general intelligence. Recently sequenced genotypes of 1400 individuals with IQ > 170 possessed a slightly reduced number of rare and probably deleterious alleles, but no specific "high-intelligence" alleles. This is consistent with the simplest hypothesis that the inherited value of general intelligence depends primarily on genotype contamination. However, any score that can be currently attributed to genotypes explains only ~5% of variation of the general intelligence. This degree of ignorance should not be surprising, because we could neither identify all deleterious alleles in a genotype, nor ascribe a coefficient of selection to each of them.

Further Reading

Di Rienzo, A & Hudson, RR 2005, An evolutionary framework for common diseases: the ancestral-susceptibility model, *Trends in Genetics*, vol. 21, pp. 596–601.
A hypothesis that many protective alleles are derived.

Guillaume, L 2014, Rare and low-frequency variants in human common diseases and other complex traits, *Journal of Medical Genetics*, vol. 51, pp. 705–714.
On the importance of rare alleles.

Hrabě de Angelis, M, Nicholson, G, & Selloum, M 2015, Analysis of mammalian gene function through broad-based phenotypic screens across a consortium of mouse clinics, *Nature Genetics*, vol. 47, pp. 969–978.
Over 80% of murine loss-of-function alleles cause identifiable phenotypes.

Kingsolver, JG & Diamond, SE 2011, Phenotypic selection in natural populations: what limits directional selection?, *American Naturalist*, vol. 177, pp. 346–357.
In nature, directional selection is more common than stabilizing selection.

Krapohl, E, Euesden, J, & Zabaneh, D 2016, Phenome-wide analysis of genome-wide polygenic scores, *Molecular Psychiatry*, vol. 21, pp. 1188–1193.
Partitioning of human behavior into 50 traits.

Pierpont, EP, Basson, CT, & Benson Jr, DW 2007, Genetic basis for congenital heart defects, *Circulation*, vol. 115, pp. 3015–3038.
Comprehensive review of birth anomalies of the heart.

Plomin, R & Deary, IJ 2015, Genetics and intelligence differences: five special findings, *Molecular Psychiatry*, vol. 20, pp. 98–108.
Intelligence as a quantitative proxy for wellness.

Sankaranarayanan, K 1998, Ionizing radiation and genetic risks IX. Estimates of the frequencies of Mendelian diseases and spontaneous mutation rates in human populations: a 1998 perspective, *Mutation Research*, vol. 411, pp. 129–178.
Review of frequencies of Mendelian diseases.

Stahl, EA, Wegmann, D, & Trynka, G 2012, Bayesian inference analyses of the polygenic architecture of rheumatoid arthritis, *Nature Genetics*, vol. 44, pp. 483–489.
Alleles with larger effects on a two-state trait tend to be rarer.

Yang, J, Bakshi, A, & Zhu, Z 2015, Genetic variance estimation with imputed variants finds negligible missing heritability for human height and body mass index, *Nature Genetics*, vol. 47, pp. 1114–1120.
Complex genetic architecture of human height and body mass index.

13

Mutational Pressure on Our Species

... the optimum mutation rate from the standpoint of human welfare for the foreseeable future is zero.

James F. Crow, 1986.

De novo deleterious mutations substantially impair human wellness. The combined mutational pressure on all Mendelian diseases is ~0.01. The combined mutational pressure on all complex diseases is known with lesser certainty, but appears to be ~0.03. All current estimates of the mutational pressures on human quantitative traits were obtained indirectly, through paternal age effects. Many of them imply that mutations which occur in the course of one generation shift the average values of some important quantitative traits by 0.01–0.03 of their current values; however, a lot of question remain unanswered. If deleterious mutations keep accumulating in the population, the mutational pressures on complex traits may increase. Unfortunately, we still do not know to what extent deleterious mutations will affect wellness of future generations. Direct data on the relationship between complements of *de novo* mutations in genotypes of individuals and their phenotypes will likely clarify this issue in the next 10–20 years.

13.1 Mutational Pressure on Diseases

In the previous two chapters, we considered imperfection of human fitness and wellness caused by deleterious alleles that are already present in the population. Effects of these alleles, produced by mutation in the past, constitute the first part of the Main Concern. Here, we deal with its second part – deleterious effects of *de novo* mutations. These effects are substantial even during a single generation. If selection is relaxed, accumulation of *de novo* mutations in the course of many generations will eventually cause a meltdown of fitness and wellness.

An average *de novo* mutation is more deleterious than a pre-existing mutant allele, because the pervasiveness of a mutant allele within the population is inversely proportional to the coefficient of selection against it (see Chapter 9), so that alleles that reduce fitness strongly are underrepresented among the pre-existing deleterious alleles. In the extreme case of dominant mutations that are inconsistent with reproduction, all of them must be *de novo* ones. Still, a majority of even *de novo* deleterious mutations are mild, at least when heterozygous (see Chapter 6), and qualitatively their effects are similar

Crumbling Genome: The Impact of Deleterious Mutations on Humans, First Edition. Alexey S. Kondrashov.

to that of pre-existing mutant alleles. Thus, for a number of generations, even a free accumulation of *de novo* mutations is unlikely to lead to any novel phenomena. Instead, it will increase imperfection and variation of genotypes which are present in any population at the state of the mutation–selection equilibrium (see Chapter 9).

Quantitatively, the impact of accumulating *de novo* mutations on a phenotypic trait is characterized by the mutational pressure (see Chapter 6). In the case of a two-state (normal versus disease) trait, the mutational pressure is the per generation rate of a steady-state increase of the disease frequency due to *de novo* mutations, if the environment does not change and there is no opposing selection. This pressure is equal to the frequency of cases of the disease that would not occur without a contribution from *de novo* mutation.

One can expect most autosomal recessive Mendelian diseases to be caused by relatively old alleles, unless inbreeding is common in the population (see Chapter 6). In contrast, a substantial proportion of dominant and X-linked recessive diseases should be due to *de novo* or very young mutations, because a mutant allele causes such a disease and becomes exposed to selection immediately or soon after its inception and, thus, does not persist for too long. Data obtained by comparison of genotypes of persons having Mendelian diseases and of their parents show that these expectations are correct. Such comparisons do not require next-generation sequencing (NGS), because only one, causative gene needs to be sequenced, as long as the disease has been correctly diagnosed. Due to their relative technical simplicity, studies of the role of *de novo* mutation in Mendelian diseases began over 20 years ago.

The total mutational pressure on single-gene Mendelian diseases is ~0.005. This figure is not surprising: dominant and X-linked recessive Mendelian diseases can be caused by mutations of ~2000 genes, and the typical per gene rate of loss-of-function mutations is $\sim 3 \times 10^{-6}$. Here, summation is appropriate, because only very rarely a person suffers from more than one Mendelian disease. The total mutational pressure on aneuploidies is very close to their birth frequency ~0.002 (Table 12.1), because almost all their cases are due to *de novo* mutations. The same is apparently true for poorly studied rare diseases caused by large-scale mutations that affect several adjacent genes, so that the corresponding mutational pressure is ~0.001. Thus, ~8 newborns per 1000 suffer from a genetically simple disease due to a *de novo* mutation.

It has long been suspected that *de novo* mutations are also to some extent responsible for at least some cases of complex diseases. There were two main reasons for this suspicion. First, many complex diseases reduce fitness severely but nevertheless persist in the population. Secondly, old fathers transmit more *de novo* mutations to offspring (see Chapter 6), and frequencies of a number of complex diseases are known to increase with the paternal age (see below). In 2002, Dolores Malaspina and colleagues reviewed the available indirect evidence for the causative role of *de novo* mutations in schizophrenia and concluded that this role is substantial.

However, to directly measure the mutational pressure on a complex disease is much more difficult than on a Mendelian diseases, because this requires analyzing many loci. In fact, at the current level of knowledge, we cannot confidently exclude any locus from the target of a complex disease. Thus, ideally, complete genotypes need to be sequenced, which was impossible before the advent of NGS. Starting from 2007, several studies revealed a role of large-scale *de novo* mutations in autistic spectrum disorders (ASDs) and schizophrenia. In 2010, a paper titled "A *de novo* paradigm for mental retardation" reported a comparison of all known protein-coding genes in 10 individuals with unexplained intellectual disability with that of their parents. A small-scale *de novo* mutation which likely contributed to the disease was found in 6 individuals.

Since these pioneering studies, a lot of data has accumulated. Currently, studies of mutational pressures on complex diseases often involve sequencing genotypes of multiple "quartets", each consisting of mother, father, their affected child, and an unaffected sibling, and comparing the complements of *de novo* mutations in affected versus unaffected offspring. Table 13.1 summarizes what is currently known. All the diseases studied so far lead to a substantial reduction of fitness, and a large contribution of *de novo* mutations to such a disease is not surprising. Absolute mutational pressures are the highest for neurodevelopmental diseases that have the largest targets (see Chapter 12) and population frequencies.

Mutational pressures presented in Table 13.1 must be underestimated, because all the studies performed so far only took into account *de novo* mutations that directly affect the encoded proteins. This would be acceptable for Mendelian diseases, which are mostly caused by such alleles. In contrast, complex traits are also strongly affected by variation within non-coding DNA (see Chapter 12).

Still, even if all the relevant *de novo* mutations were taken into account, I do not think that they would be found in 100% of cases of a complex disease. Indeed, several cases of incomplete penetrance of causative mutant alleles have been described. In such a case, an allele that, when produced by *de novo* mutation, likely triggers the appearance of a disease, is, nevertheless, present in genotypes of some clinically healthy people. Also, occurrence of complex diseases is correlated between even distantly related people, demonstrating that not all causative alleles are very young. Finally, common derived alleles, which must be quite old, are known to make a substantial contribution to susceptibility to complex diseases, although an average *de novo* mutation makes a larger contribution than an average pre-existing allele. Thus, a complex disease can appear if recombination of pre-existing alleles produces a high-contamination genotype, or due to an environmental insult, even in the absence of any relevant *de novo* mutations.

Perhaps, severe intellectual disability is an exception, because even the analyses performed so far detected *de novo* causative mutations in 60% of cases, although many genes that constitute the target of this condition must still be unknown. Also, in contrast to mild intellectual disability and to other neurodevelopmental disorders, most cases of severe intellectual disability appear to be sporadic, without any affected close relatives. If so, this condition may, in fact, mostly represent a collection of phenotypically similar fully penetrant Mendelian diseases.

Hopefully, mutational pressures on a wide variety of important diseases, including major depression, asthma, and tuberculosis will be soon studied by sequencing multiple quartets. Indeed, *de novo* mutation can play a role even when a disease has a clear environmental trigger. This is illustrated by a substantial mutational pressure on cerebral palsy, which is thought to result from environmental insults on the developing brain before, during, or soon after birth. Still, it seems that the genotype of a child determines their resistance to such insults, so that under similar circumstances a child that carries a relevant *de novo* mutation becomes affected with a higher probability. The same may be the case for an infectious disease. Relative contribution of *de novo* (germline) mutations to the diseases of old age, such as stroke, dementias, and cancers is likely to be low, because selection against alleles that can cause them is probably weak.

Table 13.1 Mutational pressures on complex diseases.

Disease	Frequency of the disease	Proportion of cases with a contribution from *de novo* mutations	Mutational pressure	Comment
ASD	0.011	0.25	0.0028	Cases of ASD to which *de novo* mutations contributed are more severe and often involve reduced IQ. The role of *de novo* mutation is larger in girls, who have ASD more rarely than boys
Schizophrenia	0.005	0.15	0.0008	Some large-scale mutations contributing to schizophrenia occur at high individual rates and are severely deleterious
Bipolar disorder	0.01	0.1	0.001	Cases of bipolar disorder to which *de novo* mutations contributed have an earlier onset
Severe intellectual disability	0.005	0.6	0.003	Sequencing complete exome makes it possible to establish genetic diagnosis for a majority of patients with severe intellectual disability
Epileptic encephalopathy	0.001	0.2	0.0002	Pathogenic *de novo* mutations were found in 10% of 65 genes studied so far
Cerebral palsy	0.002	0.15	0.0003	Apparently, only ~10% of cases of cerebral palsy can be explained by birth asphyxia
Congenital heart disease	0.007	0.2	0.0014	Only severe cases of birth anomalies of the heart are regarded as congenital heart disease
Neural tube defects	0.001	0.2	0.0002	Supplementing the diet of pregnant women with folic acid reduces the frequency of neural tube defects by 50–70%
Congenital diafragmal hernia	0.0003	0.15	0.00005	Even with modern medical care, this birth anomaly leads to early mortality of ~30%

ASD, autism spectrum disorder; IQ, intelligence quotient.

However, because many of these diseases are very common, the absolute mutational pressures on them can be high. It will also be very interesting to estimate mutational pressures on overall characteristics of wellness such as lack of disability and excellent health, in people of different ages.

The sum of mutational pressures on all complex diseases listed in Table 13.1 is ~0.01. Because these pressures must be underestimated, and because many complex diseases have not yet been studied from this perspective, I believe that the total mutational pressure on the health of young people, due to contributions of *de novo* mutation to both Mendelian and complex diseases, is between 0.02 and 0.05.

The mutational pressure on a disease can also be assayed indirectly, by comparing its frequencies in the offspring of old versus young fathers. This approach exploits a rapid increase of the mutation rate with the paternal age: a 20-year-old father transmits to his child ~25 *de novo* mutations, and a 50-year-old father transmits ~85 *de novo* mutations (see Chapter 6). Because the expected number of maternal *de novo* mutations in an offspring is always ~15, offspring of 50-year-old fathers receive ~2 times more *de novo* mutations than offspring of 20-year-old fathers. Thus, the mutational pressure on a disease in the offspring of 20-year-old fathers should be close to the difference between its frequencies in the offspring of 50- and 20-year-old fathers.

Studies of the effects of the paternal age on diseases and quantitative traits in offspring do not depend on any advanced technologies and, therefore, have a long history. Some patterns have been firmly established, while others remain controversial. Advanced paternal age clearly increases the risks of neuropsychiatric conditions, including ASD, schizophrenia, and bipolar disorder. The same trends have been reported for the probability of dying before the age of 5, the probability of remaining childless, congenital heart defects, and prostate cancer.

Moreover, three recent studies reported adverse effects of even grandparental age. Data on individuals born in Sweden since 1932 indicate that men who fathered a child after the age of 50 are significantly more likely to have a grandchild (by this child) with ASD or schizophrenia. Similarly, data on preindustrial Finns show that individuals whose fathers, grandfathers and great-grandfathers fathered their lineage on average under age 30 are ~13% more likely to survive to adulthood than those whose ancestors fathered their lineage at over 40 years. Such findings can be naturally explained by effects of *de novo* mutation.

Still, some of the observed paternal age effects are suspiciously strong. Doubling of the disease frequency with the doubling of the complement of *de novo* mutations seems to imply that all its cases are due to *de novo* mutations, which cannot be true for a complex disease. However, a number of studies found that the frequency of a disease in children of 50-year-old fathers is ~2 (or even more) times higher than that in children of 20-year-old fathers. For example, it has been reported that sons of fathers who were older than 38 years at the time of conception have 1.7 times higher risk of developing prostate cancer than sons of fathers who were younger than 27 years. Could there be other explanations for the effects of the paternal age? The answer is "yes, several of them". First, the socioeconomic environment of a child obviously depends on the age of their father (and mother). Secondly, some inheritance can be epigenetic, that is, mediated not by the nucleotide sequence of DNA but by the pattern of its methylation (see Chapter 1) and of attachment of RNA and protein molecules to it. There are data suggesting that epigenetic factors can substantially affect phenotypes of offspring, and that advanced paternal age can lead to adverse epigenetic effects. Thirdly, advanced paternal age often also means advanced maternal age, which can also increase the frequency of disease in offspring, apparently through mostly non-genetic mechanisms. Fourthly, advanced paternal age could simply reveal that something is wrong with the father himself, because a healthy male is likely to reproduce young.

It may be possible to distinguish the last two explanations from the effects of *de novo* mutation. For example, in some studies children of older fathers still had higher disease frequencies even when only those fathers who had their first child early were taken into account. In contrast, it is hard to distinguish genetic and epigenetic effects, without directly looking at the DNA sequence. Thus, although adverse effects of advanced paternal age suggest substantial mutational pressures on a number of human diseases, I do not think that they can alone produce definite quantitative estimates.

A valuable approach to measuring mutational pressures in humans would be to study *de novo* mutations and phenotypes of people whose parents (in particular, fathers) carried a dominant mutator allele in the heterozygous state and who did not themselves inherit this allele. The trouble is, strong human germline mutators have not yet been described. Still, it is very likely that alleles of replicative DNA polymerases that lack proof-reading activity and loss-of-function alleles of mismatch repair, both known to increase somatic mutation rate and to cause cancer-predisposition syndromes (see Chapter 5), are also germline mutators. If this is indeed the case, and a child of a father who carries such an allele receives, say, 500 *de novo* mutations instead of the usual 100 (see Chapter 6), measuring frequencies of diseases among such children would produce extremely important data.

Advances in medicine led to a dramatic relaxation of selection against many diseases. In particular, many birth anomalies can now be surgically repaired, and the affected people can lead mostly normal lives and reproduce. If so, could we view human populations enjoying advanced health care as mutation accumulation (MA) experiments (see Chapter 9) and infer mutational pressures on treatable diseases from the per generation changes in their frequencies? Unfortunately, the answer is "no" because one cannot standardize human environments between generations, which is always done in MA experiments. In fact, frequencies of a number of birth anomalies decline in many countries, due to improved environments. Near-elimination of rubella, reduced exposure of pregnant women to lead, and folic acid supplements played a substantial role in this trend.

Still, frequencies of some other diseases, including even those that reduce fitness despite the best currently available care, are rising. In particular, the frequency of ASD increased by a factor of 5 in the course of 50 years in some countries. The causes of this striking phenomenon remain obscure. Accumulation of mutations could play a role in it, if alleles that contribute to ASD also affect other traits (pleiotropy, see Chapter 3), and selection acting on these traits became weaker. The increased average paternal age could also be responsible, due to higher rates of *de novo* mutations and other effects. However, other causes that are unrelated to mutation, such as better diagnosis and so far unidentified adverse changes of the environment could also explain this phenomenon, partially or even completely. Indeed, there is a strong correlation between frequencies of ASD and of birth anomalies (mostly, in different individuals) across local populations. This correlation suggests that some common, currently unknown, environmental factors are partially responsible for both kinds of conditions.

I believe that, in order to be definite, any measurement of the mutational pressure on a human disease must be based on a direct ascertainment of all *de novo* mutations in the genotypes of multiple individuals with and without the disease. Better and cheaper NGS will likely lead to many such measurements in the coming years.

13.2 Mutational Pressure on Quantitative Traits

Mutational pressures on diseases have so far been studied exclusively in humans. The only exception are data on rates of visible mutations in *Drosophila* (see Chapter 6), which produce phenotypes that are mostly analogous to human Mendelian diseases. In contrast, mutational pressures on quantitative traits were measured in a number of non-human species (see Chapters 6 and 9). All these measurements involved determining the rate of changes of phenotypes, from generation to generation, in MA experiments, which cannot be performed on humans.

A direct, single-generation approach to measuring the mutational pressure on a quantitative trait is to relate the values of this trait in many individuals to the numbers and properties of *de novo* mutations in their genotypes. Unfortunately, this approach, which in the last several years produced extremely interesting data on human diseases, has not yet been applied to quantitative traits such as general intelligence, height, or blood cholesterol level. This will likely be done soon, because sequencing genotypes of many mother–father–offspring trios is not a big deal anymore. Doing the same on non-human species, to supplement data obtained in MA experiments, would also be worthwhile.

Currently, all the available estimates of the mutational pressures on human quantitative traits have been obtained indirectly, through paternal age effects. It has been reported that children of older fathers have higher body mass index (BMI), reduced facial attractiveness, delayed early development, and reduced intelligence, although in this latter case different studies produced conflicting results. In a study of over 30 000 children in the USA, scores of several neurocognitive tests were reduced by 0.05 in those whose fathers were 50 years old, in comparison with those whose fathers were 20 years old. This result implies a very high per generation mutational pressure of 0.05 on cognition (see above). In contrast, data on the intelligence quotient (IQ) of Swedes born between 1951 and 1976, as well as on 170 000 Danish conscripts, did not reveal any adverse effects of advanced paternal age.

Of course, such estimates suffer from the same problems as paternal age-based estimates of the mutational pressures on diseases. In fact, these problems may be even more serious in the case of quantitative traits, because of a likely larger influence of the socioeconomic environment. Thus, I believe that, similarly to the case of diseases, the only way to obtain a solid estimate of the mutational pressure on a human quantitative trait is to relate its values in many individuals to their complements of *de novo* mutations. Again, studying children of fathers with strong mutator genotypes could be the easiest way of obtaining such data. In non-human species, individuals carrying very many *de novo* mutations can be produced experimentally, by using artificial mutagenesis. We used this approach to study mutational pressures on *Drosophila* (see Chapter 6), and it makes sense to perform analogous experiments on mice, as the results would be more directly relevant to humans.

Average values of a number of quantitative traits in human populations slowly change in the "desired" directions, opposite to what accumulation of mutations is likely to produce. Salient examples are increased life span, height, and intelligence, which are obviously due to better environments. Average scores of tests measuring intelligence have been rising in the course of the last several decades in many countries. This trend is known as the Flynn effect. Many environmental factors, including malnutrition, iron deficiency, childhood diseases, and a long list of pollutants, all have lasting adverse effects on intelligence, and these factors are generally becoming less severe. It is not surprising that improved environment more than compensated for any possible adverse effects of selection for lower intelligence (see Chapter 11) and relaxed negative selection (see Chapter 9) in industrialized countries. Still, in the last ~20 years the Flynn effect all but disappeared in some countries, indicating that their environments may already have become close to optimal (or at least that they are no longer improving rapidly).

There are also some undesired long-term trends. In particular, easy availability of high-calorie food led to a dramatic increase of average BMI in many countries. Clearly, it is impossible to infer the mutational pressure on a quantitative trait from any data on long-term trends in its average values in human populations.

So far, we have concentrated on phenotypes of young people. However, the mutational pressure on longevity and phenotypes of old people are also of great interest for two reasons. First, aging is a process of exceptional importance for human wellness, as over 80% of deaths in industrialized countries are aging-related. Secondly, deleterious mutations are very likely to contribute to aging. There is a general agreement that aging evolves if organisms, even in their best shape, have no chance to live long. Indeed, there is no point in making a mouse that can live for 10 years if predators kill almost every mouse before it reaches the age of 2 years. Still, aging can evolve by two distinct mechanisms. One is fixation of alleles that increase performance of an organism early in life, at the expense of impairing its late performance, perhaps due to reallocation of resources from maintaining the body to reproduction. The other is accumulation, due to lack of negative selection, of mutant alleles which impair performance of organisms only at advanced ages that they are unlikely to reach regardless of their genotypes, without providing any benefits. These two mechanisms are not mutually exclusive, and, apparently, both of them played a role in the evolution of aging in a number of species.

Thus, late-acting wellness-impairing alleles are likely to be important for human aging. Unfortunately, no direct measurements of the mutational pressure on human longevity have been performed so far, and data on the paternal age effects on longevity are contradictory. Studies of complements of *de novo* mutations in genotypes of very old people are likely to produce important results, although the availability of genotypes of their parents can be a problem.

13.3 Possible Increase of the Mutational Pressure

So far, we implicitly assumed that the mutational pressure on a phenotypic trait does not change, as long as the mutation rate remains constant. However, this is not necessarily the case, and, when mutations accumulate, the mutational pressure may increase, due to synergistic epistasis. In the case of a quantitative trait, epistasis is absent if the trait landscape is exponential, and present if the relative expected value of the trait declines faster when the genotype contamination increases (Figure 9.3). Such acceleration means that deleterious alleles reinforce effects of each other (see Chapter 9) and the effect of a complement of *de novo* mutations on the trait increases with genotype contamination (Figure 13.1).

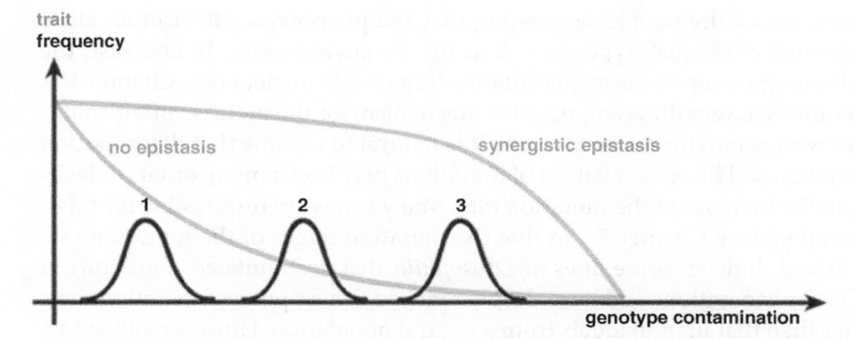

Figure 13.1 The effect of epistasis on the mutational pressure on a quantitative trait. If a quantitative trait depends on genotype contamination exponentially (no epistasis), going from its distribution 1 to 2 leads to the same relative decline of the average value of the trait as going from distribution 2 to 3. In contrast, if deleterious alleles affect the trait synergistically, its relative decline is larger in the second case. Of course, an exponential dependence is linear on the logarithmic scale.

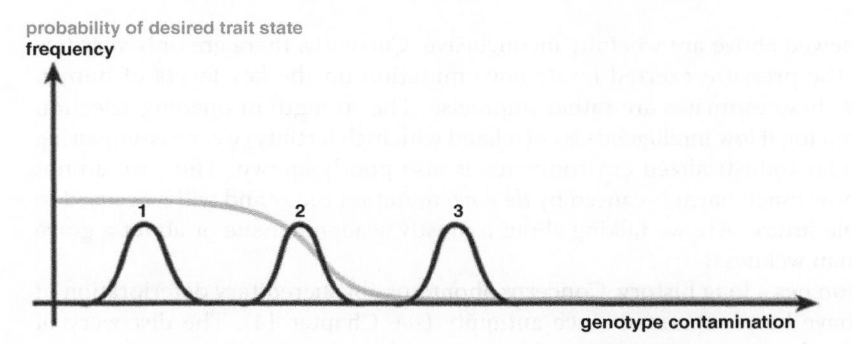

Figure 13.2 The effect of epistasis on the mutational pressure on a two-state trait. Going from the distribution of genotype contamination 2 to 3 leads to a larger decline of the probability of the desired state of a two-state trait than going from 1 to 2.

In the case of a two-state trait, synergistic epistasis appears automatically, as long as genotype contamination plays the role of the trait potential (Figure 3.4). Thus, when deleterious alleles accumulate, we can expect the proportion of people who do not have schizophrenia or who have IQ above some cut-off value (say, 140) to decline faster and faster (Figure 13.2). Apparently, the British evolutionary biologist Julian Huxley was the first one to appreciate this effect.

Due to this mechanism, the mutational pressure exerted on health by an autosomal recessive disease increases with the frequency of disease-causing alleles. Indeed, recessivity is a result of synergistic epistasis between alleles of the same gene (see Chapter 9). A *de novo* mutation causes such a disease only in an individual that received a pre-existing disease-causing allele from the other parent. Thus, under random mating, the mutational pressure is equal to the mutation rate of the responsible gene times the population frequency of its disease-causing alleles.

There are also direct indications that synergistic epistasis is, as expected, an essential feature of the genetics of complex two-state traits. In particular, alleles that, when produced by mutation, are known to trigger a complex disease, are often also present, at low frequencies, in healthy people, suggesting that the phenotypic effect of an allele depends on the rest of the genotype, as well as on the environment. In contrast, the importance of synergistic epistasis in quantitative traits is still unclear (see Chapter 15).

There is also another, very different, possible mechanism for the increase of the mutational pressure with genotype contamination. It is natural to assume that deterioration of the phenotype caused by accumulation of mutations may lead, among other undesirable changes, to the increase of the mutation rate. Many genes are responsible for fidelity of DNA handling (see Chapter 5), so that the mutation target of the mutation rate must be non-trivial. Indeed, some lines of *Drosophila* that accumulated mutations in the course of many generations and have deeply reduced fitness possess mutation rates ~2 times higher than that in individuals from a natural population. However, other MA lines of *Drosophila* do not show any marked increase in the mutation rate. The importance of this effect in humans is unknown.

13.4 *De Novo* Mutations and Human Wellness

The data reviewed above are woefully inconclusive. Currently, there are only very few estimates of the pressure exerted by *de novo* mutation on the key facets of human wellness, and these estimates are rather imprecise. The strength of ongoing selection against (or even for, if low intelligence is correlated with high fertility) wellness-impairing mutations under Industrialized environments is also poorly known. Thus, we do not really know how much harm is caused by *de novo* mutation today and will be caused in the foreseeable future. Are we talking about a mostly academic issue or about a grave threat to human wellness?

This question has a long history. Concerns about possible hereditary deterioration of our species have been expressed since antiquity (see Chapter 14). The discovery of mutation (see Chapter 4) made it possible to formulate them in the proper genetic terms. Here, a crucial role was played by Hermann J. Muller, one of the founders of genetics. In 1950, he published a classical paper titled "Our load of mutations", which introduced many ideas discussed in this book. In particular, Muller emphasized that a population can survive a constant influx of deleterious mutations only at the state of mutation–selection equilibrium (see Chapter 9), because unlimited accumulation of mutations will inevitably cause its eventual demise. Thus, any relaxation of negative selection could only be temporary, and eventually selection must resume, to establish a new mutation–selection equilibrium.

Muller subscribed to the view that the demographic transition profoundly relaxed selection among humans (see Chapter 11). Nevertheless, he did not think that accumulation of *de novo* mutations has immediate practical consequences:

> Despite our insistence in the foregoing that indefinitely prolonged continuance of the present pattern of reproductive behavior along with a continuance of modern medical practices ..., would eventually lead to grave genetic consequences if not to complete disaster for mankind, it is not our intention to leave the impression that there is in this situation a cause for acute present alarm. For, as we have also pointed out repeatedly above, the process is a very long term one, the great genetic changes being, in terms of human affairs, very slow.
>
> H. J. Muller, 1950.

However, Muller reached this conclusion on the basis of very limited information that was available in the final years of the pre-DNA times. He assumed that the average number of *de novo* deleterious mutations per human genotype is below 0.5 (instead of the current figure of ~10), and that an average human genotype carries 8 (instead of thousands) of pre-existing deleterious alleles. By the end of the 20th century, it became clear that Muller's estimates need to be updated. In 1997, James F. Crow published a review "The high spontaneous mutation rate: is it a health risk?", which took into account the data produced since 1950. Still, he reached essentially the same conclusion as Muller:

> I do regard mutation accumulation as a problem. ... But this is a problem with a long time scale; the characteristic time is some 50–100 generations, which cautions us against advocating any precipitate action. We can take time to learn more. Meanwhile, we have more immediate problems: global warming, loss of habitat, water depletion, food shortages, war, terrorism, and especially increase of the world population.
>
> J. F. Crow, 1997.

In the 21st century, NGS has produced even more of data, in particular, revealing an important role of *de novo* mutation in the origin of many diseases. Nevertheless, the available estimates of mutational pressures remain rather vague. Thus, we do not yet know for sure whether the guarded optimism of Muller and Crow was justified.

It may be tempting to try to deduce the mutational pressure on human wellness from some general considerations. Each of us carries ~1000 protein-altering deleterious alleles, but only ~1 such mutation appears *de novo* in the course of a generation. Thus, it will take ~1000 generations of completely relaxed selection to double the number of such alleles in a genotype. Also, because the standard deviation of the per genotype number of such alleles is ~30 (see Chapter 12), their number will be incremented by one standard deviation only after ~30 generations of free accumulation of new mutations.

Such calculations seem to imply that Muller and Crow were right. However, this reasoning is misleading because the already-present deleterious alleles form a rather biased sample of fresh, *de novo* deleterious mutations, enriched with mild alleles (see Chapter 11). Thus, an extra protein-altering *de novo* mutation increments genotype contamination by much more than 1/1000 of the standing contamination due to pre-existing protein-altering alleles.

Theory of the mutation–selection equilibrium implies that genotype contamination will double after 1/Mean[s] (where s is the coefficient of selection against a *de novo* mutation) generations of free accumulation of *de novo* mutations (see Chapter 9). Unfortunately, we do not have a precise estimate of Mean[s], which can be anything between, say, 0.001 and 0.01. Moreover, we do not know how doubling of the average genotype contamination would affect the average fitness or wellness, and the effect could be disproportionately severe (see Chapter 15).

To appreciate the degree of our ignorance, the following thought experiment is useful. Suppose that we add, to the genotype of an average human, a set of mutations which are expected to appear *de novo* in the course of 10, 100, or 1000 generations. What phenotype would such a new genotype produce? A human genotype that received a 1000-fold dose of spontaneous mutations will definitely be lethal, under any environment: among ~100 000 extra mutations there will likely be at least one dominant lethal and, even more importantly, their cumulative negative impact on fitness must be overwhelming. A 100-fold dose of spontaneous mutations would also very likely lead to drastic consequences, if not to outright lethality, because most of the resulting individuals would suffer from severe Mendelian and complex diseases. In contrast, we cannot say anything definite about individuals who received a 10-fold dose: their average fitness and wellness could go down dramatically or just a bit.

Recently, a laboratory population of mice which carried a mutator allele which increases the mutation rate by a factor of 17 was studied in the course of 20 generations. After several generations, mortality increased, and pregnancy rates and litter sizes declined substantially. Selection against drastic mutations definitely took place in the population, so that it was not a MA experiment, which is alone capable to produce quantitative data. Still, these results suggest that even a 10-fold dose of spontaneous mutations would cause a readily detectable harm to humans.

We also do not know how different the mutational pressures are on human fitness and wellness. One could think of reasons why the pressure of wellness could be both lower and higher than on fitness. On the one hand, like fitness, wellness depends on adaptations that evolved through natural selection (see Chapter 7), so that a mutation that is mute fitness-wise (selectively neutral, see Chapter 8), must also be mute wellness-wise. However, many human adaptations, such as extreme physical endurance or ability to tolerate starvation (or a fear of snakes and spiders) are irrelevant in modern societies. Also, due to genotype–environment ($G \times E$) interactions (see Chapter 10) the effect of a mutation should be generally smaller under more benign environments. Thus, the mutational pressure on wellness under the Industrialized environment should be lower than on fitness under the Pleistocene environment.

On the other hand, a mutation that is manifested mostly at an old age could have only a minimal impact on fitness, but could be severely harmful wellness-wise, for example, increasing the risk of Alzheimer's disease after the age of 60. Also, mutations that substantially increase the risk of diabetes and obesity in a modern society could be only mildly deleterious thousands of years ago. If so, the mutational pressure on wellness under the Industrialized environment should be higher than on fitness under the Pleistocene environment. Which of these two considerations is more important, we do not know.

Equally important, we do not know to what extent, if any, selection against *de novo* mutations is relaxed in modern human populations, or will be relaxed in the foreseeable future (see Chapter 11). Finally, we do not know for how long the improvement of human environments will be able to compensate phenotypic effects of the possible accumulation of deleterious mutations. Obviously, this cannot proceed indefinitely, because, at best, the environment can eventually become optimal. It seems that some industrialized countries have already gone a long way towards this goal.

13.5 Optimistic and Pessimistic Scenarios

Clearly, we do not know exactly how *de novo* mutations will be affecting fitness and wellness of the future generations of humans. We can only be reasonably sure that, currently, the total mutational pressure on all clear-cut diseases that are manifested before the age of 30 is 0.01–0.05. Facing such an uncertainty, it is helpful to outline two extreme, but still consistent with the data, scenarios of what could happen, optimistic and pessimistic.

According to the optimistic scenario, the total mutational pressure on the diseases of young age is ~0.01, and the mutational pressures on quantitative facets of wellness are ~0.003 of their average values. A substantial selection against deleterious mutations keeps operating, and will operate even if the environment continues to improve, because a near-elimination of pre-reproductive mortality is compensated by a reduced random variation of fertility. As a result, there will be no accumulation of deleterious alleles in the future, so that the mutational pressure will not increase and all human populations will remain at the mutation–selection equilibrium, similar to the one that existed throughout human history. As a result, the proportion of individuals that are overtly affected, under Industrialized environments, by deleterious alleles in the course of their lives will remain around 10%. Moreover, continuing improvement of the environment, including health care, will lead to increased overall wellness. Thus, deleterious alleles will never make their way into the top 10 problems facing humanity.

According to the pessimistic scenario, the total mutational pressure on the diseases of young age is ~0.05, and the mutational pressures on quantitative facets of wellness are ~0.03 of their average values. Selection against deleterious alleles is deeply relaxed under Industrialized environments and cannot prevent accumulation of all but the most deleterious mutations. Thus, the mutational pressures on many traits will likely increase with time. As a result, frequencies of overt diseases, in particular those caused by impaired functioning of the brain, will increase rapidly, and the mean values of some key traits which characterize human wellness will decline by ~30–40% in the next 10 generations, making phenotypes that currently correspond to the bottom 10–20% of the population a new norm. Some characteristics of the population, such as the proportion of people with IQ above 140, will decline even more. Soon, improvements of the environment will become unable to mask these declines. Thus, after only ~10 generations societies will begin to crumble, and preventing this is as important as dealing with climate change and habitat loss.

The truth must be somewhere in between, and, I believe, is closer to the pessimistic scenario. Still, I freely confess, this is only a judgement call – we do not really know.

Further Reading

Crow, JF 1997, The high spontaneous mutation rate: Is it a health risk?, *Proceedings of the National Academy of Sciences of the USA*, vol. 94, pp. 8380–8386.

A summary of what was known about deleterious mutations is humans before the NGS era.

Deciphering Developmental Disorders Study 2017, Prevalence and architecture of *de novo* mutations in developmental disorders, *Nature*, vol. 542, pp. 433–438.

Flynn, JR 2012, *Are We Getting Smarter?*, Cambridge University Press, Cambridge.

Description of the Flynn effect, a substantial increase in IQ scores in many countries.

Frans, EM, Sandin, S, & Reichenberg, A 2013, Autism risk across generations: a population-based study of advancing grandpaternal and paternal age, *JAMA Psychiatry*, vol. 70, pp. 516–521.

Advanced grandpaternal age increases the risk of autism.

Lynch, M 2016, Mutation, eugenics, and the boundaries of science, *Genetics*, vol. 204, pp. 825–827.

A recent debate about the Main Concern.

Malaspina, D, Corcoran, C, & Fahim, C 2002, Paternal age and sporadic schizophrenia: evidence for de novo mutations, *American Journal of Medical Genetics*, vol. 114, pp. 299–303.

Pre-NGS evidence of a substantial mutational pressure on schizophrenia.

Muller, HJ 1950, Our load of mutations, *American Journal of Human Genetics*, vol. 2, pp. 111–176.

A classical analysis of the effects of deleterious alleles on humans.

Rzhetsky, A, Bagley, SC, & Wang, K 2014, Environmental and state-level regulatory factors affect the incidence of autism and intellectual disability, *PLoS Computational Biology*, vol. 10: e1003518.

A study which reported a strong correlation between frequencies of ASD and birth anomalies across the USA.

Saha, S, Barnett, AG, & Foldi, F 2009, Advanced paternal age is associated with impaired neurocognitive outcomes during infancy and childhood, *PLoS Medicine*, vol. 6: e1000040.

In a study of >30 000 children, advanced paternal age was associated with reduced values of several quantitative traits that assay cognition.

Vissers, LELM, Gilissen, C & Veltman, JA 2015, Genetic studies in intellectual disability and related disorders, *Nature Reviews Genetics*, vol. 17, pp. 9–18.

Up-to-date review of a major role of *de novo* mutations in severe intellectual disability.

14
Ethical Issues

Now I say that the human being, and in general every rational being, exists as end in itself, not merely as means to the discretionary use of this or that will ...

Immanuel Kant,
Groundwork for the Metaphysics of Morals, 1785.

In the 20th century, attempts to influence the genetic composition of human populations without any regard for the rights and interests of individuals led to some of the worst atrocities in the human history. In contrast, reproductive counseling, which is widely practiced these days, is aimed at preventing diseases in future children, and not at effecting changes at the population level. Ethical issues are inherently difficult to resolve. Still, if human wellness is our only concern, removal of unconditionally deleterious alleles from human genotypes is not only acceptable but morally necessary, as long as it increases wellness of future children without any side effects. However, methods available for germline genotype modification today are way too imprecise for this task. Several ethical dilemmas are relevant to the Main Concern, and advancing technologies will likely produce more of them. Because values cannot be derived from facts, scientists do not possess any special expertise on the questions of ethics, and thus should not play the leading role in deciding how the fruits of science can and cannot be applied to humans.

14.1 Lessons from History

So far, we have approached the Main Concern from the perspective of natural sciences. This would be sufficient if our only goal were to understand the harm inflicted by human deleterious alleles. However, we are also going to discuss possible ways of mitigating this harm. As history amply illustrates, ethics of any actions that involve humans is of paramount importance.

Influencing human hereditary traits has been contemplated since ancient times. In Book V of his *Republic*, Plato advocated artificial selection of humans:

- And how can marriages be made most beneficial? That is a question which I put to you, because I see in your house dogs for hunting, and of the nobler sort of birds not a few. Now, I beseech you, do tell me, have you ever attended to their pairing and breeding?

Crumbling Genome: The Impact of Deleterious Mutations on Humans, First Edition. Alexey S. Kondrashov.

- In what particulars?
- Why, in the first place, although they are all of a good sort, are not some better than others?
- True.
- And do you breed from them all indifferently, or do you take care to breed from the best only?
- From the best.
- ...
- And if care was not taken in the breeding, your dogs and birds would greatly deteriorate?
- Certainly.
- And the same of horses and animals in general?
- Undoubtedly.
- ... the same principle holds of the human species! ... the best of either sex should be united with the best as often, and the inferior with the inferior, as seldom as possible ... if the flock is to be maintained in first-rate condition.

Plato, *Republic*, 380 BC.

Later in the text, Plato proposed to foist this artificial selection on the unsuspecting population through deception by the ruling elite. However, for a long time, such ideas mostly remained academic. A weird exception occurred in the 18th century in Prussia, where king Friedrich Wilhelm I commanded his tall male subjects to marry tall women, in order to provide recruits for his elite regiment known as Potsdam Giants.

Darwinian revolution strengthened the case for artificial selection of humans. If evolution by natural selection produced all the beautiful species around us, and artificial selection led to the origin of countless useful varieties of domesticated plants and animals, why not try to achieve something similar on ourselves? The key role in developing such reasoning was played by Francis Galton, Darwin's cousin and a pioneer of studies of quantitative inheritance. In 1883, Galton coined the term eugenics, which he defined as "... the science that deals with all influences that improve the inborn qualities of a race".

In the early 20th century, the inception of genetics caused a surge of popularity of eugenics in Europe and North America. Studies motivated by eugenics led to a number of advances in human genetics, in particular, in understanding of inheritance of complex traits. However, academic pursuits were only a part of it. Soon, eugenics developed into a diverse movement that was active in many countries and incorporated a wide variety of scientific and political ideas. Still, this movement had a common aim – to manipulate human hereditary traits through artificial selection. This selection could involve discouraging procreation of those who possess "undesired" phenotypes (negative eugenics) and encouraging procreation of those who possess "desired" phenotypes (positive eugenics). In deciding what is desired some eugenicists concentrated on between-population differences and others on within-population variation.

In retrospect, it is astonishing that many eugenicists were willing to do something on humans on the basis of rather rudimentary knowledge. Indeed, the very nature of hereditary variation remained enigmatic at that time, as the discovery of the central role of DNA was still decades away. Thus, crucial issues of efficiency and possible unintended consequences of artificial selection of humans could not be then addressed properly.

Naturally, eugenicists actively discussed the possible means for achieving their goals. Galton was ambiguous on this subject. On the one hand, he believed that preferential procreation of individuals with desired phenotypes can be achieved mostly by influencing the "Popular Opinion". On the other hand, with regard to people with undesired phenotypes he stated that "... if these continue to procreate children, inferior in moral, intellectual, and physical qualities, it is easy to believe the time may come when such persons would be considered as enemies of the State, and to have forfeited all claims to kindness". The 20th century eugenics inherited this ambivalence, as some eugenicists were content to rely only on persuasion, while others were ready to resort to coercion.

On top of the common aim, all hues of eugenics possessed another fundamental unifying feature. Almost without exception, their primary goal was to influence properties of human populations and societies, and not of individuals. One might think that the welfare of the population and of individuals should be in harmony, because a fit population must consist of fit individuals, and improving the gene pool of the population at the same time prevents a lot of individual suffering. This is, of course, true; however, actions aimed at long-term welfare of the population can be inconsistent with the welfare of some of its current members.

Disregard of interests and rights of individuals expressed by some eugenicists was nothing short of grotesque. For example, in a book *Passing of the Great Race*, published in 1916, an American, Madison Grant, argued that "Mistaken regard for what are believed to be divine laws and sentimental belief in the sanctity of human life tend to prevent both the elimination of defective infants and the sterilization of such adults as are themselves of no value to the community. The laws of nature require the obliteration of the unfit and human life is valuable only when it is of use to the community or race".

Similarly, in 1921 a Russian, Nikolay Koltsov, said, in an address to the Russian Eugenical Society titled "Improving the Human Race" that "If it were proven that, in every war, a stronger, more vital, more eugenically worthy race wins, than, from the eugenic point of view, it would be possible to not protest against wars; in particular, because even the most determined opponents of wars do not deny that wars have their beneficial facets". He concluded this address by pronouncing that "Eugenics is religion of the future, waiting for its prophets".

Both between- and within-population eugenics led to actions by a number of governments. Actions concerned with between-population eugenics, grounded in ancient practices of genocide and segregation, were of two sorts. One directly sought to promote some populations at the expense of others. This infamously culminated in a murder of about 6 million Jews and about 500 000 Romas in 1941–1945 by the Nazis. The other sought to prevent interbreeding of people of different ancestry. This included mandatory sterilization of people of mixed European–African ancestry, also practiced by the Nazis, as well as laws banning marriages and criminalizing sexual relationships between individuals of different "races" that existed in many countries and in most states of the USA. Voters in Alabama removed the last such ban from the state Constitution only in 2000, by a "whopping" margin of 60% to 40%. Fortunately, in the USA all these laws became mute in 1967, after the Loving v. Virginia decision by the Supreme Court.

On top of being immoral, between-population eugenics is also irrational. No human population is genetically "superior", at least as far as deleterious alleles are concerned. Instead, the average genotype contamination is essentially the same in all human populations (see Chapter 11), which is not surprising, as all of them were subject, certainly until very recently, to strong natural selection. Also, there is no evidence of significant genetic incompatibility between even the most distant human populations. One may approve or disapprove of how the former US President, Barack Obama, a biracial person, handled his job, but he is obviously a very intelligent and physically fit individual. Moreover, genotypes of all modern non-Africans carry ~2% of alleles acquired as a result of interbreeding with the Neanderthals, which occurred soon after the great out-of-Africa migration of *Homo sapiens* about 100 000 years ago, and some human populations also carry up to 7% of alleles from the Denisovans or some other enigmatic kinds of humans. *Homo neanderthalensis* and *H. sapiens*, often regarded as different species, are several times more genetically distant from each other than any two *H. sapiens* populations. There is an ultimate irony in interspecific hybrids opposing between-population marriages. In fact, because different populations carry different complements of deleterious alleles, many of which are at least partially recessive, between-population marriages, which reduce the frequency of strongly deleterious homozygotes, should, on average, produce healthier children. This ubiquitous phenomenon, known as hybrid vigor or heterosis, is the other side of inbreeding depression (see Chapter 9).

In contrast to the case of between-population eugenics, government actions concerned with within-population eugenics had no clear historical precedents. They targeted people with "undesired" phenotypes, in particular, those suffering from neurodevelopmental disorders, in two ways. One was compulsory sterilization and the other was murder ("involuntary euthanasia"). Compulsory sterilization laws existed in many countries and many states in the USA. The first of them was introduced in the State of Indiana in 1907, and some of the last ones were rescinded in 1976 in Sweden and in 1996 in Japan. Remarkably, a number of countries with vocal eugenics movements, including Great Britain, never introduced such laws. Totally, several hundreds of thousands of people became victims of compulsory sterilization. Systematic murder was confined to Nazi Germany, where, starting from 1939, about 300 000 people were killed, including several thousands of disabled children.

Thus, the worst crimes committed in the names of between- and within-population eugenics were equally evil. However, in contrast to between-population eugenics, within-population eugenics had a rational aim. Needless to say, ends do not justify means. Still, who would object against improving wellness of further generations, if this could be done by perfectly ethical means?

Even criminal means, however, are not necessarily effective. Artificial selection imposed with within-population eugenics purposes likely led to some reduction of frequencies of the targeted phenotypes, because variation of all human traits is partially heritable (see Chapter 12). This includes even such traits as having tuberculosis, which was one of many grounds for compulsory sterilization: although the disease is caused by a bacterium *Mycobacterium tuberculosis*, the genotype does play a role in whether a person exposed to the pathogen becomes sick. However, this reduction must have been small, because, at the level of the gene pool of the whole population, even the most brutal eugenic actions led to only rather weak selection. To produce rapid changes, selection must involve a large proportion of the population.

Georg Hegel once said that "What experience and history teaches us is that people and governments have never learned anything from history, or acted on principles deduced from it." The eugenics affair proves him wrong: after World War II some lessons have definitely been learned, albeit slowly. By the end of the 20th century, a nearly universal consensus emerged. I would formulate it as follows:

1) Striving to improve the genetic composition of human populations can be extremely dangerous. Perhaps only two other grand designs, to make everybody believe in a true God and to create a perfect society, have demonstrated a comparable potential to cause evil.
2) Poor science applied to humans could be worse than no science at all. Even today, despite enormous recent advances in human genetics, we still cannot confidently predict the possible consequences of manipulating human genotypes (see Chapter 15).
3) Ethics is crucial when humans are involved. Even scientifically sound policies, applied to humans as if they were bacteria, can lead to horrendous abuses.

Still, this consensus is not comprehensive. Indeed, there is no agreement on the key issue. What means of affecting genotypes of further generations of humans could be acceptable, if any?

Opposition to most or even all such means is quite common. For example, the Charter of Fundamental Rights of the European Union broadly prohibits "eugenic practices, in particular those aiming at the selection of persons". In 1989 the Nobel laureate biologist Salvador Luria asked sarcastically, in a letter to the journal *Science* concerned with the plans to sequence the human genome: "Will the Nazi program to eradicate Jewish or otherwise 'inferior' genes by mass murder be transformed here into a kinder, gentler program to 'perfect' human individuals by 'correcting' their genomes in conformity, perhaps, to an ideal, 'white, JudeoChristian, economically successful' genotype?". However, many other people, including myself, do not view every attempt to control human genotypes as inherently evil.

14.2 Modern Practices

In fact, various means of influencing what genotypes future children will possess have been widely practised in many countries for the last several decades. In contrast to early 20th century eugenics, the aim of the modern practices is not to effect changes at the population level, but to enable individuals to control their reproduction. Thus, individuals, and not governments, decide which actions to take, from those that are technically and legally available. Crucially, an individual must always have an option to do nothing. This paradigm is sometimes called liberal or libertarian eugenics, although, due to negative connotations of the word eugenics, this term is not used widely.

Instead, the process of evaluating probable outcomes of reproductive decisions, and of conveying the conclusions to individuals who contemplate these decisions, may be called reproductive counseling (or simply counseling, for brevity). It can occur before or after conception, and can utilize data on both genotypes and phenotypes. Currently, data on complete genotypes are not routinely used for counseling and, instead, only one or several genes are investigated. This practice is known as genetic screening. Preconception counseling can be either "premarital", offered before the potential parents have any children, or "family", offered to parents that already have at least one child. Currently, no government mandates any actions based on the results of any form of counseling, at least legally.

Deliberate choice of partners for reproduction is an important adaptation for having more numerous and fit offspring. It is ubiquitous within the animal kingdom and, of course, humans are no exception. However, natural mate choice is based only on phenotypes. The rationale for premarital counseling is that phenotypes can be misleading, if deleterious alleles remain phenotypically cryptic. Thus, data on the genotypes of would-be parents may help to predict phenotypes of their future children.

The simplest and the most practically important situation of this kind appears when both parents are healthy heterozygous carriers of pathological recessive alleles of the same autosomal gene, because their child will have a 25% chance of being either a homozygote, if these alleles are identical, or a compound heterozygote, if they are different (Figure 2.3 and Figure 2.17). Either way, the child will have the corresponding Mendelian disease, at least in the case of fully penetrant loss-of-function alleles. Similarly, if a woman is a heterozygous carrier of a pathological recessive allele of an X-chromosome gene, 50% of her sons will have this allele on their only X chromosome and, thus, will be affected, regardless of the genotype of the father.

Current guidelines recommend "pan-ethnic" premarital screening of all people in the USA for only two particularly common autosomal recessive diseases, cystic fibrosis and spinal muscular atrophy, for which the overall frequencies of carriers are 1/30 and 1/80, respectively. However, in some populations recessive deleterious alleles of one or several autosomal genes are anomalously common, apparently because they offer, when heterozygous, some protection against a locally prevalent disease (see Chapter 10). Salient examples are alleles which, when homozygous, cause sickle-cell anemia (several African and South Asian populations, in some of which the frequency of carriers is 1/7), thalassemia (a number of Mediterranean and Middle Eastern populations; e.g., in the Republic of Cyprus the frequency of carriers is 1/7), and Tay–Sachs and several other diseases (e.g., Ashkenazi Jews, in which the frequency of Tay–Sachs disease carriers is 1/25). In such populations, screening of prospective parents for the locally common disease-causing alleles can make a big difference. In several countries premarital counseling that involves genetic screening for such alleles is a condition of legal marriage. Still, a prospective couple can marry regardless of the outcome of this screening.

Genetically, the choice of a sperm donor is analogous to the choice of a mate. However, each donor can potentially become a biological father of over 100 children. Also, rejecting a particular anonymous donor is not involved with any emotional cost, unlike a decision to cancel a planned marriage. Thus, genetic screening of potential donors makes a lot of sense. Nevertheless, there is no consistent approach to this issue. In particular, the USA does not mandate any screening, and cryobanks that provide donor sperm follow rather different practices. Some of them screen all donors only for disease-causing alleles of cystic fibrosis and spinal muscular atrophy genes, and others screen for alleles that can cause over 20 recessive Mendelian diseases.

Family counseling is particularly important if something is seriously wrong with at least one of the children that the couple already has, because this increases the odds of the same disease in their future children. The first, crucial step is to establish the diagnosis for the sick child. If a child of any sex has an autosomal recessive Mendelian disease, any future child has a risk of 25%, and if a boy has an X-linked recessive disease, only future boys are at 50% risk. In the case of a dominant disease, inherited from one of the parents but not manifested in her or him due to incomplete penetrance, the risk is equal to half the penetrance of the pathological allele.

Until recently, it was thought that if a child has a Mendelian disease due to a *de novo* mutation that occurred in one of the parents (see Chapter 13), the risk of the same disease for future children is not elevated, because the mutant allele should be absent in other gametes produced by the same parent. Now we know that this is not true, because the *de novo* mutation that caused the disease is often a part of a cluster (see Chapter 4), especially if inherited from the mother. Empirically, the risk for other children can be as high as several percent.

If a child of the couple has a genetically complex disease, the risk for future children also has to be established empirically. For example, siblings of a child with autistic spectrum disorder (ASD) have 6–8% chance of also having this condition. In this case, the risk is higher if the affected child is a girl, apparently because it takes a larger complement of ASD-causing alleles to produce symptoms in a girl than in a boy.

A principal limitation of preconception counseling is that the genotype of the would-be child is still unknown, so that all predictions regarding their phenotype must be based on Mendel laws and, thus, can only be probabilistic. Postconception counseling is free from this limitation. The genotype of a would-be child may be investigated either in the course of reproduction by *in vitro* fertilization (IVF) before an embryo is implanted ("preimplantation diagnosis") or during pregnancy ("prenatal diagnosis"). Usually, several embryos are produced in preparation for IVF, and preimplantation diagnosis may help to determine which of them will be implanted. Prenatal diagnosis may lead to abortion of the affected fetus or, rarely, to *in utero* therapy. Recently, non-invasive prenatal diagnosis, based on fetal DNA that is normally present in the maternal blood, became possible starting from the 10th week of pregnancy. Prenatal diagnosis by means of ultrasound imaging can also detect serious anomalies of the fetal phenotype.

In most industrialized countries, testing fetal phenotypes is advised for all pregnancies. In contrast, current guidelines do not recommend any genetic screening of embryos or fetuses, unless their risk of having a Mendelian disease is known to be elevated. If a mother is older than 35 years, this substantially increases the risk of a *de novo* aneuploidy. In all other cases, *de novo* mutations are individually too rare to justify genetic screening of all fetuses at the current level of knowledge (see Chapter 15). Thus, this screening is performed only if genotypes of the parents imply a high risk due to transmission of pre-existing pathological alleles, for example, if both of them are known to be heterozygous carriers of the same autosomal recessive disease.

If preconception counseling established a high disease risk, the would-be parents generally have the following options: (i) take chances and do nothing; (ii) do not have children together; or (iii) use postconception counseling. The second option can take many forms. The bride and groom can call off a planned marriage, or marry but refrain from having biological children and, perhaps, adopt, or marry and procreate by using donor sperm for *in utero* insemination (IUI) or donor eggs for IVF. However, by far the most commonly used option is the third one. Because IVF is expensive and invasive, postconception counseling usually relies on prenatal diagnosis.

For example, in the Republic of Cyprus, although its population overwhelmingly belongs to the Eastern Orthodox Church which generally opposes abortion, near-complete eradication of thalassemia was achieved almost entirely by genetic screening of fetuses and abortion of those affected. Universal premarital screening plays only a secondary role, alerting carrier couples (about 1 in 50) to the 25% risk. The same is true for eradication of Tay–Sachs disease among secular Jews in Israel.

Even in traditionalist Saudi Arabia, where marriages are arranged and abortion of affected fetuses is illegal, only a fraction of marriages are called off after premarital counseling establishes that the prospective spouses are both carriers of thalassemia. Apparently, only in several Ultra-Orthodox Jewish communities premarital counseling alone was almost 100% effective in eradicating Tay–Sachs and several other autosomal recessive diseases common among Ashkenazi Jews, because almost all at-risk marriages are called off.

Abortion due to genetic diseases or phenotypical anomalies of the fetus is legal in almost all industrialized countries with a few exceptions, such as Ireland. For example, in France ~0.5% of pregnancies are now aborted for these reasons, which led to a drastic reduction of birth frequencies of a number of severe conditions. Even in the case of Down's syndrome, 80% of affected fetuses are aborted there, despite this condition being not as grave as many other genetic diseases.

Although the aim of reproductive counseling is wellness of individuals, it, nevertheless, affects the gene pool of the population. When carriers of an autosomal recessive disease abstain from marrying each other and choose other partners, this leads to an increase in the frequency of the disease-causing allele, relative to what would happen if they intermarry and produce sick children who would not reproduce. In contrast, if a woman who carries an X-linked recessive allele abstains from reproduction, this purges this allele from the gene pool. However, such effects are small and apparently do not guide individual reproductive decisions.

All the practices described above are passive, in the sense that they do not involve deliberate changes of germline genotypes (germline genotype modification, GGM). So far, there is only one GGM technique that can be applied to humans, being approved in Great Britain in 2015. This technique exploits the fact that eukaryotic mitochondria have their own genotypes (see Chapter 1). When these genotypes carry a disease-causing mutation, it can be replaced, together with mitochondria, by donor mitochondria with normal genotypes. One way of doing this is to transfer the maternal nuclear genotype ("spindle") from the mother's egg into the donor egg with healthy mitochondria in its cytoplasm, whose own spindle was removed, and to fertilize the resulting egg with the father's sperm, in preparation for IVF (Figure 14.1). Then, the child develops from a zygote in which only the nucleus belongs to their parents, and 16 569 nucleotides

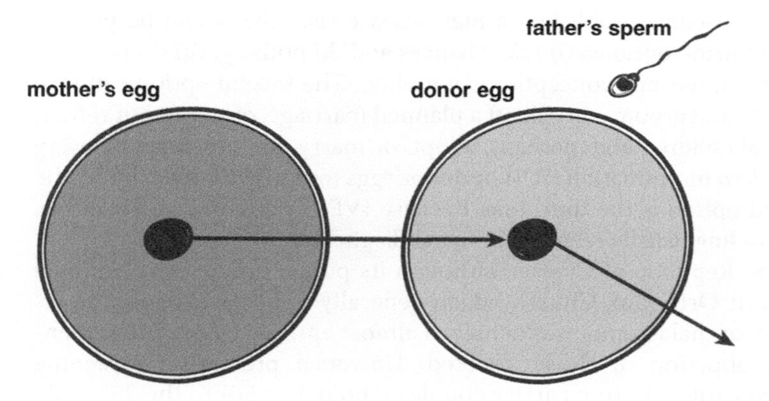

Figure 14.1 Replacement of the nuclear genotype of the donor egg with that from the mother's egg.

of their genotype, out of 3.2 billion, come from the woman who donated her egg (in mammals, the sperm does not transmit mitochondria). This complicated procedure would be unnecessary if the pathological allele (usually, just one wrong nucleotide) could be reverted to its normal state in the genotypes of the maternal mitochondria; however, this is currently impossible (see Chapter 15).

14.3 Humanist Ethics and the Main Concern

Modern practices described above are legal in all or almost all democratic countries, where they are viewed by the majority as ethically acceptable. On what grounds? Unfortunately, ethical problems are fundamentally different from scientific problems, considered in previous chapters. Revealing mysteries of nature may take a lot of time, effort, and knowledge. Still, contrary to what some philosophers would say, definite truths about nature exist and can be discovered. The Earth is (approximately) spherical and not flat, a molecule of water consists of two hydrogen and one oxygen atoms, and there are 23 chromosomes in the haploid human nuclear genome. These claims eventually gained the status of well-established facts, which will never be revised and which no reasonable person should dispute.

In contrast, there is no general agreement on even the basic principles of ethics. When we state that "It is wrong to torture a child", is this an expression of objective, absolute truth – a moral fact – or just a social convention? Are we moral beings endowed with free will and thus responsible for our actions, or just aggregates of atoms, molecules, and cells, so that our experience of responsibility and of good and evil is an illusion? All this is still debated.

Pure logic is incapable of offering guidance on ethical principles, and natural sciences cannot provide any help either. Of course, science can, or at least should strive to inform us what would be the consequences of our possible actions (or inactions), but it is mute on what is right and what is wrong. As David Hume emphasized, one cannot derive values from facts. Only an undisputed authority, such as a Holy Scripture, can provide a firm foundation for ethics. The trouble is, different people accept different undisputed authorities, and many accept none.

Thus, the only consistent approach appears to be the one espoused by the American Founders, who opened the Declaration of Independence by simply stating "We hold the following truths self-evident…". But, again, different truths may be self-evident to different people at different times. As a striking example, a number of industrialized countries evolved, in the course of several decades, from criminalizing homosexuality to legalizing same-sex marriages. Such evolution is hard to predict, and one cannot be sure that a currently predominant moral opinion will never change. Ethical consensuses are notoriously difficult to reach, and well-meaning people may hold irreconcilable views on important issues.

Still, in order to proceed, we need some point of departure. I will adhere to what can be called humanist ethics, which is widely – although by no means universally – accepted in modern industrialized societies. Its key premise has been formulated by Immanuel Kant (see epigraph). An almost universal rejection of this premise by early 20th century eugenics, which treated individuals not as ends, but as means of producing populations that would conform to this or that design, was its fundamental flaw. In contrast, the standards of decency of our time demand that we accept Kant's premise in its strongest form and attribute the same dignity and rights to every human being (one may add individuals of some other species to those who are so protected – here, this serious issue is irrelevant) who is aware of their existence – young or old, smart or ignorant, healthy or sick, rich or poor.

Two complementary facets of humanist ethics are essential for us. On the one hand, any act that does not harm any person is acceptable. This does not mean that it is OK to blow up ruins of ancient cities or to pollute the atmosphere – but only because we humans lose as a result. No evil could be committed on Earth if it were not inhabited (and will never be visited) by sentient beings. A humanist does not accept that an act can be wrong only because it allegedly contradicts God's will, is unnatural, violates a tradition, is inconsistent with the beliefs of the majority, and so on. For example, voluntary sterilization, by tube ligation in women and vasectomy in men, as well as any means of safe preconception birth control is definitely acceptable for a humanist. In contrast, teachings of the Roman Catholic Church which prohibits using condoms – even for married couples, even when one of the spouses is infected with HIV – look monstrous from this perspective (recently, Pope Francis relented a bit).

On the other hand, infringing on the rights and interests of a human being is prohibited, with the only possible exception being when it is done to protect other human being(s). This premise, known as the harm principle, has been formulated by John Stuart Mill, who said that "the only purpose for which power can be rightfully exercised over any member of a civilized community, against his will, is to prevent harm to others". According to modern standards, this harm must be substantial and immediate, so that remote dangers of "deteriorating gene pool" or "degeneration of the human race" cannot trump liberties of an individual. Thus, mandatory sterilization as a means to check the growth of the human population or the accumulation of deleterious alleles in it is out of the question for a humanist, at least before people are about to start dying of starvation or the society is on the verge on unraveling under the burden of mass hereditary disability.

On top of prohibitions, any ethics also implies demands. The most fundamental demand of humanist ethics is to act for the benefit of humans, as long as nobody is harmed in the process. In particular, withholding needed medical care is immoral and may be criminal. In most civilized countries, with the notable exception of the USA, access to health care is recognized as a fundamental human right.

This demand can be extended to children who are not yet born. In 2001, the Australian bioethicist Julian Savulescu formulated a principle of Procreative Beneficence. He considered a standard situation in IVF, where several embryos are produced, but only one is to be implanted. According to Savulescu, "couples (or single reproducers) should select the child, of the possible children they could have, who is expected to have the best life, or at least as good a life as the others, based on the relevant, available information".

Thus, Procreative Beneficence justifies influencing genotypes of children through artificial selection. One who believes that human life begins from conception may object to selection and, thus, destruction of embryos. However, selection is only one of several means that can potentially address the Main Concern (see Chapter 15). Let us for a while concentrate on GGM, which, in my opinion, can be ethically uncontroversial for a humanist, at least in principle.

I believe that Procreative Beneficence can be extended to children not yet conceived, assuming that GGM eventually becomes technically feasible. The only difference is that, in the case of IVF one of the embryos must be selected, while in the case of GGM there is a default option of doing nothing. Thus, for this case I would reformulate the Procreative Beneficence principle as follows: "If this is possible, couples (or single reproducers) should modify their germline genotypes in such a way that the child they will conceive is expected to have a substantially better life than a child conceived without any modification, and the best life, or at least as good a life as a child conceived after any other feasible modification, based on the relevant, available information".

Imagine, for the sake of argument, that there is a pill that reverts some, or even all, clearly and unconditionally deleterious alleles (such as loss-of-function alleles, see Chapter 2) in my germline cells to their normal alleles, without any side-effects. Such reversion would definitely reduce the risk of overt disease and improve wellness of my children quantitatively, possibly by a lot (see Chapter 15). Then, Procreative Beneficence implies that I not only may but even have a moral obligation to take this pill before reproducing. From the point of view of humanist ethics, I do not see how this conclusion could be avoided, because nobody is harmed by this so far unavailable pill. It would be irrational and wrong to reject a treatment that could benefit a child because of crimes committed by the Nazis. Durability of improvements that such a pill could cause, which will be passed on to further generations of my descendants, is its extra benefit.

Let us briefly examine, from the point of view of humanist ethics, two arguments that are often advanced against all forms of GGM. First, sometimes it is asserted that eliminating deleterious alleles is demeaning to those people who carry them. I do not think this is true, although the situation may be viewed as an existential paradox. Once there is a human being, they have the same amount of humanity, regardless of the quality of their genotype. However, this is perfectly consistent with wishing that future humans are deleterious alleles-free. For example, parents who love their son with Duchenne dystrophy can hardly wish their next son to have this tragic condition, and routinely use prenatal diagnosis to rule this out. The same is true, of course, for diseases without a hereditary cause. As Savulescu argued, laws that require seatbelts do not mean disrespect for paraplegics. Humans cannot be undesirable, but impaired wellness can and is. Also, the genotype of every human suffers from a profound weak imperfection (see Chapter 10) and can, therefore, benefit from GGM (see Chapter 15).

Secondly, it is often said that GGM is immoral because it will likely produce genetic haves and have nots. Indeed, it least initially, GGM probably will not be available to everybody. This is sad, but not new. Rejecting a beneficial treatment because it is not available to every person on Earth would mean shutting down all health care in industrialized countries.

Blanket opposition to GGM in humans often goes hand-in-hand with irrational antiscience attitudes. People and organizations who oppose it as a matter of principle often also oppose genetic modification in agriculture or all forms of birth control. Paradoxically, such attitudes are common at the opposite fringes of the ideological spectrum. Usually, they boil down to a feeling that genotypes are "natural" or "sacred" and therefore should not be messed with.

However, I do not see why unconditionally deleterious alleles that reside in my genotype are any more natural or sacred than pathogenic bacteria that reside in my throat. Thus, reverting these alleles, if it becomes possible, should not be any more controversial that using antibiotics against *Streptococcus*. Also, I see no intrinsic ethical differences between preventing the appearance of new mutant alleles and reverting the already-present ones. A would-be-father obviously should not introduce extra deleterious alleles into his germline – and thus must avoid exposure to radiation and chemical mutagens (see Chapter 4). If efficient passive antimutagens were available, it is hard to think of any rational reasons for him to not use them. If so, what is inherently wrong with active antimutagenesis?

Many breakthroughs in medicine were initially met with opposition, which, in hindsight, often looks callous and plain stupid. Vaccination, antiseptics, and anesthesia are famous examples. I hope that not reverting to their normal states hundreds of broken or altogether missing genes in gametes from which children are to be conceived will eventually become as unacceptable as not vaccinating children against polio and measles. We are content to entrust precious genotypes of our offspring to a Russian roulette of random mutation, Mendelian segregation, and meiotic recombination only because currently there are no other options.

14.4 The Main Concern and Ethical Dilemmas

Even if we agree that GGM can be acceptable and even morally necessary under some conditions, this does not mean that dealing with the Main Concern never raises difficult ethical issues. It does. Unfortunately, accepting humanist ethics (or any other ethical principles) is not always sufficient to determine the right course of action. Instead, there may be situations, known as ethical dilemmas, where, even on the basis of the same ethical principles, well-meaning people may reach opposite conclusions. In the framework of humanist ethics, such dilemmas readily appear when interests of different humans clash. At least four ethical dilemmas are relevant to the Main Concern.

First, is it always ethical for an individual to reproduce? This question arises if one knows that their child is at an elevated risk of a genetic condition. Should a couple keep procreating after having an autistic child, despite the probability of ~7% that a later child will be similarly affected? Should a couple conceive a child without an intention to use prenatal diagnosis if they know that both of them are heterozygous carriers of a severe autosomal recessive disease, so that with a probability of 25% their child will be tragically ill? Rarely, a healthy woman nevertheless carries a deleterious mitochondrial mutation in most or even all of her germline cells (due to a very large cluster of mutations, Figure 4.7), in which case all her offspring may be condemned to suffering and early death. If the situation is known to her, should she have children? Contemplating this dilemma, one needs to keep in mind that even in the absence of any obvious risk factors, a child will be born with a serious disease with a probability above 1% (see Chapter 12). Thus, short of declaring all procreation immoral, we perhaps need to draw a line – but where?

Secondly, when is it ethical for an individual to influence genotypes of their children? Because safe preconception GGM, a preferable option, is not yet possible, let us concentrate on postconception selection. Infanticide is now almost universally viewed as murder, but abortion is the subject of endless debates. Obviously, from the moment when the fetus becomes a human being, abortion is unacceptable. But when does this happen? Traditional teachings of Judaism, Christianity, and Islam claim that a fetus acquires a soul 40 days, immediately, and 120 days after conception, respectively. Currently, many people, Christians and not, believe that a zygote is already a human being. Against this view, one can argue that a zygote does not always develop into a child even under the best conditions – most pregnancies terminate spontaneously at very early stages (see Chapter 11), a brainless anocephalus can be born, and in ~1% of cases a zygote gives rise to two or even more identical twins. At the other extreme is a belief that abortion is acceptable at any moment until birth. An obvious counterargument is that a viable child can be delivered by a Cesarean section starting from the ~35th week of pregnancy. In the USA, abortion is legal until the fetus becomes viable outside the womb, which is as arbitrary a criterion as any other. If, at some point in the future, it will become possible to support the whole course of development of a human zygote artificially, would this make all abortions illegal?

Even if abortion may be acceptable, what features of the would-be child can be moral grounds for it? Inevitable painful early death? Grave disability? Mild disability? Undesired sex? In particular, many people find abortion due to the last reason at least morally suspect. Still, selecting fetuses by sex is widespread in a number of countries, and led to a large excess of boys over girls among the newborns in India and China.

This dilemma is going to become even more relevant. Determining the complete genotype of a fetus after the 10th week of pregnancy by non-invasive means will soon become routine. Our understanding of the relationship between genotype and phenotype is also going to improve. Thus, there will be women seeking abortion because they believe – correctly or not – that their child will possess too many deleterious mutations, a low intelligence, or an undesired skin color.

Thirdly, is it ethical for governments to regulate any reproductive decisions by individuals? This issue is different from that of ethics of these decisions, because even immoral does not necessarily mean illegal. The government may want to do two things, to prevent conception or birth of children who would be condemned to grave suffering, and to protect fetuses.

Currently, governments of industrialized countries do not interfere with individual decisions to conceive children. For example, the Health Code of California states: "The extremely personal decision to bear children should remain the free choice and responsibility of the individual, and should not be restricted by the state." This hands-off approach can be supported by a number of arguments. One is that any interventions are fraught with abuses, as the history of eugenics abundantly demonstrates. Another is that a law aimed at preventing birth of children with inherited diseases would be impossible to enforce, without draconian measures. Are we going to criminalize sex between carriers of the same autosomal recessive disease or to mandate prenatal diagnosis and abortion of affected fetuses conceived by two such carriers? Thirdly, laws of such kind can be counterproductive. It is widely accepted, for example, that criminalizing exposure of the fetus to illegal drugs discourages addicted women from seeking treatment. Who would benefit from jailing parents who produced their second child with Tay–Sachs disease?

However, there could also be serious arguments against this approach. In particular, there is no consistency in how governments deal with this issue and with some other issues. Incest involving consenting adults is a felony in many jurisdictions, but unprotected sex between carriers of the same autosomal recessive disease is never a crime, although in the second case the conceived child will be at a greater danger. Also, there is a striking contrast between how society deals with vertically (genetic) versus horizontally (infectious) transmitted diseases. Having sex is also an extremely personal decision, but, nevertheless, it is a crime to knowingly infect a partner with a sexually transmitted disease. In some jurisdictions even informing a partner, in advance of consensual sex, that you are infected does not absolve you from criminal responsibility. If transmitting syphilis to a consenting partner is a crime, why is transmitting a fatal allele to a child, who cannot give their consent before being conceived, not?

In contrast to the prevailing hands-off approach to individual decisions to conceive children, governments do try to regulate abortion. However, the relevant laws vary widely. Early abortion is legal on demand in a majority of industrialized countries but criminalized in some (with or without exceptions, such as for saving the pregnant woman's life). Even in countries where abortion is legal, a number of governments try to prevent selection of fetuses on the basis of their sex, by withholding this information from the would-be parents. However, this will soon become impossible, when early determination of the sex of a fetus using a simple blood test, instead of ultrasound investigation, becomes routine. Similarly, banning abortions "solely" due to disorders diagnosed in the fetus, as it has recently been done in Indiana, will likely have little effect, except to make women who seek abortion for this reason to conceal their motives.

Anyway, this third dilemma is only marginally relevant to harmful effects of numerous mildly deleterious alleles, which are the leading cause of the Main Concern. Although cumulatively they cause a lot of harm, such alleles usually do not lead to severe impairment of individual children (see Chapter 12). As long as this remains the case, humanist ethics prohibits any restriction of individual liberties with the aim of controlling such alleles.

Fourthly, could it be ethical for governments to try to address the Main Concern using soft power, for example by giving financial incentives to procreate only to people with high intelligence, such as university graduates? This proposition, reminiscent of positive eugenics and briefly tried in Singapore over 30 years ago, looks morally suspect, and could be a slippery slope on the way to explicit coercion. Still, it seems that humanist ethics does not prohibit such measures outright. And are they really worse than forcing workers who earn the legal minimal wage to choose between buying food and medicines?

How are the modern practices of reproductive counseling related to these dilemmas? Preconception counseling does not involve any of them and, therefore, few people object to it as a matter of principle. At least from the point of view of humanist ethics, it is certainly right to warn prospective parents who ask for medical advice that their children would be at a grave risk. In contrast, postconception counseling and, in particular, prenatal diagnosis, are inherently controversial. Nevertheless, even many people and organizations who oppose abortion in most situations believe that it is acceptable to prevent the birth of a severely ill child.

If and when GGM becomes technically possible, it will surely lead to new ethical dilemmas. Two kinds of them can be anticipated.

First, GGM cannot be completely safe, and a balance between its benefits and risks will need to be found. Would it be ethical to perform GGM which increases the expected intelligence quotient of a child by 15 points and their life expectancy by 10 years if it introduces an unintended drastic deleterious mutation with probability 10%, 1%, or 0.1%? Of course, any medical intervention is involved with this problem, and a risk which is acceptable for a life-saving coronary bypass (~1% death rate) rules out a cosmetic surgery.

Secondly, there will be an issue of choosing the target of GGM. Safely reverting an unconditionally deleterious allele may be uncontroversial, but what about an allele that is deleterious with probability 95% but beneficial with probability 5%? And will it be moral – and should it be legal – for a person to modify their germline with the aim that their children have a particular skin or eye color? The current trend is to give people more reproductive choices – but how far should it go?

With the only exception of mitochondrial replacement, modifying genotypes of future children has never been attempted, and is illegal in many countries. As long as the available techniques remain imprecise, this is justified. Because GGM cannot be used to cure an already-existing sick person, using it cannot be ethical unless the risk of serious complications is, at the very least, not higher than that involved with natural procreation. If and when suitable techniques are developed, and if GGM is shown to substantially increase the wellness of offspring (see Chapter 15), its blanket prohibition will become untenable in modern democracies. Of course, for governments and societies that do not respect human rights, GGM may provide a yet another opportunity to violate them.

Advances of science not only open new possibilities but also create new ethical dilemmas. Dramatic improvements in medical care created dilemmas concerned with the definition of death, euthanasia, and continuing life support. One does not need to be a prophet to predict bitter disputes over what can and cannot be done in order to address the Main Concern, using tools that will likely become available in the future.

14.5 Role of Scientists

What is the proper role of natural scientists in the ongoing and future debates about ethical issues and public policies relevant to the Main Concern? I believe that this role should not be central. A claim about properties of nature can be true or false, but not good or evil, and these properties can be discovered by making mathematical models, observations, and experiments. In contrast, there is no rational way to resolve an ethical issue. Appellations to beliefs and emotions play the key role in ethical debates, which makes them, to say the least, frustrating for a scientist, because in our profession the search for truth works completely differently.

Also, the history of eugenics provides yet another lesson, specifically for scientists: be extremely careful trying to apply your knowledge to human affairs. Indeed, many biologists tainted their legacies by making naive, tasteless, or even outright disgusting statements and proposals in this regard. As an example, let us compare two American biologists – a well-known but mediocre conservative, Charles B. Davenport, and a genius, leftist, Hermann J. Muller. Ironically, their activities aimed at improving the human gene pool were similarly flawed.

A successful zoologist, Davenport was a professor at Harvard and later director of the Cold Spring Harbor Laboratory. He became one of the driving forces of the American eugenics movement, and in 1916 formulated the "Eugenics Creed". It reads:

> I believe in striving to raise the human race to the highest plane of social organization, of cooperative work and of effective endeavor.
>
> I believe that I am the trustee of the germ plasm that I carry; that this has been passed on to me through thousands of generations before me; and that I betray the trust if (that germ plasm being good) I so act as to jeopardize it, with its excellent possibilities, or, from motives of personal convenience, to unduly limit offspring.
>
> I believe that, having made our choice in marriage carefully, we, the married pair, should seek to have 4 to 6 children in order that our carefully selected germ plasm shall be reproduced in adequate degree and that this preferred stock shall not be swamped by that less carefully selected.
>
> I believe in such a selection of immigrants as shall not tend to adulterate our national germ plasm with socially unfit traits.
>
> I believe in repressing my instincts when to follow them would injure the next generation.

For a modern reader like myself, this Creed sounds ugly. A human being is, first and foremost, a person, and not a trustee of their germ plasm. A supremacist belief that those who subscribe to this Creed have particularly valuable genotypes is self-serving, unfounded, and dangerous. And immigrants, adulterating national germ plasm, sounds racist. Davenport's activities likely contributed to the persistence of racial segregation in the USA.

However, even if every article of this Creed were agreeable in substance, I still think that a scientist who makes such pompous pronouncements embarrasses themselves. Who is Davenport (or me, or any other scientist) to tell people how many children to have? Who cares what we believe regarding how people should behave? We better leave pontificating about such matters to political leaders, religious gurus, and cultural icons.

In contrast to Davenport, Muller was not a part of the American establishment. In 1932, five years after discovering that X-rays are mutagenic (this work eventually earned him a Nobel Prize), he resigned from his job at the University of Texas. He left the USA first for Germany but soon, after Hitler came to power, moved to the USSR. In 1936, Muller wrote a long letter to the Soviet tyrant Joseph Stalin, one of the worst mass murderers in history. Mixing science with Marxist demagoguery, Muller argued that the USSR should institute a large-scale program of artificial insemination of the willing women by the sperm from "transcendently superior" donors, each of which could thus sire tens of thousands of children. Ludicrously, he claimed this proposal to be unrelated to "bourgeois" eugenics.

If Stalin had followed Muller's advice, this would likely have led to violations of human rights at a scale much larger than eugenics caused in the USA, because it was not in the nature of the communist dictatorship to allow people to make choices. However, Stalin was hostile to all things genetic. This soon led to a pogrom of Soviet genetics and to the murder or imprisonment of a number of scientists, but, ironically, also saved the Soviet people from yet another abuse and spared Muller the disgrace of being its intellectual driver. In 1937, Muller managed to escape from the USSR, thoroughly disillusioned with communism. He eventually returned to the USA, secured a permanent job at the University of Indiana, won a Nobel Prize in 1946, and kept doing groundbreaking science until his death in 1966.

True, not all ventures of scientists into ethical and political minefields were so misguided. For example, in 1913 the co-discoverer of evolution by natural selection, Alfred R. Wallace, expressed strong skepticism towards eugenics and wisely cautioned that under the then current state of the society it will likely lead to abuses and coercion. Still, such wisdom may appear to be more the exception than the rule.

I do not take sides on an ethical issue, unless the basic tenets of humanist ethics are sufficient to resolve it. I see this as a sensible compromise between entertaining even troglodyte opinions and promoting my own views. In fact, I do not have a definite view on some ethical dilemmas described above. However, even when I do have one, expressing it here would be inappropriate and misleading. As a geneticist, I have a special expertise in the subjects that were discussed in the previous chapters, and speak about them with authority. In contrast, my moral and political judgements are of no particular interest.

A scientist can fearlessly advocate a political position only if it follows directly from knowledge about nature and from the most basic ethical considerations. I feel comfortable to argue that creationism should be banished from schools because it is a lie. Similarly, I am ready to express my support for germline reversion of clearly and unconditionally deleterious alleles, as long as it is guaranteed that nobody could be harmed by it. In contrast, I am not willing to argue whether a person should be allowed to modify their germline to ensure that their children have a desired eye color. When an issue is not ethically transparent, it is suitable for a scientist to play only a limited, although still very important, role of an expert and try to predict what would happen after every possible course of action. Unfortunately, our knowledge is often insufficient even for this, as the issue of reverting deleterious alleles illustrates (see Chapter 15).

Thus, our long-term contribution to society is the greatest if we do exactly what we are trained and paid to do – produce new knowledge. On this road a biologist who studies humans has to deal with two kinds of obstacles that are unfamiliar, for example, to a physicist. One of them is taboos, imposed on research by society. Recently, the journal *Nature* listed four subjects in human genetics that are widely viewed as taboo, which include the nature of within-population variation in intelligence and the nature of phenotypic differences between populations ("races").

I believe that such taboos are misguided, and that only a clear and present danger can be grounds to refrain from research. Artificial creation of a lethal pathogen and development of weapons of mass destruction for a dictatorship are obvious examples. In contrast, to avoid a problem only because somebody may not like the results of its investigation is fundamentally wrong. We have a moral duty to resist such pressures and to pursue truth, which is easier said than done, because our very ability to do expensive research is controlled by society at large. Scientists who are brave enough to work on taboo problems often encounter difficulties and their motives may be viewed with suspicion.

The second kind of obstacles are the ideological biases of scientists themselves. This may sound weird: by the same token by which a property of nature cannot be good or evil, it also cannot be liberal or conservative. Is Mendelian inheritance liberal or conservative? However, this truth is sometimes forgotten when humans are the subject. For example, due to a variety of reasons that have nothing to do with science, the claim that genotypes play a major role in neurodevelopmental disorders and in within-population variation in human intellectual abilities became associated with conservative ideology. In contrast, the claim that the genotype is essential for the sexual orientation of an individual is often perceived as liberal. Both claims are, in fact, true – and this is the only thing that should matter for a scientist. In an ideal world a bleeding-heart liberal scientist and a hardline conservative scientist should reach exactly the same conclusions on the basis of the same data. Unfortunately, scientists who investigate human biology sometimes allow their ideologies to interfere with their work. Of course, this is a mortal professional sin.

Further Reading

Campbell, IM, Stewart, JR, & James, RA 2014, Parent of origin, mosaicism, and recurrence risk: probabilistic modeling explains the broken symmetry of transmission genetics, *American Journal of Human Genetics*, vol. 95, pp. 345–359.
Due to clustering, the presence of a mutation in a child increases the risk for siblings, especially for mutations of maternal origin.

Clayton, EW, McCullough, LB, & Biesecker, LG 2014, Addressing the ethical challenges in genetic testing and sequencing of children, *American Journal of Bioethics*, vol. 14, pp. 3–9.
An ethical dilemma concerned with determining genotypes of children.

Cowan, RS 2009, Moving up the slippery slope: mandated genetic screening on Cyprus, *American Journal of Medical Genetics*, vol. 151, pp. 95–103.
Description of screening for beta-thalassemia on Cyprus.

Gillham, NW 2001, Sir Francis Galton and the birth of eugenics, *Annual Reviews of Genetics*, vol. 35, pp. 83–101.
A brief biography of the founder of eugenics.

Glad, J 2003, Hermann J. Muller's 1936 letter to Stalin, *The Mankind Quarterly*, vol. 43, pp. 305–319.
A cautionary tale for social activist scientists.

Ramsden, E 2009, Confronting the stigma of eugenics: genetics, demography and the problems of population, *Social Studies of Science*, vol. 39, pp. 853–884.
A description of early attempts to draw lessons from the history of eugenics.

Reed, SC 1974, A short history of genetic counseling, *Social Biology*, vol. 21, pp. 332–339.
Reproductive counseling before DNA sequencing.

Savulescu, J 2001, Procreative beneficence: why we should select the best children, *Bioethics*, vol. 15, pp. 413–426.
Principle of Procreative Beneficence.

Sleeboom-Faulkner, ME 2011, Genetic testing, governance, and the family in the People's Republic of China, *Social Science & Medicine*, vol. 72, pp. 1802–1809.
Description of laws and practices concerned with reproduction in China.
Turda, M 2010, *Modernism and Eugenics*, Palgrave Macmillan, Basingstoke.
A concise history of European eugenics from a novel perspective.

15

What to Do?

Do not withhold good from the one who needs it when you have power in your hand to do it.

The Proverbs of Solomon 3:27.

It is likely that most non-neutral mutations are deleterious unconditionally. Reverting all deleterious alleles in a human genotype may produce a substantial improvement of wellness, although we do not know this for sure. Artificial selection in humans is ethically problematic and unrealistic. Thus, it seems that the only possibility to get rid of unconditionally deleterious alleles in human genotypes is through deliberate modification of germline genotypes. Unfortunately, techniques suitable for this task will likely appear not earlier than 10–50 years from now. Data on sequences of complete genotypes may soon become useful for predicting severe diseases in offspring. Cryopreserving your own sperm at a young age can be a wise individual decision. There will be a rapid progress in our understanding of phenotypic effects of deleterious mutations, contingent on a strong support of the relevant research by governments and society.

15.1 Conditionally Beneficial or Unconditionally Deleterious?

The genotype of every human harbors a large number of deleterious alleles, produced by mutation in the past generations (see Chapter 11). These derived, mutant alleles substantially impair wellness (see Chapter 12). Thus, replacing them all with ancestral alleles looks like a worthy goal.

Still, to be sure that this is, indeed, the case we need information on phenotypes that correspond to weakly perfect, deleterious alleles-free genotypes. Would individuals with such genotypes be much better off, in terms of fitness, wellness, or any other important traits, than all of us? Because weakly perfect genotypes are too improbable to appear within any human population (see Chapter 10), we have to resort to indirect inferences. First, we will consider effects of mutant alleles on fitness.

Let us start from the fundamental, qualitative question: how often is a mutant allele deleterious unconditionally? Getting rid of such an allele could do only good, and the more of them that are replaced with normal alleles, the better. True, magnitudes of

Crumbling Genome: The Impact of Deleterious Mutations on Humans, First Edition. Alexey S. Kondrashov.

deleterious effects are very different for different alleles. Nevertheless, an allele which increases the probability of a heart attack by only 0.1%, without ever providing any benefits, is still unconditionally deleterious, albeit not as strongly as an allele which increases this probability by 10%, and we would be better off without either of them. In contrast, replacing an allele that can occasionally be beneficial is not necessarily a good idea, even if its average effect is deleterious.

An allele can be deleterious only conditionally due to two phenomena. The first is sign epistasis, a situation when not only the strength but the very direction of selection at a locus depends on the genetic background (see Chapter 3). Sign epistasis is definitely common at the scale of profound between-species differences. An allele which works well as a part of the octopus genome, being involved in the formation of its non-inverted eye, would not work in the human genome. Conversely, a human allele involved in the development of our inverted eyes would not be fit for an octopus (see Chapter 10). Horse alleles or donkey alleles work well together, but being brought into the same diploid genotype they produce a sterile mule. Thus, there must be loci where the horse allele is superior within the horse genotype, and similarly the donkey allele within the donkey genotype. In other words, at such a locus an allele from one species is incompatible with the rest of the genotype from the other species. Low fitness of hybrids between dissimilar enough forms of life is due to this phenomenon, known as Bateson–Dobzhansky–Muller incompatibility (this is the same Muller as in earlier chapters). In particular, population genetic data suggest that mild incompatibility between some alleles of *Homo sapiens* and *H. neanderthalensis,* which are sometimes regarded as different species, was manifested after their interbreeding, which eventually gave rise to modern non-African human populations.

In contrast, at the level of variation within populations and species incompatibilities between alleles at different loci are rare. Indeed, we attribute different individuals to the same species precisely because of overall compatibility of their genotypes, which allows them to produce viable and fertile offspring together. This rule is not without exceptions: according to some estimates, two *Drosophila melanogaster* genotypes harbor, on average, one or two pairs of substantially incompatible loci. Still, it seems that only a minority of polymorphic loci is involved in within-species incompatibility with other, unlinked loci.

Within-species incompatibility can also involve two polymorphic loci that reside within the same gene (Figure 15.1). In this case, due to tight linkage, alleles at these loci almost always stay in the original, compatible combinations even after meiosis. Thus, such within-gene incompatibilities, and the resulting sign epistasis, are only rarely manifested. Still, they are known for several human genes, although the proportion of polymorphic loci participating in within-gene incompatibilities appears to be small.

Incompatibility and sign epistasis may also appear due to interactions between alleles at the same locus. In humans, the most practically important intralocus incompatibility occurs at locus Rh that encodes the so-called rhesus factor, because, if carried by a rhesus-negative mother (homozygous by Rh^- allele), a rhesus-positive fetus (Rh^+Rh^- heterozygote) may suffer from hemolytic anemia. As a result, alleles Rh^+ and Rh^- are sometimes incompatible with each other.

Perhaps, the most plausible mechanism for sign epistasis is stabilizing selection on quantitative traits (see Chapter 7). In this case, an allele that increases the value of the trait would be beneficial on a genetic background that leads to too small a value of the trait, but deleterious on a background causing too high a value of the trait, relative

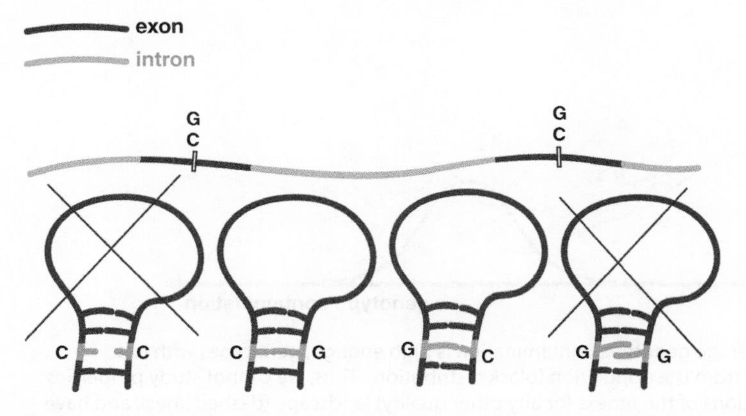

Figure 15.1 Within-gene incompatibility. Two loci (nucleotide sites) within exons of the same gene each harbor alleles G and C. These sites interact within the RNA molecule encoded by this gene, so that Watson–Crick (see Chapter 1) pairs G:C and C:G confer higher fitness and non-Watson–Crick pairs G:G and C:C confer lower fitness (crossed out). Thus, G (C) at one locus is superior when the other locus carries C (G).

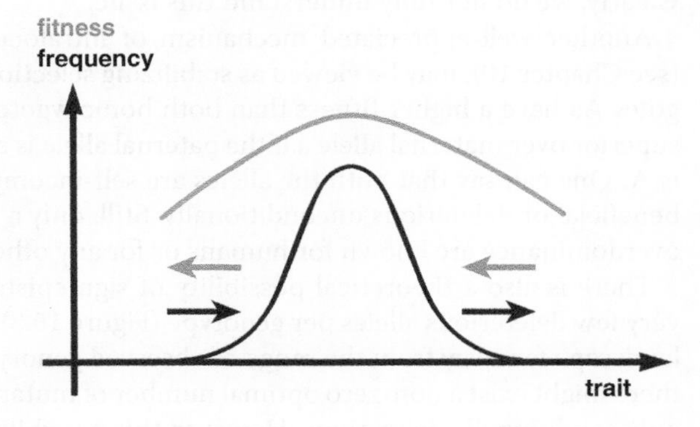

Figure 15.2 Sign epistasis is an unavoidable result of real stabilizing selection. An allele which increases the value of the trait (black arrows) moves the genotype towards the fitness optimum if other alleles produce a small value of the trait and away from it in the opposite case. Of course, the opposite pattern is present if an allele decreases the value of the trait (gray arrows).

to the fitness peak (Figure 15.2). Thus, an allele with a small effect on the trait will be beneficial and deleterious with the same probability of 50%. However, comparison of genomes of different species shows that most derived alleles that are not functionally mute are deleterious at least most of the time, if not always, because they mostly have low frequencies. In fact, it seems that by far the most common effect on the molecular function of a derived allele is its impairment (and not an enhancement or an alteration), and if an impairment were beneficial with probability 50%, all functions would eventually collapse.

A possible explanation is that stabilizing selection is usually not real but only apparent (Figure 7.14), produced by directional selection on correlated, hidden variables, most likely, on genotype contamination (see Chapter 9). Still, it seems that at least sometimes stabilizing selection is real. Both too high and too low a blood pressure is bad *per se*. For a quantitative trait such as the level of expression of a protein-coding gene, real stabilizing selection seems likely, because both too little and too much of the protein must reduce fitness. Sophisticated analysis of some population-level data also implies that at least some stabilizing selection is real. And yet genotype-level

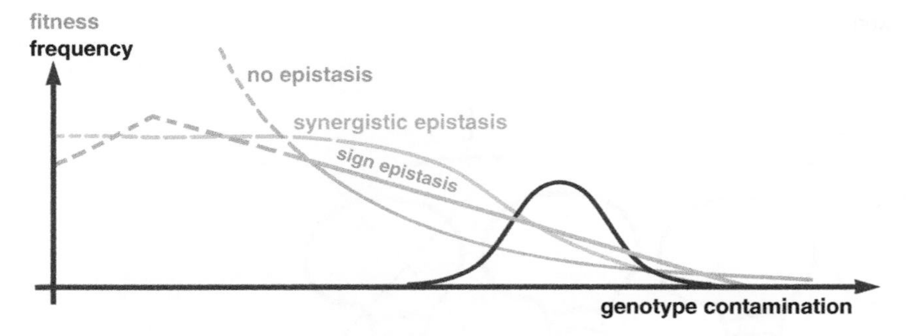

Figure 15.3 When the average genotype contamination is high enough, genotypes with low contamination are absent from the population (black distribution). Thus, we cannot study properties of the corresponding portions of the fitness (or any other quality) landscape (dashed lines), and have to make guesses. Without epistasis, the quality of the mutation-free genotype is the highest, in comparison with all other genotypes, and is very high; under narrowing, synergistic (but not sign!) epistasis it is the highest, but not very high; and under sign epistasis it is not even the highest.

analyses fail to detect any common mode of selection, except a negative one. Clearly, we do not fully understand this issue.

Another well-appreciated mechanism of intralocus sign epistasis, overdominance (see Chapter 10), may be viewed as stabilizing selection within a locus. When heterozygotes Aa have a higher fitness than both homozygotes AA and aa, maternal allele A is superior over maternal allele a if the paternal allele is a, and inferior if the paternal allele is A. One can say that both the alleles are self-incompatible, so that neither of them is beneficial or deleterious unconditionally. Still, only a small number of definite cases of overdominance are known for humans or for any other species.

There is also a theoretical possibility of sign epistasis in the uncharted territory of very few deleterious alleles per genotype (Figure 15.3). Indeed, extrapolating the fitness landscape far away from the range of observed genotype contaminations is risky. Thus, there might exist a non-zero optimal number of mutant alleles, which would make them only conditionally deleterious. However, this possibility appears to be rather remote, as there is no plausible biological mechanism for it.

To summarize, I do not believe that within-species sign epistasis is a common phenomenon. If so, the same allele at a non-neutral locus must confer a higher fitness when present within (almost) every genotype that can be encountered within the species. Admittedly, this conclusion is based only on a preponderance of evidence, and has not been proven beyond reasonable doubt.

The second phenomenon that could make an allele only conditionally deleterious is the existence of multiple fitness landscapes such that the allele is deleterious under some of them but beneficial under others (Figure 10.1), without sign epistasis under any particular landscape. Multiple fitness landscapes may appear due to different controllable environments or different states of the population. At some human polymorphic loci, such as those affecting skin color, alleles are conditionally beneficial, depending on the environment. Frequency-dependent landscapes, under which rare alleles are advantageous, are experienced by polymorphic loci responsible for variation of immune response and, remarkably, of facial features: rare facial features, which make an individual recognizable, were favored in the course of human evolution. We know this because selectively neutral variation of genome regions around these loci is elevated, which is a signature of long-term balancing selection.

Still, such signatures are quite uncommon in the genome. Thus, it seems that different directions of selection under different fitness landscapes is a property of only a small proportion of polymorphic loci. Indeed, most non-neutral polymorphic loci carry a derived allele that simply impairs the function of the affected gene, and it appears that most such alleles cannot be beneficial under any environment or the state of the population.

If, instead of fitness, we consider human wellness, the most intriguing evidence of conditional beneficence of some generally deleterious alleles come from recent reports that at least some alleles that contribute to autistic spectrum disorder (ASD) are associated with higher intelligence in those people who are not clinically autistic, and that alleles that contribute to schizophrenia are associated with creativity in healthy people. Also, according to some data, relatives of people with neuropsychiatric conditions have more children. Thus, it could be that a small number of alleles that cause such conditions when too many of them are present in the genotype can increase wellness and fitness of an individual.

However, these findings need to be confirmed and, anyway, this situation looks more like an exception than the rule. In particular, recent data on people with very high intelligence do not reveal any complex interactions between alleles, which could result in sign epistasis. Instead, concordance between monozygotic twins in belonging to the top 5% of the intelligence quotient (IQ) distribution is 0.45, and for dizygotic twins this concordance is 0.25. Thus, doubling the degree of genotype similarity only doubles the concordance. If exceptionally high intelligence was due to lucky combinations of alleles that individually reduce it, concordance between monozygotic twins would be much higher than twice the concordance between dizygotic twins.

15.2 Mutationless Utopia: What Could It Be?

Let us accept a plausible hypothesis that most non-neutral derived alleles are deleterious unconditionally, so that "the fewer mutant alleles, the better" principle is generally true. Then, the next question is quantitative: how large is the potential benefit for fitness of replacing all deleterious derived alleles in a genotype with the corresponding ancestral alleles? In 1950, Muller guessed that "... the average man must be in one way or another, all told, at least 20% below the par of the fictitious all-normal man". Today we know that Muller substantially underestimated contamination of human genotypes. However, we still do not know the magnitude of human weak imperfection (see Chapter 10) because, in the absence of individuals with "all-normal" genotypes, it can be estimated only in several indirect ways, none of which is really satisfactory.

The principle flaw of these estimates is that weak imperfection strongly depends on the shape of the fitness (or wellness) landscape far to the left from the relatively narrow range of genotype contaminations that are actually present in populations (Figure 15.3). Thus, studying the available genotypes does not shed any direct light on properties of the weakly perfect genotype. If further reduction of even an already low genotype contamination increases fitness substantially, which would be the case if epistasis is absent and selection is exponential (see Chapter 9), weak imperfection can be huge.

In contrast, if fitness landscape is nearly flat where contamination is low, which would be the case under strong synergistic epistasis between deleterious alleles, weak imperfection would be much lower. There are a lot of conflicting data on epistasis, and no clarity has been reached.

Let us start from briefly considering three indirect approaches to estimating the magnitude of weak imperfection that are based on properties of genotypes with typical contaminations. First, we can use data on selection, assuming that all the variation in fitness is due to deleterious alleles. In contrast to imperfection I, evolvability of fitness E can be measured directly, and theory shows that $I \geq E$ (see Chapter 7). A number of measurements of E have been performed on different species, including humans, and some of them produced values as low as ~0.1. In humans the value of E was only ~1 or even less even under Pleistocene and Preindustrial environments (see Chapter 11). Thus, the lowest possible values of weak imperfection are modest. For example, $I = 1$ means that fitness of the weakly perfect genotype is two times above the average fitness (see Chapter 7). However, $I = E$ only under the strongest form of synergistic epistasis, truncation selection, and in general E does not impose any upper limit on I. Thus, data on E underestimate weak imperfection, and we do not know by how much.

Secondly, we can rely on the average genotype contamination C. If human populations are at a deterministic mutation–selection equilibrium, and our estimate of the genomic deleterious mutation rate U (see Chapter 8) is basically correct, $C = U = 10$ (see Chapters 9 and 11). Then, if epistasis is absent, $I = \exp(C)$, implying that a weakly perfect genotype would have fitness ~10 000 times above the population average. However, if a sufficiently strong synergistic epistasis is present and effects of individual mutations are small enough, an arbitrarily low I is consistent with any given U and C, because truncating an arbitrarily small proportion of a population can be sufficient to produce any required selection differential (see Chapter 9).

Thirdly, we can approach our problem from the perspective of within-population genetic variation. Because contamination of human genotypes is mostly due to very numerous alleles with small individual effects, its standard deviation is at least 10 times, and probably more, smaller than its average (see Chapter 11). It seems plausible that weakly perfect genotypes, whose contamination is many standard deviations below the average, possess fitness much above the mean one. However, as before, we cannot be quantitative without knowing the shape of the fitness landscape.

Alternatively, we can try to estimate weak imperfection by measuring fitnesses of genotypes with increased contamination. This approach was proposed in 1956 by Morton and colleagues in the same paper in which they showed that $C = U$ at the mutation-selection equilibrium (see Chapter 9). In contrast to genotypes with substantially reduced contamination, genotypes with increased contamination are not difficult to obtain. Thus, one can attempt to infer properties of the fitness landscape to the left from the range of actual genotype contaminations from those to the right from this range (Figure 15.4). Indeed, if doubling of the genotype contamination reduces fitness, say, by the factor of 5, this implies that eliminating genotype contamination altogether would cause its 5-fold increase, as long as epistasis is absent (see Chapter 9). Genotypes with increased contamination can be produced in three ways, through free accumulation of spontaneous mutations, artificial mutagenesis, and inbreeding.

Each generation of free accumulation of spontaneous mutations increments the average genotype contamination by $U \times \mathrm{Mean}[s]$, where $\mathrm{Mean}[s]$ is the average coefficient

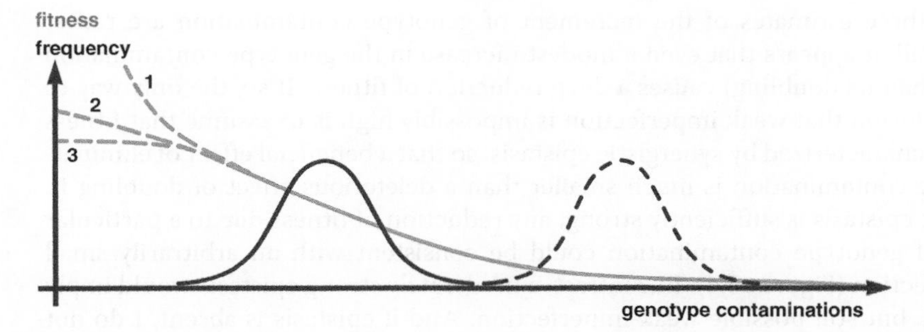

Figure 15.4 Estimating weak imperfection of the population with some distribution of genotype contamination (solid line) by studying a population with incremented genotype contaminations (broken line). Unfortunately, the magnitude of weak imperfection implied by a particular decline of fitness due to increased genotype contamination depends strongly on epistasis (1, no epistasis; 2, moderate epistasis; and 3, strong epistasis).

of selection against a *de novo* mutation. Thus, it takes 1/Mean[s] generations to double the average equilibrium genotype contamination U. Let us assume that Mean[s] = 0.001, so that doubling of the average genotype contamination takes 1000 generations of completely relaxed selection. Our data (see Chapter 9) show that after only 50 generations (almost) without selection the average fitness of *Drosophila melanogaster* population declines by a factor of 2. If so, under exponential selection, after 1000 generations fitness should decline by a factor of $2^{1000/50}$, and we have to conclude that the equilibrium mean population fitness constitutes only $(1/2)^{20} = 0.000001$ of that of the weakly perfect genotype.

A single treatment of *D. melanogaster* males with a powerful mutagen, ethyl methanesulfonate (see Chapter 4), produces as many mutations as ~100 generations of spontaneous mutation. Data obtained by Hsiao-Pei Yang and colleagues suggest that this treatment produces genotypes with fitness reduced by a factor of ~3. If so, the equilibrium mean fitness must constitute $(1/3)^{1000/100} = 1/60\,000$ of the maximal one. Of course, this estimate, as well as the previous one, is very sensitive to Mean[s], which is not known precisely.

The above two methods cannot be applied to humans. In contrast, inbred humans, born to consanguineous couples, are available for study. Children whose parents are first cousins carry identical by descent alleles at 1/16 of their loci (see Chapter 2). Thus, such a child carries 1/16 of their deleterious alleles in homozygous state, and a rare deleterious allele which was heterozygous in the outbred population remains heterozygous and becomes homozygous with probabilities 15/16 and 1/32, respectively. Suppose that an average contribution of a homozygous deleterious allele into genotype contamination is 4 times larger than that of a heterozygous one. Then, the average genotype contamination of children from first cousin marriages is higher by a factor of $4(1/32) - (1/16) = 1/16$. Data on populations living under less-than-optimal conditions suggest that before Industrialization fitness of such children was reduced by ~10%. This implies that the average equilibrium fitness is $0.9^{16} = 0.03$ of the maximal one. Of course, this estimate is highly sensitive to the poorly known degree of recessivity of mutant alleles: if we assumed that the average contribution to the genotype contamination of a homozygous allele is, say, 10 times that of a heterozygous allele, the implied weak imperfection would be much smaller.

All these three estimates of the increment of genotype contamination are rather imprecise. Still, it appears that even a modest increase in the genotype contamination (much less than its doubling) causes a deep reduction of fitness. If so, the only way to avoid a conclusion that weak imperfection is impossibly high is to assume that fitness landscape is characterized by synergistic epistasis, so that a beneficial effect of eliminating genotype contamination is much smaller than a deleterious effect of doubling it. If synergistic epistasis is sufficiently strong, any reduction of fitness due to a particular increment of genotype contamination could be consistent with an arbitrarily small weak imperfection (Figure 15.4). In contrast, a moderately strong epistasis would imply a substantial, but still possible weak imperfection. And if epistasis is absent, I do not know how to explain the above data.

To summarize, the data on fitnesses of existing or of overcontaminated genotypes are insufficient to estimate fitnesses of undercontaminated genotypes and weak imperfection, as long as the strength of synergistic epistasis remains unknown (Figure 15.4). It is plausible, but not certain, that getting rid of all individually rare deleterious alleles would lead to a large increase in fitness. I believe that this issue can be resolved only empirically by artificially creating organisms with deleterious alleles-free genotypes and studying their phenotypes. It is already technically possible to do this on bacteria, as well as on baker's yeast, a unicellular eukaryote.

So far, we have considered fitness – but what about imperfection of human wellness? Would a mother want her child to have a weakly perfect genotype and, if yes, how strongly? We know even less about this issue, because population genetic data shed light on how alleles affect fitness, but not wellness (see Chapter 11).

Still, getting rid of pre-existing deleterious mutant alleles would surely lead to elimination of conditions which cannot develop without them. These conditions definitely include Mendelian diseases, except cases of mostly dominant diseases caused by *de novo* mutations (see Chapter 13). Also, it seems very likely that a weakly perfect genotype would be nearly immune to complex diseases which are strongly affected by genetic factors, including schizophrenia and ASD, but probably excluding severe intellectual disability, which appears to be essentially Mendelian (see Chapter 13). Indeed, as long as a disease is truly complex and cannot be caused by mutant alleles of just one gene, *de novo* mutation in the course of one generation would be unlikely to produce enough damage to push a weakly perfect genotype beyond the disease threshold (see Chapter 3). Birth anomalies would probably not be completely eliminated, because the environment or chance may be alone sufficient to produce some of them but their frequency would certainly go down.

In contrast, we know nothing definite about how weak perfection would affect quantitative facets of wellness. Still, it is plausible that genotypes with contamination 10 or even more standard deviations below the current average would produce much better phenotypes. A clue to the magnitude of this improvement can be provided by the extent to which extraordinary abilities are caused by genotypes. Many geniuses are clearly inborn: even under the best possible environment, only an exceptional child can perform and compose music like young Mozart. Such inborn gifts could result, at least partially, from happy developmental accidents, instead of genotypes. However, data on people with very high IQ indicate a major role of genotypes, because close relatives tend to have (or not have) very high IQ together.

The simplest assumption is that the genetic cause of very high intelligence is an unusually low genotype contamination. Indeed, direct sequencing of genotypes of very smart people showed that they carry smaller than average numbers of deleterious alleles. However, this effect was small, perhaps because we still cannot precisely assay the contamination of a genotype (see Chapter 12). Similarly to the case of fitness, only empirical studies of the effects of unusually low genotype contamination on wellness can settle this issue. An extra obstacle to obtaining such data is that there are no animal models of composing music or programming.

Above, we considered imperfection due to only one of its five evolutionary causes – rare, substantially deleterious alleles (see Chapter 10). Let us now expand out mutationless utopia and include drift imperfection due to fixed slightly deleterious alleles (SDAs), the only other evolutionary mechanism of imperfection which is relevant to the Main Concern (see Chapters 8 and 10). The substantial contribution of relatively common alleles, with frequencies > 5%, into a number of complex diseases suggests that alleles capable of producing important adverse effects can, nevertheless, be only slightly deleterious. If many susceptibility alleles can reach high frequencies, many others are likely to be fixed in the population, at any particular moment.

SDAs can involve all kinds of changes of the DNA sequences. In particular, Boissinot and colleagues demonstrated that the human lineage accumulated a lot of slightly deleterious transposable elements from the LINE-1 family. Collectively, fixed SDAs could be responsible for up to 1000 mullers of contamination in every genotype (see Chapter 9). However, as we have already seen, this figure cannot be translated into reduction of fitness without knowledge of synergistic epistasis, if any. Thus, direct data are again needed to reliably assay imperfection due to fixed SDAs. Such data will be particularly difficult to obtain, because this will require identifying and reverting millions of fixed SDAs. Still, genome-wide purging of SDAs can potentially produce a very large benefit, both fitness- and wellness-wise.

On top of harboring fixed deleterious alleles, the human genome also contains a lot of features which facilitate mutagenesis, while being apparently useless, if not harmful, *per se*. Two classes of such features are particularly prominent. The first are hypermutable CG contexts within protein-coding exons, which are responsible for up to 50% of pathogenic missense mutations causing Mendelian diseases. These contexts could likely be safely replaced, because any amino acid sequence can be encoded without any CGs. Secondly, the human genome, as well as other mammalian genomes, carries a lot of dead transposable elements and other repetitive sequences, and illegitimate recombination between them is responsible for a majority of large-scale mutations (see Chapter 5), which are particularly dangerous. Thus, removing mutagenic features from the human genome would substantially reduce the genomic rate of spontaneous deleterious mutations.

15.3 Mutationless Utopia: Is It Ever Going to Happen?

If it were technically possible, would we want to turn mutationless utopia into reality? In my opinion, a likely deep reduction of the frequency of severe diseases with a substantial genetic component, which together affect ~5% of young people and the majority of old people, is alone a sufficient reason to consider this idea very seriously.

The possibility of a substantial improvement of quantitative facets of wellness is another reason. Finally, prenatal mortality in a population of deleterious alleles-free people would likely be way below its staggering current level of 70%.

In Chapter 14, we considered some ethical objections against this radical way to address the Main Concern. From the biological perspective, I could think of only one sensible objection – possible undesirable side-effects at the population level. Are we ready to sacrifice genetic diversity for weak perfection? Do we want a brave new world of "designer babies" in which all humans are effectively identical twins with almost the same phenotype (see Chapter 10)? Are not we scared by a society populated by uniform walking perfections who are all, say, dark-skinned, blue-eyed, and harmoniously built, and who all understand physics better than Albert Einstein and play tennis better than Serena Williams?

I do not think that anything like this is going to happen, even if parents acquire unlimited power to modify genotypes of their children, because there are a lot of loci where no allele would be universally preferred. First, parents want their children to be similar to themselves – as long as their traits are not outright bad. I am happy that my children did not inherit my scoliosis, but also that they did inherit my light skin. Only if I lived in the tropics, I would prefer my children to be dark-skinned, to make them less dependent on sunscreen and to reduce their risk of melanoma. In contrast, a dark-skinned father is likely to want his children to look like him, and not like me.

Secondly, even under a uniform environment, some facets of fitness and wellness are frequency-dependent. In particular, as it was the case in the course of human history, fitness-wise, it is still advantageous, wellness-wise, to have a recognizable face and, therefore, to possess a unique (or, at least, a very rare) combination of alleles at a number of polymorphic loci that control variation of facial morphology. Sensible parents would understand this and would not want their children to be copies of a popular sex symbol. Even some genotypes that lead to recognizable conditions, such as Asperger's syndrome, may still be preferred by some parents, because this condition is often associated with high intelligence or unusual talents, despite the impaired ability to interact with others.

Still, while diversity of skin colors, facial features, and talents and temperaments is welcome, diversity of vision acuities or blood pressures is not. Everybody should have 20/20 eyesight, and a blood pressure close to 115/75, perhaps with small individual variations, but definitely not 160/100 or 80/50. Unless the current tentative conclusions about the nature of human genetic variation are fundamentally wrong, only a small proportion of it is responsible for desirable diversity, and most of it is unconditionally deleterious.

Finally, humanist ethics does not allow population-level concerns, even legitimate, to trump liberties of individuals and interests of their future children. Thus, to prohibit voluntary germline genotype modification (GGM) only because it can make humans too uniform would be wrong. Indeed, marriages between individuals from different populations also reduce diversity of genotypes, phenotypes, and cultures. Still, we are not going to prescribe individuals from a small minority to marry exclusively within their kin, in order to preserve their gene pool and language.

Of course, I am not arguing that parents should, morally or legally, have absolute control over genotypes of their children. Discussing particulars of application of technologies that are probably decades away does not make sense. I only claim that reverting even all unconditionally deleterious alleles would not diminish valuable diversity of human populations.

Still, even a worthy goal should not be pursued, unless it can be reached by means that are both technically feasible and ethically acceptable. Deleterious alleles can be removed in two ways – by selection or through active antimutagenesis. Negative natural selection did a fair job preventing unlimited accumulation of substantially deleterious alleles in the course of billions of years. However, this job was far from perfect, because each genotype still carries many such alleles. Could negative artificial selection be more efficient?

This is not impossible. Natural selection assays genotype contamination indirectly, through phenotype, and random phenotypic variation diminishes its efficiency. Thus, it would be very interesting to directly select flies or mice for lower contamination of their genotypes. If, in an experimental population, only individuals with below-average genotype contaminations are allowed to reproduce, this should lead to rapid removal of rare deleterious alleles. Currently, the key obstacle to performing such an experiment is our still very limited ability to assess quantitatively the contamination of a genotype, or even to identify all alleles that are substantially deleterious.

Eventually, this will change, and it will become possible to determine the contamination of a human genotype. However, artificial selection of humans is bound to raise ethical issues, because selection is antithesis to equality (see Chapter 14). Even in the absence of any coercion by the government, aborting a fetus due to too high a genotype contamination, which, nevertheless, is unlikely to cause any serious diseases, is at least morally suspect. In contrast, voluntary sexual or fertility selection could be acceptable, but neither of them looks like a viable option.

Artificial sexual selection could be most painlessly implemented by using sperm from donors with exceptionally good genotypes. Using cryopreserved sperm requires *in utero* insemination (IUI; see Chapter 14), which is only minimally invasive, and leads to pregnancy with probability up to 25% per cycle. Such sexual selection was advocated, as a eugenic panacea, by many people, including Muller and, as late as 1962, by Julian Huxley, the grandson of "Darwin's bulldog" Thomas Huxley (and the person who gave Muller his first faculty job in 1916, at Rice University).

Biologically, this idea makes sense. In many socially monogamous species, females routinely engage in extra-pair copulations with particularly attractive males, and offspring from such matings often have substantially higher fitness than that sired by social mates of females. However, it does not look likely that donor insemination is ever going to have a substantial impact on the human gene pool. In the USA, only about 1 child per 1000 is conceived through IUI. The much more invasive *in vitro* fertilization, which allows a couple that cannot conceive naturally still have a child together, is several times more common.

Moreover, even these relatively rare instances of IUI involve only rather modest selection. Women choose donor insemination when they do not have a male partner, when their partner is sterile, or when conceiving a child with the partner carries a known high risk of a genetic disease. When natural conception is not involved with a drastically elevated risk, which is by far the most common situation, it is rather unlikely that many women who have a male partner will ever choose IUI. Indeed, all humans have rather similar overall genotype contaminations, and using donor sperm is not going to make a big difference for an individual child. "Transcendently superior" sperm donors, hailed by Muller (see Chapter 14), do not exist in real populations. Even if it eventually becomes possible to reliable identify the best available sperm donors, how many women will choose IUI only because it can increase the expected IQ of a child by, say, 10%? Very few, I suppose, because even when there is a high known risk, couples often prefer preimplantation or prenatal diagnosis, in order to have their shared children, over donor insemination.

Artificial selection against deleterious alleles through differential fertility also does not look realistic, due to the same reason. A person, or a couple, may refrain from reproduction if their children carry a high risk of a serious genetic disease. Fortunately, such situations are not very common. In all other cases, few people would determine the number of their children on the basis of a quantitative prediction about their generally normal wellness.

Thus, antimutagenesis appears to be the only realistic means for removing deleterious alleles from human genotypes. Passive antimutagenesis by storing your own sperm at a young age could be a useful palliative, as it can reduce the mutation rate by a factor of ~1.5, if widely used. Still, the mutationless utopia can be achieved only by active antimutagenesis via GGM. But will the necessary technologies ever be developed?

Muller claimed in 1936 that "... it is not possible artificially to change the genes themselves in any particular, specified directions. The idea that this can be done is an idle fantasy, probably not realizable for thousands of years at least". The current progress in dealing with DNA suggests that he may have been way too pessimistic. The CRISPR/Cas9 system (see Chapter 4), although far from perfect, already makes it possible to edit existing genotypes, which is the natural way to revert a small number of alleles. Technologies of genotype editing are likely to progress rapidly.

However, for achieving mutationless utopia *de novo* non-template genotype synthesis could make more sense. In 2010, a team led by one of the DNA sequencing pioneers, Craig Venter, produced a bacterium *Mycoplasma mycoides*, whose genotype, with a length of 1 million nucleotides, was synthesized in this way. The sequence of this genotype was a copy of the genotype sequences from natural bacteria, with only a small number of functionally mute changes that were introduced deliberately.

This project took many years to complete and involved assembling longer DNA segments from shorter pieces. Clearly, much better technology would be needed to synthesize the very long genotype of a mammal (see Chapter 1). Ideally, we need molecular machines, perhaps consisting of many interacting human-designed proteins, which can precisely manufacture very long DNA molecules with prescribed nucleotide sequences. Such technology would not violate any laws of nature, and no one should be surprised if it emerges in 30–50 years from now. Recent work on synthetic yeast chromosomes showed that manufactured DNA molecules can be incorporated into a eukaryotic genome.

Even rather crude methods could be useful for modifying genotypes of somatic cells, which can be used for the treatment of genetic diseases. Indeed, a therapeutic effect is possible even if spurious modifications are common and a desired modification occurs only in a small proportion of cells. GGM in non-humans, used to produce genetically modified organisms, also does not need to be precise. For example, recently the laboratory of George Church at Harvard used CRISPR/Cas9 to remove a number of retroviruses from a porcine genotype, in order to produce pigs that can be safer organ donors for humans. In contrast, GGM in humans must be precise, to make the associated risk of unintended and harmful genetic changes very low (see Chapter 14).

Of course, getting rid of unconditionally deleterious alleles would be impossible without first identifying them. We can already do this for alleles that destroy molecular function (see Chapter 2). In contrast, identifying deleterious missense alleles, deleterious non-coding alleles, and SDAs that are fixed in the population (see Chapter 8) is progressively more difficult (see Chapter 11). Also, it is much easier to study the effect of an allele on fitness than on wellness, although these are likely to be tightly correlated. Still, a rapid progress in this area is likely.

Naturally, no attempts to eliminate deleterious alleles in humans should be made before experimenting on other species. Weakly perfect bacteria, flies, and mice must be created first. Results of such experiments will surely clarify our thinking about the mutationless utopia.

An already available technique of mitochondrial replacement (see Chapter 14) provides an opportunity to initiate experimental investigation of weak imperfection. So far, only donor mitochondria with their native genotypes have been used. However, it might be possible to replace these genotypes with artificially synthesized DNA molecules. Because mammalian mitochondrial genomes consist of only ~16 000 nucleotides, this synthesis is not difficult, and the only problem is to deliver artificial genotypes into mitochondria and to make sure that they begin to function. An average mammalian mitochondrial genotype carries several rare deleterious alleles and, perhaps, hundreds of fixed SDAs. If weakly perfect mitochondria substantially improve phenotypes of experimental mice, this would stimulate research on other means of GGM.

15.4 What Can I Do Without Germline Genotype Modification?

It seems unlikely that a technology suitable for replacing even one wrong nucleotide in a human germline genotype will appear in the next 10 years. Can a person, nevertheless, do something, today or in the near future, to ensure that their children have good genotypes, conducive to their wellness? Modern practices, described in Chapter 14, are aimed against only relatively common, genetically simple, severe impairments. Is it possible to do more, in particular, by utilizing information on complete genotypes?

To put this question into perspective, let us first examine whether an individual may benefit from knowing their own genotype. Some information that can be gleaned from a genotype can surely be amusing, revealing the proportion of Neanderthal blood in the veins of its owner and the geographical origins of their more recent ancestors. But can this information be useful?

The answer is “yes” if a person has a clear-cut medical condition. Sequencing the genotype of an affected individual can be essential for precise diagnosis of some conditions, helping to choose the best available treatment. Still, currently one or more disease-causing alleles can be identified in the genotype of only 1 out of 4 patients, and even such identification informs medical decisions only in a proportion of cases. Establishing genetic causes of a complex, as opposed to Mendelian, disease is particularly hard.

Benefits of knowing the whole genotypes are the largest in the case of cancer. The presence of somatic mutations in a relatively small number of genes in malignant cells is often crucial for choosing the best treatment. Soon, sequencing the tumor will be a routine step in cancer care.

In contrast, if there is no specific reasons for concern, the current answer is more on the "no" side. Indeed, the genotype of every healthy individual carries ~100 alleles that have been reported to contribute to a disease. Some of these alleles are, in fact, benign, and reports of their adverse effects are mistaken. Other alleles, although potentially pathogenic, are incompletely penetrant (see Chapter 3). On top of this, every human genotype also carries hundreds of "variants of unknown significance", many of which are likely to disturb the molecular function, but whose medical relevance remains obscure. Thus, at the present level of understanding of the connection between genotypes and phenotypes (see Chapter 1), knowing your own complement of potentially disease-causing alleles can do more harm than good, by causing fruitless anxiety and encouraging unnecessary tests, without providing any medically actionable information.

Eventually, this situation will change, and knowing the genotype of a healthy young person, or even of a newborn, will make sense medically, helping, in particular, to anticipate problems in the future. Ongoing clinical trials MedSeq and BabySeq attempt to determine if this time is already here. Sequencing the genotype of a newborn will at some point become the most economical method of early screening for a number of treatable Mendelian diseases. Positive findings will be rare (~1 in 1000), but very important. Otherwise, I believe that the benefits of knowing your own genotype, in the absence of specific medical indications, will outweigh the drawbacks no sooner than in 10-20 years, and probably not by much.

Let us now go back to reproductive decisions. Currently, preconception analysis of genotypes of prospective parents is recommended only very restrictively, as screening for disease-causing recessive alleles of a small number of genes (see Chapter 14). Would it be better to look for such alleles in complete genotypes? This may sound like a good idea, because the overall risk of a Mendelian disease in a newborn is ~3% (see Chapter 12) and most healthy humans carry at least one allele which is capable of causing a Mendelian disease. Most of these diseases are recessive, although even a dominant disease inherited from a parent may not be visible in his or her phenotype, due to its late onset, incomplete penetrance, or variable expressivity (see Chapter 3).

In theory, data on complete genotypes of would-be parents can reveal risks of all Mendelian diseases in their children, except those caused by *de novo* mutations. Nevertheless, collecting such data is not yet practiced widely, and for a good reason: for (almost) every pair of genotypes, there will be genes that can be responsible for Mendelian diseases in which both genotypes carry "variants of unknown significance" or even alleles that were reported to be pathogenic. A potential child may receive such alleles from both parents and, thus, be at risk of a recessive Mendelian disease. In other words, no prospective marriage would get a clean bill of health. However, most of these alarms will be false, because the suspicious alleles are either benign or have low penetrance. At the same time, risks of not yet described (and likely rare, see Chapter 12) Mendelian diseases will remain undetected.

The first attempt to analyze complete genotypes of prospective parents was made by a company GenePeeks in 2014. So far, the only goal of this analysis is to help selecting a sperm donor which is suitable for a particular woman, and ~20% of potential donors get rejected because of a high predicted risk of a Mendelian disease.

Such predictions are going to become more and more precise: when millions of complete genotypes are known, we will have a much better idea of which alleles can cause Mendelian diseases, and with what probabilities. Still, currently analyzing complete genotypes of all prospective parents cannot be recommended. However, this conclusion may be reversed in the next several years.

Knowing the genotype of an embryo or a fetus can produce a more precise prediction of the phenotype of the future child, because this prediction does not involve any uncertainty due to Mendelian segregation and recombination (see Chapter 2). Also, ~0.5% of newborns suffer from a dominant single-gene Mendelian disease caused by a *de novo* mutation, absent in the parents (see Chapter 13). Nevertheless, sequencing complete genotypes of all fetuses also cannot be recommended at this time for the same reason: every genotype will raise some suspicions, mostly unfounded.

What about complex diseases? We know that genotypes are to a large extent responsible for them (see Chapter 12). However, the path from genotype to a complex disease is, by definition, much more complex, and much less understood, than that to a Mendelian disease. I doubt that information on complete genotypes, of parents or of fetuses, will become very useful for predicting complex diseases in children in the next 20–30 years.

Finally, predicting the values of quantitative traits of a future child is not only currently impossible, but also appears to be worthless, as far as reproductive decisions are concerned. Indeed, for a sensible person, decisions of such importance as whether to marry, conceive, or abort could be informed only by risk of serious diseases, but not by the expected height or intelligence of a child, as far as they are within normal limits. Perhaps, the only feasible exception is the choice of a sperm donor, which needs to be made anyway.

Save replacing mitochondria with faulty genotypes (see Chapter 14), the only means of (passive) antimutagenesis that is already available is to store your own sperm. This idea was first alluded to by Muller in 1950 and elaborated by Crow in 1997. Data on accumulation of mutations in aging male germline (see Chapter 6) and on adverse phenotypic effects of advanced paternal age (see Chapter 13) strongly suggest that using sperm produced at a young age can substantially increase wellness of children. If a woman is younger than 35, IUI makes it possible to achieve pregnancy with probability 60–70% after 6 months. Commercial cryobanks currently charge ~ US$300 per year for storing an unlimited amount of sperm, but this price will likely become lower for long-term storage.

Storing sperm makes sense only if it is collected before the age of 30 (the earlier, the better), and used at least 10, if not 20, years later. In other words, if you store your sperm today, there will be no reason to use it until, at least, 2027. By this time, a much clearer picture of possible benefits and risks of using stored sperm will emerge. In particular, there are conflicting evidence on whether IUI involves an increased risk of birth anomalies, and more data are needed on this. Also, some data indicate that chromosomes in sperms of young men have shorter telomeres, which may be a disadvantage. Crucially, effects of the paternal age on the phenotypes of children will be quantified much better. Thus, to store his sperm now would be a wise individual decision for a young man. It would also be wise for governments to treat sperm storage as a public health issue and to start encouraging it.

15.5 Prognosis

Save a global calamity, a variety of issues related to the Main Concern will be actively investigated in the next 20 years. At the level of genotypes, millions of mother–father–offspring trios will produce nearly perfect data on spontaneous mutation from the perspective of DNA sequences and molecular function. The dependence of the mutation rates on the paternal age and other factors will also be known with high precision. However, I would be very surprised if this led to a radical revision of the already available estimates.

In contrast, the current understanding of phenotypic effects of deleterious alleles is still rudimentary, which leaves room for dramatic progress. The following fundamental problems will be addressed:

1) Phenotypic effects of pre-existing genetic variation. What are the properties of complex trait loci affecting various phenotypic traits, in particular, fitness and wellness? Are wellness-impairing alleles really almost exclusively derived and deleterious, at least under harsh environments? Is sign epistasis really rare and synergistic epistasis really common? How precise is the estimate of ~10 mullers of contamination in an average human genotype? Brute force sequencing of ~1 billion of genotypes and relating them to the phenotypes of their owners will likely shed light on these questions.
2) Mutational pressure on human wellness and its facets. How large are mutational targets of the key phenotypic traits and average effects of an individual *de novo* mutation on them? How strong are mutational pressures? These questions will be addressed by relating complements of *de novo* mutations in genotypes of many humans to their phenotypes. Hopefully, the gap between optimistic and pessimistic scenarios that are consistent with the data (see Chapter 13) will be much narrower in 20 years' time than it is today.
3) Strength of selection against deleterious alleles in modern human populations. Is this selection still substantial? If "yes", does it push our key traits, including general intelligence, in the direction that is conducive or adverse to wellness? Studying dynamics of alleles that affect these traits in human populations using very large samples may help to clarify these issues.
4) Phenotypes produced by deleterious alleles-free genotypes. By how much is their fitness and wellness superior over ours, under different environments? Here, progress may be the slowest, and the degree of our imperfection may remain obscure for quite a while.

An equally important issue is the development of technologies that are precise enough for GGM in humans. It seems plausible that in 20 years' time methods of DNA editing will be good enough to make it possible to revert individual small-scale alleles. In contrast, precise synthesis of arbitrary human genotypes is probably not going to happen that soon. Thus, even if the new data show that the pessimistic scenario is closer to the truth, it is unlikely that active antimutagenesis will have been already implemented at a large scale by 2037. In contrast, we can expect clarity regarding benefits and dangers of passive antimutagenesis by sperm freezing.

Of course, these advances are contingent on the support by governments and the society at large. Research of all facets of deleterious mutation should be encouraged and funded. It is time to realize that germline mutation poses no less a threat to human wellness than cancer-causing somatic mutation. In 1971, the then US President, Richard Nixon, initiated "War on Cancer". Due to the enormity of the problem, it was not a fast, across-the-board practical success. However, the resulting progress in understanding cancer biology was transformative, and any cures that will be developed in the future will be grounded in it. I hope that "War on Mutation" will be declared soon. The human genome consists of parts, and it is not in our powers to prevent it from crumbling. Still, diligent efforts could eventually mitigate the harm of mutation.

Further Reading

Charlesworth, B 2013, Why we are not dead one hundred times over, *Evolution*, vol. 67, pp. 3354–3361.
Imperfection due to fixation of many very slightly deleterious alleles can be very high.

Clarke, T-K, Lupton, MK, & Fernandez-Pujals, AM 2016, Common polygenic risk for autism spectrum disorder (ASD) is associated with cognitive ability in the general population, *Molecular Psychiatry*, vol. 21, pp. 419–425.
Evidence that alleles which increase the risk of ASD may increase cognitive abilities.

Corbett-Detig, RB, Zhou, J, & Clark, AG 2013, Genetic incompatibilities are widespread within species, *Nature*, vol. 504, pp. 135–137.
A study of sign epistasis in *Drosophila melanogaster*.

Eisenberg, DTA, Hayes, MG, & Kuzawa, CW 2012, Delayed paternal age of reproduction in humans is associated with longer telomeres across two generations of descendants, *Proceedings of the National Academy of Sciences of the USA*, vol. 109, pp. 10251–10256.
Genotypes of sperms produced by older men have longer telomeres, which may be beneficial.

Gerlach, NM, McGlothlin, JW, & Parker, PG 2012, Promiscuous mating produces offspring with higher lifetime fitness, *Proceedings of the Royal Society B*, vol. 279, pp. 860–866.
In a dark-eyed junco (a songbird), both male and female offspring produced by extra-pair fertilizations have higher lifetime reproductive success.

Morton, NE, Crow, JF, & Muller, HJ 1956, An estimate of the mutational damage in man from data on consanguineous marriages, *Proceedings of the National Academy of Sciences of the USA*, vol. 42, pp. 855–863.
A classical paper which elucidated some key properties of the mutation–selection equilibrium and introduced indirect estimates of weak imperfection.

Qin, J, Sheng, X, & Wang, H 2015, Assisted reproductive technology and risk of congenital malformations: a meta-analysis based on cohort studies. *Archives of Gynecology and Obstetrics*, vol. 292, pp. 777–798.
Assisted reproduction appears to increase the risk of birth anomalies.

Sheehan, MJ & Nachman, MW 2014, Morphological and population genomic evidence that human faces have evolved to signal individual identity, *Nature Communications*, vol. 5: 4800.
Evidence of frequency-dependent selection on polymorphic loci that determine human facial features.

Spain, SL, Pedroso, I, & Kadeva, N 2016, A genome-wide analysis of putative functional and exonic variation associated with extremely high intelligence, *Molecular Psychiatry*, vol. 21, pp. 1145–1151.
So far, the genetic basis of extremely high intelligence remains poorly understood.
Yang, Y, Muzny, DM, & Xia, F 2014, Molecular findings among patients referred for clinical whole-exome sequencing, *JAMA*, vol. 312, pp. 1870–1879.
Currently, determining sequences of all protein-coding genes makes it possible to establish diagnosis for ~25% of patients with hard-to-diagnose diseases.

Index

Crumbling Genome: The Impact of Deleterious Mutations on Humans, First Edition. Alexey S. Kondrashov.

l

m

n

o

Printed and bound by CPI Group (UK) Ltd, Croydon, CR0 4YY
16/04/2025
14658429-0001